OUR ACOUSTIC
ENVIRONMENT

ENVIRONMENTAL SCIENCE AND TECHNOLOGY

A Wiley-Interscience Series of Texts and Monographs

Edited by ROBERT L. METCALF, *University of Illinois*

JAMES N. PITTS, Jr., *University of California*

WERNER STUMM, *Eidgenössische Technische Hochschule, Zurich*

PRINCIPLES AND PRACTICES OF INCINERATION
Richard C. Corey

AN INTRODUCTION TO EXPERIMENTAL AEROBIOLOGY
Robert L. Dimmick

AIR POLLUTION CONTROL, Part I
Werner Strauss

AIR POLLUTION CONTROL, Part II
Werner Strauss

APPLIED STREAM SANITATION
Clarence J. Velz

ENVIRONMENT, POWER, AND SOCIETY
Howard T. Odum

PHYSICOCHEMICAL PROCESSES FOR WATER QUALITY CONTROL
Walter J. Weber, Jr.

ENVIRONMENTAL ENGINEERING AND SANITATION
Joseph A. Salvato, Jr.

NUTRIENTS IN NATURAL WATERS
Herbert E. Allen and James R. Kramer, Editors

pH AND pION CONTROL IN PROCESS AND WASTE STREAMS
F. G. Shinskey

INTRODUCTION TO INSECT PEST MANAGEMENT
Robert L. Metcalf and William H. Luckmann, Editors

OUR ACOUSTIC ENVIRONMENT
Frederick A. White

OUR ACOUSTIC ENVIRONMENT

FREDERICK A. WHITE

Professor of Nuclear Engineering
Professor of Environmental Engineering
and Industrial Liaison Scientist
Rensselaer Polytechnic Institute

A WILEY-INTERSCIENCE PUBLICATION

John Wiley & Sons

New York / London / Sydney / Toronto

Library of Congress Cataloging in Publication Data:

White, Frederick Andrew, 1918-
  Our acoustic environment.

  (Environmental science and technology)
  "A Wiley-Interscience publication."
  Includes bibliographical references and index.
  1. Sound. 2. Noise pollution. 3. Noise
control. I. Title.

QC225.15.W49   534   75-8888
ISBN 0-471-93920-X

Printed in the United States of America

10 9 8 7 6 5 4 3 2 1

SERIES PREFACE

Environmental Sciences and Technology

The Environmental Sciences and Technology Series of Monographs, Textbooks, and Advances is devoted to the study of the quality of the environment and to the technology of its conservation. Environmental science therefore relates to the chemical, physical, and biological changes in the environment through contamination or modification, to the physical nature and biological behavior of air, water, soil, food, and waste as they are affected by man's agricultural, industrial, and social activities, and to the application of science and technology to the control and improvement of environmental quality.

The deterioration of environmental quality, which began when man first collected into villages and utilized fire, has existed as a serious problem since the industrial revolution. In the last half of the twentieth century, under the ever-increasing impacts of exponentially increasing population and of industrializing society, environmental contamination of air, water, soil, and food has become a threat to the continued existence of many plant and animal communities of the ecosystem and may ultimately threaten the very survival of the human race.

It seems clear that if we are to preserve for future generations some semblance of the biological order of the world of the past and hope to improve on the deteriorating standards of urban public health, environmental science and technology must quickly come to play a dominant role in designing our social and industrial structure for tomorrow. Scientifically rigorous criteria of environmental quality must be developed. Based in part on these criteria, realistic standards must be established and our technological progress must be tailored to meet them. It is obvious that civilization will continue to require increasing amounts of fuel, transportation, industrial chemicals, fertilizers, pesticides, and countless other products and that it will continue to produce waste products of all descriptions. What is urgently needed is a total systems approach to modern civilization through which the pooled talents of scientists and engineers, in cooperation with social scientists and the medical profession, can be focused on the development of order and equilibrium to the presently disparate segments of the human environment. Most of the skills and tools that are needed are already in existence. Surely a technology that has created such manifold environmental problems is also capable of solving them. It is our hope that this Series in Environmental Sciences and Technology will not only serve to make this challenge more explicit to the established professional but that it also will help to stimulate the student toward the career opportunities in this vital area.

Robert L. Metcalf
James N. Pitts, Jr.
Werner Stumm

PREFACE

Courses on ecology and the environment have recently become part of almost every college curriculum. However, a survey course on the environmental aspects of sound is rare, despite its relevance and educational merit. This situation can be partially attributed to the lack of textbooks suitable for the nonspecialist. Books are indeed available on such subjects as physical acoustics, underwater sound, and architectural acoustics, and highly technical works on sound and vibration have been written for electrical and mechanical engineers. But there are few single-volume publications for the student who desires an overview of the diverse topics related to environmental acoustics. Of course, many timely and useful publications deal with noise. However, noise refers only to the *unwanted* portion of our sonic environment. The objective, here, is to provide an educational stimulus that is somewhat broader in scope. Hopefully, long after some of the specific noise problems of our society are resolved, the reader of this book will retain an interest in the *desired* sounds of our environment, that is, the sounds of communication and music. These sounds have reflected the thought and culture of every civilization, even though the technology for preserving and reproducing sonic signals did not appear until the twentieth century.

Another objective is to present acoustics for the increasing number of students with interdisciplinary interests. Few subjects provide a broader base than acoustics for revealing the interdependence of the physical and biological sciences, and for demonstrating the connection between science, engineering, and the arts. Acoustics affords a link between art and architecture, electrical engineering and physiology, physics and biology, psychology and the field of medicine, linguistics and computer technology. Even if we limit our attention to topics that can clearly be classified as environmental, we observe a broad overlap of disciplines. The control of loud or intrusive noise, for example, is important because such noise interferes with speech and music perception, in addition to being a potential hazard for hearing loss. Also, legal aspects of noise abatement cannot be divorced from city planning and other societal problems. Hence, I hope the book will be useful to anyone whose professional assignments require some knowledge of acoustics and acoustic terminology.

The organization of this book into three parts provides options for its use in seminars, minicourses, independent study, or formal presentations. Part I contains an introductory chapter, followed by a discussion of the three p's of sound: (1) sound production, (2) sound propagation, and (3) sound perception. Sound is characterized in simple terms, and the decibel scale is defined and explained. Sound propagation in both outdoor environments and enclosed spaces is discussed in relation to relevant physical phenomena. Sound perception is dealt with in several chapters in order to survey both the physiological and subjective aspects of human hearing, and to identify the instrumentation that permits objective and quantitative sound measurements. Thus, Part I provides the foundation for understanding the applied topics of Parts II and III.

Part II focuses on the contemporary problem of noise. Chapter 8 presents a general overview of this problem, and traffic and aircraft noise are surveyed in Chapters 9 and 10. Chapter 11 includes several topics on industrial noise control. The legal aspects of noise abatement at the federal, state, and local levels are reviewed in Chapter 12. This chapter also summarizes international efforts toward noise control. Considerations of noise in city planning, condensed from the periodical literature, are presented in Chapter 13.

Part III introduces topics that should be of interest to students, not just for a semester, but for a lifetime. Chapter 14 deals with the sounds of speech, the parameters that affect speech intelligibility in usual speaking environments, and the voice spectrograph. Fundamentals of musical acoustics are introduced in Chapter 15, and the special environments provided by auditoriums and concert halls are discussed in Chapter 16. Entire books have been devoted to the subject matter of Chapter 17; the physiological and psychological effects of sound. Hopefully, some students will seek out these more detailed monographs after reading this introductory presentation.

Instructors using this book for a course in environmental acoustics may wish to arrange for complementary demonstrations or field studies. Sound level meters are now available at a reasonable cost, and recordings of environmental sounds can be obtained that permit one to both hear and observe sound spectra via an oscilloscope. Furthermore, there is no real substitute for the experience of making a few actual acoustic measurements, such as the sound pressure level of traffic noise or the reverberation time of an auditorium. The material of this book may also be augmented through formal assignments, in which students select specific topics of personal concern, pursue extensive reading in the periodical literature, and prepare updated oral reports or term papers.

FREDERICK A. WHITE

Troy, New York, 1975

ACKNOWLEDGMENTS

It is a pleasure to acknowledge both the direct and indirect contributions to this book from many organizations and individuals. My personal concern with environmental acoustics was increased substantially during a sabbatical leave that I spent at the Bell Telephone Laboratories, through the many interdisciplinary publications of Dr. L. L. Beranek and, more recently, through the professional stimulus provided by many engineering and research groups throughout the United States. I would like to especially mention the following organizations or staff members: Acoustical and Insulating Materials Association; Acoustical Society of America; American Petroleum Institute, N. K. Weaver, M.D.; B and K Instruments, Inc.; Bell Laboratories, Dr. F. L. Flanagan, Dr. W. M. Gibson, Dr. M. R. Schroeder; Boeing Company, A. L. Wood; Bolt Beranek and Newman, Inc., Dr. L. L. Beranek, Dr. T. J. Schultz; State of California, Office of the Attorney General, N. C. Yost; Central Institute for the Deaf, St. Louis, Dr. J. D. Miller; Columbia University, Professor Cyril M. Harris; Consolidated Edison, Dr. R. A. Bell; Dames and Moore, Dr. F. M. Kessler; Dartmouth College, Professor W. L. Gulick; Detroit Edison, R. A. Popeck; DuPont deNemours & Co.; Eastman Kodak Co., Dr. J. D. Walling; Eastman School of Music, Professor Wayne Barlow; Eckel Industries; Edison Electric Institute; Ford Motor Co.; General Electric Co., D. C. Graham, M. W. Schulz, Dr. H. Nagamatsu, R. B. Tatge, M. Weiss, R. J. Wells; General Radio Co.; General Motors Corp., E. A. Ratering, R. K. Hillquist; L. S. Goodfriend and Associates; Goodyear Atomic Corp., C. P. Blackledge, Jr.; Harvard University, Professor E. N. Hiebert; Hewlett-Packard Corp., J. S. Lands; Industrial Acoustics Co., Inc.; International Business Machines Corp., Dr. N. R. Dixon; Jaffe Acoustics, Inc., Christopher Jaffe; Lockheed-California Co.; McDonnell-Douglas Corp.; Mechanical Technology, Inc., Dr. O. Shinaishin; Michigan Bell Telephone, W. J. Jend, M. D.; National Aeronautics and Space Administration, H. K. Holmes, H. H. Hubbard, Dr. G. M. Wood; National Research Council, Canada, Dr. N. Olson; New York State Department of Environmental Conservation, Office of Noise Control, Dr. F. G. Haag; New York State Department of Transportation; New York State Public Service Commission, R. D. Vessels; Rensselaer Polytechnic Institute, G. Krycuk, Professor H. E. Rodman; Pratt and Whitney Aircraft; Princeton University, Professor F. A. Geldard, Professor E. G. Wever; Purdue University, Professor M. J. Crocker; RCA Laboratories, Dr. H. F. Olson; Rochester Gas and Electric Corp., C. J. Deinhardt, D. K. Laniack; Schantz Organ Co., J. A. Schantz; Society of Automotive Engineers; Spectral Dynamics Corp.; Standard Oil Company of California, Dr. J. G. Seebold; Stanford Research Institute, Dr. K. D. Kryter; Streck's Inc., F. Streck; U. S. Air Force, Lt. L. S. White; U. S. Army (Watervliet Arsenal) F. Wentink; U. S. Department of Transportation, (Federal Aviation Administration), Dr. J. V. Tobias; U. S. Environmental Protection Agency, C. Caccavari; U. S. Gypsum Co.; Westinghouse Electric Corp., Dr. R. S. Musa; Wyle Laboratories. I am also indebted to hundreds of other authors and acousticians who have developed the material in this book, because collectively, they are its true authors.

I thank also the many journals and publishers who kindly gave permissions to reproduce materials that are referenced throughout the work, to my secretary Evelyn Cook for preparing the typescript, and to the staff of Wiley for the invitation to write this type of book. Finally, I am grateful to my wife, Janet, who has helped me with this and all previous undertakings.

F. A. W.

CONTENTS

Part II   *Noise Control*

Part III   *Speech, Music, and Physio-psychological Effects of Sound*

OUR ACOUSTIC
ENVIRONMENT

I

SOUND AND HEARING

1

OUR SONIC ENVIRONMENT

We live in a world of sound. No other form of energy so pervades every facet of our living. It is the primary medium for communication and without sound the continuous transfer of knowledge of the world around us would almost disappear. Sonic signals in fact, impinge upon our local environment in such an unending stream that we are usually unconscious of most of them. We *hear* only the signals for which we are listening, even though we may be enveloped in a sound field that is produced from dozens of sources. On a city street we may be engaged in a conversation with a friend. Superimposed upon these speech signals may be the noise of traffic, the roar of a jet flying overhead, the sounds of construction, the siren of an emergency vehicle, or the output from a transistor radio. In the relative isolation of a rural environment we are recipients of a quite different variety of sound waves: the myriad sounds of wildlife, rain, wind, waterfalls, and ocean surf.

There are other *characteristic* sounds that we have learned to associate with detailed types of human activity or that reveal the nature of objects. We *recognize* the twang of a bowstring, the crack of a bat, the report of a rifle, the closing of a door, the sound of footsteps, and the pouring of liquid into a glass. In a few instances our unaided ear can even tell us who the person is, when the glass is full, when water is at 212°F, or when a watermelon is ripe. Almost solely through sound we are able to determine when an automobile engine is ready to start; indeed, starting a car would be very difficult without hearing characteristic engine sounds. Through reflected sound a blind man can estimate the size and shape of a cathedral, and with a stethoscope a physician determines the presence or absence of life itself.

SOUNDS IN NATURE

Nature furnishes us with an abundant variety of sound sources and the several media whereby sound is transmitted. Frequently portions of the earth's surface are also covered with one of the most effective and esthetically pleasing sound absorbing materials known to man, that is, a freshly fallen blanket of snow. The loudest sounds in nature usually accompany the eruption of volcanoes, earthquakes, avalanches, and lightning discharges. When the Krakatoa volcano erupted in the Netherland Indies on August 27, 1883, about a cubic mile of rock was hurled into the air and the event was detected on an island in the Indian Ocean about 3000 miles away. On rare occasions meteors traveling at supersonic speeds produce a sonic boom when they enter the earth's atmosphere. More common sounds that add greatly to human experience are the interactions of wind with natural and man-made objects.

The Greeks ascribed certain musical sounds to *Aeolus* (god of the winds): hence the term *aeolian tones,* which we associate with high winds passing over slender objects, such as pine needles, or small branches. In some instances we can hear sounds having a definite musical pitch. The pitch or tone depends on the relative speed between air and object—the higher the speed, the higher the tone. This phenomenon can be demonstrated easily by revolving a thin rope through still air. Such a tone results from eddies that are formed similar to those observed on the surface of water when a rapidly moving object traverses it. The fluttering of a flag also gives evidence of such eddies in an air stream. The flagpole serves as the obstacle, eddies then form on each side, and the flag flutters with a frequency that depends on the rate of eddy formation. It has been found that the frequency of such vortex or eddy formation (Fig. 1), and hence the frequency associated with an aeolian tone is approximately $f = v/5d$ where v is the velocity of the airstream and d is the diameter of the object. For this reason we hear the low-pitched roar of wind through tree branches of large diameter, and the higher-pitched whistling of wind over a grassy meadow or through pine trees.

For example, a wind blowing at 40 mph through branches of one-half inch in diameter will produce a "wind whistle" note whose frequency is close to middle "C" on a piano.

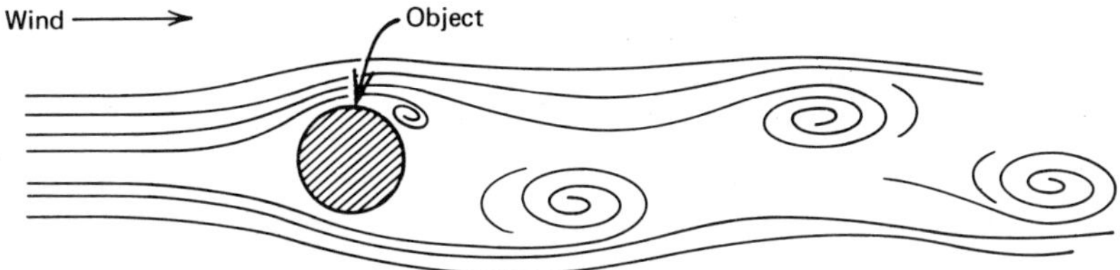

Figure 1   The formation of eddies due to wind moving rapidly over a slender object.

Other interesting sounds in nature include bird calls, whose function is to stake out territorial rights. In a forest that is sparsely populated, there is little problem with the ownership of real estate. When an area is densely populated, however, the male finch actually establishes territorial rights by means of a "claim song" characteristic of his species. The bird perches on the highest tree in his domain and establishes "squatters rights" to the immediate zone within the sound of his call. In this manner he creates an *acoustic fence* that warns would-be trespassers. Other bird and animal species have developed their own specific signals for the imminence of danger, courtship, or the availability of food. A beaver, for example, will warn other beavers to dive by slapping his broad tail on the surface of a pond. Bats even have a highly developed sound-ranging system for navigation and food procurement, but much of their acoustic signal lies in the ultrasonic range.

Humans too respond to some of nature's sounds when made by mosquitoes, hornets, and rattlesnakes. We have even named birds for their distinctive sounds: mourning dove, woodpecker, and killdeer, and we become aware of the changing of the seasons by certain sounds of wildlife.

SOUNDS IN THE OCEAN

The sound of ocean turbulence is dramatic, and this type of sound includes frequencies comprising most of the audio band. Surface waves are generated by a superposition of forces that result from winds, earth movements, and solar and lunar tides. Even a limited observation of the sea surface will show that the wind generates waves that are somewhat trochoidal rather than sinusoidal in shape. A sustained wind will cause waves to build up in height faster than it will increase the distance between adjacent crests. A condition of instability is reached when the height of a wave compared to the distance between successive waves attains a critical ratio of about 1 to 7 (Fig. 2). The wave crest breaks forward in the direction of the wind and "whitecaps" are produced. At appropriate wind velocities this cycle can continue. The whitecap grows again in height with a much greater distance occurring between successive waves. In some parts of the world, high winds can build up such surface waves to crests 60 ft high. Along a coastline, wave formation is more complex, but the sound of surf is usually one for which we have a pleasurable association.

Rather little airborne sound penetrates an ocean surface. Because of the widely different physical characteristics of air and water only about 0.1% of sound energy traverses the air-water interface, when normally incident upon it. Nevertheless, the sea itself is filled with sound. Marine life whose survival depends upon sonic signals and their interpretation abounds. Porpoises, dolphins, and some whales seek their prey, aided by their highly developed sound-sending and echolocation systems.[1] Marine biologists even find that porpoises in different parts of the world emit sounds in different dialects.

Icebergs are not even silent, for as they drift southward from the polar ice cap to warmer waters they give off continuous acoustic emissions. This characteristic noise results from the explosion of tiny air bubbles trapped within the ice, as melting occurs. This sound is to be distinguished from the rather loud cracking that we commonly hear when liquids are poured over ice cubes in a glass. The latter is due to large thermally induced stresses arising from the substantial temperature difference between the ice cubes and the warmer liquid.

The ocean is also a complex medium of sound transmission, and often it is a more favorable medium than air for transmitting weak signals.

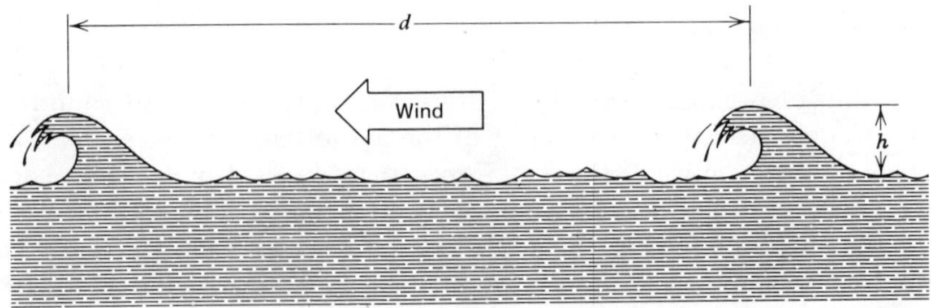

Figure 2   The generation of "whitecaps" resulting from wind on the ocean surface. A condition of instability is reached when the height to wave separation ratio , h/d, is approximately 1 to 7; the resultant sound is broad band noise.

Several factors can be cited. First, for sound that originates within the ocean, transmission losses will be less because of reflections from the air-water interface and the ocean floor. Second, for equal wavelengths, sound absorption in water is some hundred times less than in air. In addition, the sound velocity in the sea is a variable, dependent upon the temperature, salinity, and depth—resulting in unusual transmission paths.

The acoustic model of the Atlantic Ocean, for example, is three layered.[2] The top layer or surface channel is thin (0–400 ft) and the velocity of sound increases slightly as a function of depth. The middle layer (400–4000 ft) is a zone wherein the sound velocity has a large negative gradient. Within the bottom layer (4000–15,000 ft in most of the ocean) the velocity increases with depth, as in the top layer. Such acoustic inhomogeneities and the shape of the ocean floor can provide "sound channels" along which acoustic energy can be propagated great distances. In an experiment performed in 1960, an explosive charge detonated near Australia sent a sound wave along a "channel" to the Cape of Good Hope, then northerly to a listening station in the Bermuda Islands, through a distance of 13,000 miles.

One of the earliest references to this remarkable property of the ocean to transmit sound can be found in a notebook of Leonardo da Vinci. In 1490 he wrote:

> If you cause your ship to stop, and place the head of a long tube in the water and place the outer extremity to your ear, you will hear ships at a great distance from you.

The reflection and transmission of underwater sound has, of course, formed the basis of modern navigational aids. One scheme for obstacle avoidance submitted by an Englishman, L. F. Richardson, reached the British Patent Office on April 20, 1912, just six days after the Titanic's fateful collision with an iceberg in the North Atlantic. Subsequent engineering advances in underwater techniques occurred during World War II when the word SONAR (SOund NAvigation and Ranging) was coined to describe the method for determining the location and nature of submerged objects by reflected sound. Today, oceanographers also use sound as a probe to map the shape of the ocean floor. They have found trenches or chasms having dimensions of 10 to 15 miles wide and one mile deep, which are comparable to those of the Grand Canyon.

COMMUNICATION AND LANGUAGE

Communication in acoustics implies an information exchange resulting from the sending and receiving of meaningful sounds. This definition may be expanded to include sounds that are received from nature or inanimate objects but that convey a definite meaning. Speech is a more restricted and structured form of communication that man has built up from human sounds, that is, the phonetics of vowels and consonants. These sounds of speech are the building blocks for the words that comprise language. Prehistoric men probably communicated with each other by means of a few primitive sounds and visual signals. The sounds were undoubtedly associated with danger, the search for food, and activities having family or tribal significance. The advent of more advanced societies was accompanied by a need for communicating ideas and concepts rather than the signals required for survival.

A language was required to remove the barriers of human isolation, and it enabled groups of individuals to attain common goals. Through language, societies were able to undertake the building of cities and large-scale explorations. This ability to communicate through speech is the most distinctive characteristic of the human race—a faculty that we take for granted. Any child with normal hearing will quickly acquire a language, whereas even the smartest of chimpanzees, brought up with children in human families, never acquires this power. And even the most articulate parrot does not possess the ability to rearrange words and convey a new concept; it simply parrots.

A modern argument is even advanced that we *think rationally only because we possess a language.* This argument is hard to refute. We "think" or "talk to ourselves" in words or groups of words that emerged from a spoken language. Admittedly, contemporary engineers may "think" using the symbols of mathematics or computer language. For the most part, however, we recall the past, live the present, and plan the future within a formal framework of a language of words.

If we adopt our extended meaning of "communication," we note that countless sounds convey highly specific and useful information. Chimes or carillons mark the hour in villages and on college campuses. Firefighters are often summoned to a specific location by means of a predetermined code of horn or siren blasts. We can recognize a motorcycle, automobile, or truck, determine whether the vehicle is approaching or receding, and even estimate its approximate speed—solely by sound. In this sense, the warning qualities of traffic noise have replaced the primeval danger signals of earlier civilizations. We immediately recognize a person by hearing his voice, and through speech or music we are able to sense the entire spectrum of human emotions and feelings.

Sound requires no line-of-sight contact, and unlike our eyes, our ears are never closed. Sound is literally *broadcast,* and this fact alone accounts for the many occasions in which sound is superior to other communication forms. Indeed, to the trained ear, the distinction between the direct and indirect sound waves reaching the ear furnish other data. The presence of obstacles can be detected by reflected sound patterns and binaural hearing provides directional information through our perception of very small differences of sound intensity and sound arrival time.

The fact that sound travels differently in different media has also been useful in communicating. In earlier days, people often put their ears to the steel railroad track so they could obtain an early warning of the approach of a distant train.* Sound travels much faster in steel than in air; more important, the rails provided a sound transmission channel so that the signal was not attenuated as much as for airborne sound.

ACOUSTICS IN ANTIQUITY

The early history of acoustics comes to us through music and architecture.[3] A piece of sculpture (circa 4000 B.C.) in a Bagdad museum shows several musicians playing harps and flutes, and there is evidence that music as an art form was well developed by Egyptians, Hindus, Chinese, and Japanese by about this date. By the time that Egypt was building its great pyramids and sphinxes, it also had organized choruses of 12,000 voices and orchestras of 600 pieces. From engravings on buildings and tombs, on plaques and coins, it can be surmised that music was an integral part of the life of most ancient nations. Musicians were on the staffs of royal courts; they sounded the trumpets for the hunt, summoned armies to battle, and played for all festive occasions. Plato insisted that music was an essential ingredient of education and that it provided a unique means for expressing the inner self.

We are especially indebted to the Greek culture for some of the earliest scientific observations on musical scales and the nature of sound. The word *acoustics,* adopted in the seventeenth-century to apply to the "science of sound," is also derived from the Greek word meaning *hearing.*

Pythagoras, a Greek mathematician, astronomer, and geographer (sixth century B.C.) became interested in determining whether any unique relationship existed between the pitches or "notes" that seemed pleasing to the human ear. His rudimentary apparatus was a monochord consisting

---

* At one early Austrian railroad, dachshunds (called Zugsvormeldehund) were assigned at station and refueling stops. They could detect approaching trains before people could hear them and announced a train's arrival by their loud barking.

of a stretched string, passing over two wooden "bridges," that could be plucked or bowed. For identical strings under equal tension, he found that the lengths of string segments were in a ratio of exactly two to one for notes corresponding to the fundamental, and octave. This octave is the basis of our modern musical scale. It was not until the seventeenth century, however, that a further frequency dependence was discovered for stretched strings in terms of their length, mass, and tension.

Aristotle, the philosophical giant of Greek culture, also said something about acoustics. In discourses titled *Sound and Hearing* he advanced a theory for the propagation of sound in air:

> Sound takes place when bodies strike the air, not by the air having a form impressed upon it, as some think, but by it being moved in a corresponding manner; the air being contracted and expanded and overtaken, and again struck by the impulses of the breath and strings. For when the air falls upon and strikes the air which is next to it, the air is carried forward with an impetus, and that which is contiguous to the first is carried onward; so that the same voice spreads every way as far as the motion of the air takes place.

We are in possession of other interesting writings of this period. Aristoxenus, a pupil of Aristotle, wrote three short works under the caption *Elements of Harmony* and the famed mathematician, Euclid, authored an *Introduction to Harmonics*. Typical of most of the Greek philosophers, these men had little interest in undertaking physical experiments in support of their theories.

From the writings of Marcus Vitruvius Pollio* (circa 20 B.C.) we learn about the design of Greek theaters. Vitruvius was a Roman architect and engineer during the time of Caesar Augustus, and he demonstrated substantial insight into the acoustical properties of buildings. He wrote concerning interference phenomena and echoes, and his reference to a "reinforcement factor" was the antecedent to our term *reverberation,* which plays such an important role in the acoustics of rooms and auditoria. For the most part, however, the acoustic properties of the Greek amphitheaters were arrived at intuitively, in the same sense that the great medieval Gothic cathedrals of Europe were built long before the laws of mechanics were formalized by Galileo and Newton.

The Greek amphitheater, however, exhibited a recognition of the primary acoustic objectives associated with auditoria: to provide an environment (1) free from noise, (2) satisfactory for speech intelligibility, and (3) suitable for retaining the richness of music. Figure 3 shows a typical

---

* Commonly known as Vitruvius.

Figure 3   Photograph of early Greek amphitheater at Epidauros designed to provide an optimum visual and aural link between audience and stage. (Courtesy of R. V. Schoder, S.J., Loyola University of Chicago.)

Greek structure purporting to achieve these goals. The site was selected away from the noise of the city, and the theater made use of the natural topography of the land. Using an appropriate hillside for the seating area, the theaters permitted good visual and aural contact with the stage. A structure in back of the stage, an early form of the modern "shell," was added to the original amphitheater when it was found that the audience could benefit markedly by receiving both direct and reflected sounds.

We should also mention the first "noise ordinance." Julius Caesar is said to have expressed great displeasure at having his sleep disturbed by the noisy chariots traversing the streets of Rome at night. He therefore issued orders to have such activity restricted—at least in the area of the city surrounding his court. There is also some evidence that straw was sometimes laid over city streets to attenuate traffic noise.

EARLY MEASUREMENTS RELATING TO SOUND

It was not until the seventeenth-century that the nature of sound was critically examined with carefully planned experiments.

The development of the vacuum pump permitted definitive tests to be performed, which showed that sound waves could not traverse a vacuum, but required a material medium. Measurements were also made of the velocity of sound in air. One of the first such measurements was made by Marin Mersenne, a Franciscan friar (1588-1648), a contemporary of Galileo and a friend of Descartes. Mersenne also was the author of a comprehensive treatise on sound and music titled *Harmonie Universelle,* and he is often called the "father of acoustics." He used two principal methods to determine the speed of sound in air, circa 1636, in a series of experiments near Paris. The first consisted of measuring the interval between the observed visual flash and the sound detection of distant cannon. Five separate experimental stations were employed, and observations were made from 3000 to 18,000 meters. His second method consisted of the precise timing of echoes that were reflected over known distances. As a result of these experiments and using the best available clocks, Mersenne arrived at a value of 1038 ft/sec for the velocity of airborne sound.

Some two decades later, in 1657, one of the first scientific societies was formed in Florence, the *Accademia del Cimento* (Academy of Experiment). One of its first tasks was to remeasure the velocity of sound. Using Mersenne's first method, but with improved apparatus, the Academy approached the modern value of 1087 ft/sec (331 m/sec) at 0°C. Later methods, of course, utilized electrical timing and about the time of World War I, even more refined sound-ranging measurements were conducted in the Bavarian mountains, using microphones. The temperature dependence of the velocity of sound was measured in one of the first Arctic expeditions.

Newton's contributions were primarily theoretical. In Book II of his famed *Principia* (1687) he provided a derivation for the expression for sound velocity $c = (E/\rho)^{1/2}$ where E is the modulus of elasticity of the propagating medium and ρ is the density. Newton calculated theoretically that airborne sound should travel at about 968 ft/sec, substantially below the experimentally observed value. The reason for this discrepancy was later pointed out by the eminent French mathematician Pierre Laplace (1749-1827) who suggested in 1816 that Newton's derivation assumed isothermal compressions in the propagating medium rather than the adiabatic conditions that actually occur.*

---

* The expression $c = (E/\rho)^{1/2}$ is, of course, correct if E, the bulk modulus of elasticity, is computed from the adiabatic expansion relation $PV^{\gamma} =$ constant, which yields $E = \gamma P$.

Measurements of the velocity of sound in solids and liquids were not made until early in the eighteenth century. An experimental determination of the velocity of sound in iron was made by J. B. Biot, a French physicist, in 1808. He mounted a bell at the end of a long linkage of cast iron pipes (950 meters). When the bell was struck by a hammer, signals were received at the other end, the first signal arriving via the iron, the second through the air. The value obtained for iron in this first test was about 3500 m/sec. The velocity of sound in water was determined by Daniel Calladon, a Swiss physicist who conducted a series of tests in Lake Geneva. A submerged bell was struck and a quantity of gunpowder was simultaneously exploded (above the surface of the water). An observer on the shore was then able to measure the time interval between the flash and first signal traversing the body of water. A value of about 1435 m/sec at 8°C was recorded, which is close to the modern value.

In June 1845 another classic measurement was made by Christian Doppler, a professor who was later appointed the Director of the Physical Institute at Vienna. The "Doppler effect," which explains the apparent change in frequency of an approaching or receding source, was first quantitatively explained by him. He used the Rhine Railroad in Holland, a locomotive plus one car, and secured the services of musicians and musically trained observers. His experiment consisted of two measurements. In the first, the observers, stationed at three points along the railroad track, listened to the sound of trumpets from the car as it sped past them. In a corollary measurement, musicians played their trumpets by the side of the track, while observers in the railroad car were transported rapidly along the route. As described in Chapter 3, the apparent frequency change is directly related to the relative velocity of observer and sound source.

Modern experiments, of course, permit both the velocity of sound and the Doppler effect to be determined for all media within a laboratory, with an exceedingly precise monitoring of time and temperature.

TELEPHONY

In the lobby of one of the world's great research centers, the Bell Telephone Laboratories in Murray Hill, New Jersey, there is the bust of the man who was granted U. S. Patent No. 174,465. The man was Alexander Graham Bell, scientist, businessman, and one-time Professor of Vocal Physiology, School of Oratory, at Boston University. Starting from his basic patent, a communications system emerged that permitted human speech to be transmitted at great distances. This system has changed our civilization. Today, in the United States alone, about 100 billion telephone calls per year provide the audio link for commerce, government, and personal dialogue.

Telephony marked the culmination of man's continuing search for improvements in distant signaling. Semaphores and various flash, flag, and visual signals had been used for centuries. As early as the sixth century B.C., Cyrus the Great of Persia stationed men on hilltops who relayed messages to distant points by shouting; similar "voice towers" were allegedly employed by Julius Caesar in Gaul. But distant audio signaling was just not possible, or at least not practical until the advent of electricity.

In 1831 Joseph Henry, later a professor at Princeton and associated with the Smithsonian Institution, demonstrated electromagnetic signaling using wires; he completed a circuit between the windings of a small electromagnet that activated a bell, and a suitable switch and power source. At will, he was able to produce a succession of audible signals by completing and interrupting his bell system. This invention was the precursor of telegraphy. There was much to be said for a bell or audio system, rather than a visual system, because the receiver's attention could be instantly attracted by the sound, but his attention need not be fixed upon the instrument until a message actually commenced.

Practical telegraph systems were invented nearly simultaneously by Wheatstone and Cooke in England and by Samuel Morse in the United States. It was in October 1837 that Morse filed a caveat in the U.S. Patent Office covering (1) his signaling apparatus, (2) a system of signals or code, and (3) a register for the permanent recording of signals. By 1843, Congress was still somewhat unconvinced of the utility of this device. Nevertheless, the House of Representatives by an 89 to 83 vote, appropriated $30,000 for the construction of a telegraph line between Washington and Baltimore. The project proved successful, and in the following year Morse key-clicked his first long distance message: "What hath God wrought?"

The Washington-Baltimore demonstration was sufficient to convince scientists, government, and business, and a few entrepreneurs began envisioning bigger ventures. Cyrus Field even conceived of a trans-Atlantic link between the United States and England. This gigantic enterprise required massive financing programs in both the United States and Britain, and the involvement of many naval vessels for laying the transoceanic cable. Finally, on August 16, 1858, Queen Victoria and President Buchanan exchanged greetings by telegraph, although reliable service was not available for some years.

Telegraphy showed that messages could be transmitted over vast distances, but for Alexander Graham Bell, a code was a poor substitute for the human voice. His interest in communication arose from many sources, and throughout his lifetime he made many contributions to those whose normal hearing or speech was impaired. It was in the summer of 1874 that he conceived of a speaking phone—a device that could actually transmit characteristic voice sounds. He was aided by a talented instrument maker, T. A. Watson, and on March 7, 1876 he was awarded patent rights. There were many contributors to this new invention, and rival claimants engaged in expensive litigation. A significant contribution was made in 1877 by Edison, who engineered a variable resistance transmitter comprising carbon granules; this patent was sold to the Western Union Telegraph Company and was eventually incorporated into the commercial telephone.

In July, 1877 the Bell Telephone Company was started, and in 1885 the American Telephone and Telegraph Company was formed to install and operate the long lines for an interurban network, which were then limited to distances such as those from New York to Washington and New York to Boston. Transcontinental commercial service from New York to San Francisco was opened in 1915, providing an instantaneous voice link of East to West.

RADIO BROADCASTING

Radio broadcasting enveloped the world with speech and music. Thomas Edison had predicted the transmission of sound "without benefit of wires." It was a young Italian scientist, Guglielmo Marconi, however, who put together a number of electronic developments in applying for a patent in London, on June 2, 1896. The following year Marconi founded the "Wireless" Telegraph and Signal Co., Ltd., and he demonstrated his system to the Italian Navy in the Mediterranean. Shortly thereafter he personally supervised an exchange of messages for the United States in Atlantic waters between the battleships *Massachusetts* and *New York*.

Then, in 1906, on Christmas eve, from a small transmitting station at Brant Rock, Massachusetts, Reginald A. Fessenden, a Canadian engineer, startled operators of several radio-equipped boats of the New England fishing fleet. With amazement they heard not the staccato of the Morse Code, but the actual sounds of speech and music. By the following year he was transmitting speeches to distances of more than 200 miles along the eastern coast. In 1908, Lee deForest, a young Ph.D. graduate from Yale and inventor of the "triode," arranged a voicecast from the Eiffel

Tower in France. Two years later there was an experimental program by the Metropolitan Opera; and in 1916 David Sarnoff, an immigrant from Minsk, Russia, who was later to head the Radio Corporation of America, announced: "I have in mind a plan of development which would make radio a household utility—the idea is to bring music into the house by wireless."

His prophecy was soon to be fulfilled. In 1919, a Westinghouse engineer started some local musical broadcasts, over station 8XK. His occasional programs met with popular approval and a growing audience. On November 2, 1920 station KDKA, with a 100-W transmitter, went on the air on election night to broadcast the results of the voting that showed Warren Harding to be the victor over James Cox (whose vice-presidential running mate was Franklin D. Roosevelt). It has been estimated that about 2000 people heard these first election returns. In 1922 the British Broadcasting Company was created, and by 1923 there were 500 broadcast transmitters in the United States. Further growth in the United States was so rapid that in 1927 the Federal Radio Commission was granted legal powers to control wavelength and regulate power. Sound, through radio, was blanketing the nation.

Additional improvements in sound transmission accompanied the advent of television. In 1939 millions got their first glimpse of TV at the New York World's Fair, and in the summer of 1962, the first live TV pictures were transmitted across the Atlantic via *Telstar*. In 1964, *Syncom* 3 brought TV coverage of the Olympic games from Tokyo to Point Mugu, California, and on July 20, 1969, the world heard the voice of an American astronaut, Neil A. Armstrong speaking from the moon the historic words, "That's one small step for a man, one giant leap for mankind." From a technological point of view, the problem of voice transmission was substantially less difficult than the engineering of a complex guidance system. Nevertheless, at a distance of a quarter of a million miles, the most intimate link between man and the ground communication center at Houston, Texas, was not via a digital computer, but through the actual sound of a human voice.

Today, after only slightly more than a half century of commercial broadcasting, a signal can be beamed, via satellites, over the entire world.

SOUNDS OF MUSIC

The continuing popularity of music can be attributed to the fact that music is a simple and economical vehicle for effecting an instant change in the environment. Man seems to possess an insatiable appetite for continuing change, and music can offer this change even if all other environmental factors are invariant.

The availability of a new and "instant environment" through recorded music is indeed one of the historic hallmarks of the twentieth century. The environments of previous cultures were amenable to change through the media of sculpture, art, architecture, and music, but the possibility of "storing" musical sounds for subsequent retrieval came a full century after the recording of visual environments had been achieved through photography.

From 1500 to 1800 substantial efforts were devoted to the design and refinement of actual musical instruments. Great organs were installed in the cathedrals of Europe, many of which are still extant. Stringed, brass, and woodwind instruments evolved that were prototypes of virtually all the classical instruments of the present era. King Henry VIII of England owned some 400 musical instruments of which 150 belonged to the flute family. It is said that he himself played the recorder, and that he possessed 76 of them! The violin appeared in its present form about the middle of the sixteenth century, and by 1571 there were several violins in the royal band of Queen Elizabeth. Perfection of craftsmanship of the violin reached its zenith at the hands of Antonio Stradivarius (1644-1737) in Cremona, Italy.

The formal development of music is usually divided into three periods. The first period was the homophonic or unison music of the ancients up to the eleventh century. Music of this form still finds expression in contemporary music of the Orient and Asiatic nations. Polyphonic music, with several parts, but without any independent musical significance of the several voices, extended from the tenth to the seventeenth century. Modern polyphonic music in which many melodic lines were superposed began in the baroque period (1600-1750) and has continued to the present in various forms and styles. Bach and Handel were the giants of the baroque era and they were born in the same year, 1685. During this period music developed into a highly structured harmonic language. Bach consolidated the tempered scale, the four-part harmony, and produced music of complexity, decoration, and movement. Handel, in a musical feat without parallel, composed the complete oratorio, *Messiah*, in 28 days. Both Bach and Handel were organists and composers, and their work served as the foundation for European classical music.

It is not fortuitous that the advent of great church music accompanied the construction of the cathedrals, and the formation of great orchestras accompanied the building of great halls. The opera house, La Scala, in

Milan, Italy, opened its doors in 1778 to begin its tradition of glamour and artistry. About a century later, in 1871, the Royal Albert Hall with a capacity of 5080 was opened in London, and in 1883 the Metropolitan Opera House of New York presented its opening performance of *Faust*.

Toward the close of the nineteenth century Thomas A. Edison filed two patents that were destined to play a major role in providing new markets for acoustics. One was his invention of the phonograph (U.S. Patent No. 200,524, December 1877); another was his motion picture machine (U.S. Patent No. 589,168, August 1891). In 1899, the first magnetic recorder was invented by Valdemar Poulsen in Copenhagen, Denmark.

New sounds of music and entertainment in the twentieth century were made possible by many developments in electrical engineering. The first sound film ever shown was offered to a New York audience on August 6, 1926. This was *Don Juan,* which had music but no recorded speech, and on October 6, 1927 the first full-length film with a complete sound track *The Jazz Singer,* starring Al Jolson, was premiered. This was about the same time that radios became a fixture in every household. In the next decade the radio became one of the most sought after options of automobile owners. Shortly thereafter, the electronic organ joined the piano in many homes and public places, and sound amplification systems came into common use as stadiums were built with capacities of up to 100,000.

A vast market for recorded music built up in the 1930s, which escalated even further with the LP or $33\frac{1}{3}$ rpm microgroove records introduced commercially by *Columbia* in 1948. This was the same year that magnetic tape began to replace acetate discs for making original recordings. Ten years later another major era began with the general availability of stereo records. Within the last decade the convenience of "cassette" recordings has provided a new "portable environment." Via tape one can reproduce virtually any program of recorded music in an automobile, or in the most remote location. Furthermore, the sounds of music are piped into factories, dental offices, commercial buildings, elevators, and the cabins of jet aircraft.

INFRASONIC TO ULTRASONIC

While this book is concerned primarily with frequencies within the range of human hearing, some mention should be made of the other portions of the acoustic spectrum. Human hearing encompasses the frequency range of about 15 to 20,000 cycles/sec. Frequencies below 15 cps are termed *infrasonic* or subsonic frequencies; frequencies above 20,000 cycles/sec are called *ultrasonic*. The limits cannot be precise because the range of audible frequencies will vary with individuals. For many people, the ultrasonic will begin at about 16,000 cycles/sec. Of what use are these portions of the acoustic spectrum that the ear cannot hear?

Below 15 cycles/sec we can still sense by feeling. The sensation of feeling is complementary to hearing, and without it we would not experience the same environment. Many people are unaware that they both hear and feel the output from a symphony orchestra or a pipe organ. The sensation of feeling is derived from the complex exchange of airborne energy to the actual structure of an auditorium, and the transmission of vibrational energy to the listener is an important part of the total emotional impact. Subsonic energy also comes to us directly or indirectly from many sources: for example, thunder, vibrations from trains and heavy vehicles, sonic booms, and machinery in offices or factories.

Infrasonic acoustics is also important in mechanical engineering, and vibrational analysis is a powerful diagnostic tool for inspecting gears, bearings, turbines, and complex structures. The field of medicine also employs infrasound for diagnosis and therapy.

The application of ultrasonics covers many fields.[4] We have already mentioned SONAR, which uses ultrasonic frequencies for underwater sound ranging. During World War II ultrasonics began to be used for detecting flaws in metals, in addition to its continuing use for locating submarines. Today there are few areas of science or technology that do not benefit from the application of this tool. There are several reasons why high frequency sound is so important. First, the frequency range is extended to many megacycles (in liquids and solids). These higher frequencies have a shorter wavelength; thus ultrasound can be focused. Second, the sound power that can be generated at very high frequencies is much greater than for the sonic range. A small radio may produce an output of 10^{-8} watt/cm^2, yet it is common to achieve power levels of at least 1 watt/cm^2 in many ultrasonic applications. Third, ultrasonic power can be applied in very short pulses, making it easy to monitor sound transients and other phenomena.

Industrial uses of ultrasonic radiations include ultrasonic cleaning, drying, soldering, drilling, electroplating, and metal or plastic welding. Ultrasonics can also locate flaws in metals at depths much greater than X rays, and it has been used extensively as a thickness gauge. One could even say that it is a tool for protecting the environment because ultrasonics has ingeniously been applied as a burglar alarm. In agriculture, ultrasonics has been used experimentally to accelerate the growth of plant life, and ultrasonic energy is being considered, in very specialized situations, as a potential substitute for chemical pesticides. In chemical engineering it is employed for homogenization, emulsification, and the degassing of liquids; ultrasonic vibrations have even aided combustion processes because of its stirring and atomizing action. In some instances the molecular structure of organic liquids can be deduced. Exploratory applications to biology and medicine include localized internal heating, the destruction of bacteria, eye research, cardiology, obstetrics, and cancer therapy. For many years, of course, ultrasonics has been a useful tool in dentistry.

Within the last two decades a new technology has emerged related to "acoustic emissions" or the "low-level sounds" emitted by a material when it undergoes deformation. In some instances, the acoustic signal accompanying strain and deformation is strong enough to be audible to the unaided ear. Familiar examples include the cracking sound of timber when subjected to forces resulting in fracture, and the audible sounds produced by the breakup of rock formations. With new sophisticated instrumentation that is able to detect "inaudible" sounds over a wide frequency range, acoustic emissions now furnish information about the microscopic behavior of materials, and they provide a new means for monitoring pressure vessels, pipelines, airplanes, buildings, and bridges.

REFERENCES

1. K. S. Norris and G. W. Harvey, *J. Acoust. Soc. Am.,* **56,** 569 (1974).
2. T. Arase, *J. Acoust. Soc. Am.,* **31,** 588 (1959).
3. For an interesting historical account of early developments in acoustics, including translations of original works, see *Acoustics, Historical and Philosophical Development,* edited by R. B. Lindsay, Dowden, Hutchinson and Ross, Stroudsburg, Pa., 1973.
4. F. A. White, *American Industrial Research Laboratories,* Public Affairs Press, Washington, D.C., 1961, Chapter 10.

2

SOUND CHARACTERIZATION

SOUND SOURCES

The existence of sound implies a source of sonic energy and one or more media through which this energy is transmitted. In some instances we may possess a substantial amount of information about the source itself, in which case we can predict the nature of the emitted sound wave. In other instances we deduce the nature of the source by an analysis of the perceived wave motion. If we tried to enumerate all possible types of sound generators, we would find that most of them could be grouped into the following categories:

1. Vibrating solid bodies, for example, tuning forks, transformers, loud speaker diaphragms, stringed or percussion instruments, and sounds arising from complex components or structures.

2. Vibrating air columns, as exemplified by resonating organ pipes and wind instruments.

3. Transient forms of mechanical or electrical power, such as volcanoes, avalanches, and lightning discharges.

4. Impact phenomena, for example, tapping and hammering.

5. Sounds from rapidly expanding gases, such as from jets, rockets, and chemical explosions.

6. Complex sonic disturbances resulting from rapidly moving objects in fluids, such as fans, propellers, and shells.

For vibrating solid objects, the amount of acoustic energy that reaches the observer is often dependent upon the "coupling" of the object to its immediate environment. For example, the strings of a piano transmit an appreciable amount of energy only when connected to an appropriate sounding board, and the radiating surfaces of motors and machines generate appreciable noise via the media of their supporting structures. These secondary sound sources thus play an important role in sound amplification. Conversely, if such secondary sound sources are suitably damped or constrained from vibrating, the intensity of many unwanted sounds can be substantially reduced. Later we note that the sound pressure produced is proportional to the product of the amplitude and frequency of oscillation, that is, to the velocity of the vibratory surface.

SOUND WAVES

Sound waves result from the local atmospheric disturbance generated by the sound sources that we have mentioned above. When a pebble is dropped into a pond the disturbance of the surface of the water is clearly visible, and waves can be seen radiating in all directions from the point of impact. These waves, however, are said to be *transverse* because the water molecules comprising the medium undergo displacements that are perpendicular to the direction of wave propagation. In the case of a sound wave in air, energy is propagated by air molecules that are displaced in the same direction as the advancing wave front, and the sound wave is said to be *longitudinal*. Of course, sound waves can also be transmitted in liquids and solids, but we shall give attention initially to the manner by which sound energy is transmitted in air.

At normal atmospheric pressures the high molecular density of air allows us to classify air as an *elastic* material. By this we mean that a coupling exists between neighboring molecules. If a portion of the medium is displaced, the displacement will not be transferred instantaneously to distant parts of the medium. Rather, the original displacement will generate an elastic force in the ensemble of molecules adjacent to it; these molecules will then act on other molecules, and so on. The displacement will thus be propagated through the medium with a finite speed. As we shall see shortly, this speed will depend both upon the elasticity and the density of the air.

Figure 1 is a representation of an ensemble of molecules undergoing vibration because of excitation from the source. Molecules immediately adjacent to the source receive kinetic energy and impart some of their momentum to neighboring molecules. Figure 2 suggests the motion of a single air molecule. A restoring force will cause the molecule to return to

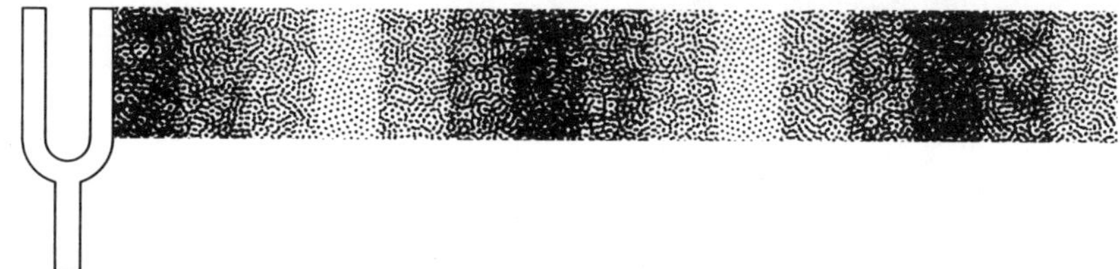

Figure 1   Instantaneous molecular density pattern associated with a plane wave.

its initial position after being displaced by acoustic excitation. Because of inertia it will also overshoot its equilibrium position, in a manner similar to the motion of a pendulum. Furthermore, because air is an elastic medium, if a loudspeaker vibrates with simple harmonic motion, the many air molecules that surround it will also undergo simple harmonic vibrations, resulting in the distributed pattern of compressions and rarefactions. Thus, in the propagation of a sound wave, the molecules do not travel great distances; they simply oscillate back and forth about a mean position and transmit their energy to distant molecules by successive collisions.

The macroscopic phenomenon that permits us to detect a sound wave is the *pressure* generated by the ensemble of molecules that are set in motion by the source. The pressure will also vary with the same frequency as the source. Furthermore, the magnitude of this sonic or dynamic pressure will always be small in comparison with the ambient or "static" pressure of the atmosphere. Normal atmospheric pressure is the weight per unit area of the column of air that extends from sea level to the top of the earth's atmosphere (Fig. 3).

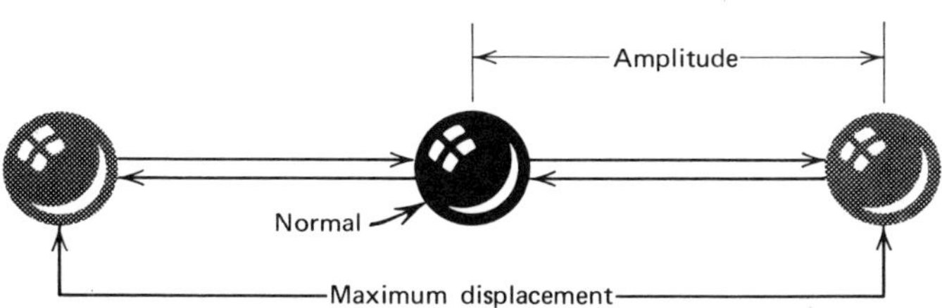

Figure 2   Displacement of an air molecule from its normal "rest" position because of acoustic excitation. The molecular motion is regular and repetitive if the sound source undergoes periodic vibrations.

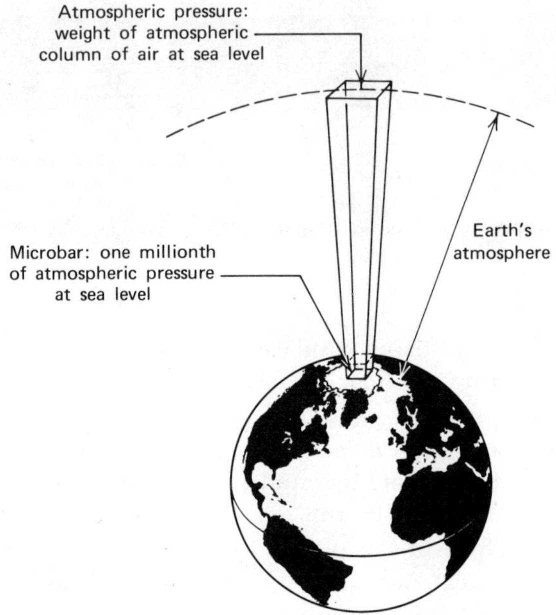

Figure 3   A microbar is a millionth of the barometric pressure at sea level, and is equal to 1 dyne/cm².

The units of pressure that are employed by meteorologists serve as the basis for other units that will be introduced later. Our ambient sea level pressure ranges from 980 to 1040 millibars,

where

$$1 \text{ bar}^* = 10^6 \text{ dynes/cm}^2$$

$$1 \text{ millibar (mb)} = 10^3 \text{ dynes/cm}^2$$

$$1 \text{ microbar } (\mu b) = 1 \text{ dyne/cm}^2$$

Figure 4*a* (a barogram) shows the time-dependent changes of the atmospheric pressure, but these changes occur at such a slow rate that they can never be sensed by the human ear. Figure 4*b* is the pressure variation of a 500 cycle/sec audible tone. The atmospheric barogram indicates pressure changes that are many times greater than that of the typical acoustic signal, and the time scale is very different. Only if pres-

* A bar is equivalent to a pressure of 0.9869 atmosphere.

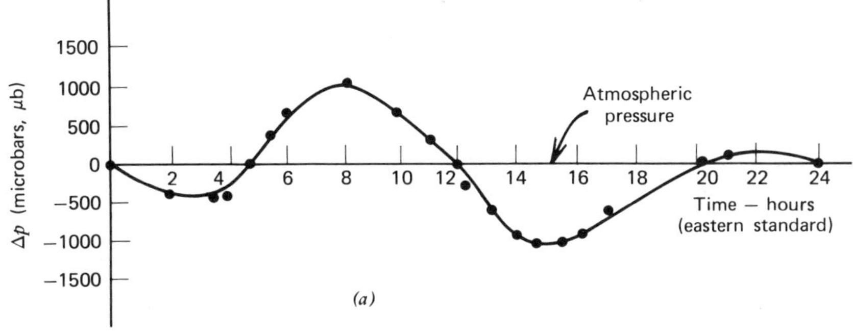

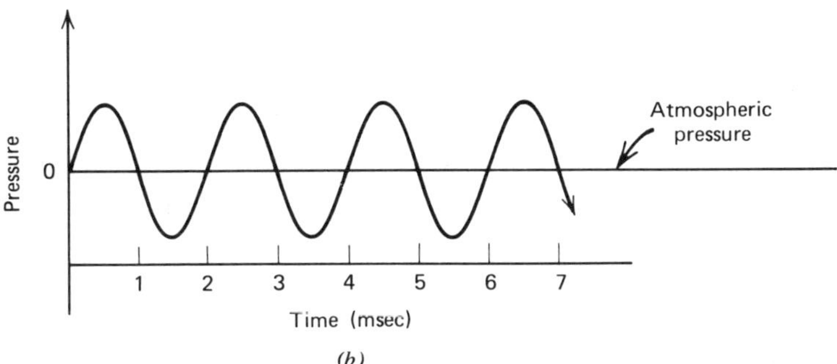

Figure 4*a*   Meterologic barogram showing a time-dependent variation from "normal" atmospheric pressure. The graph shows the average diurnal atmospheric pressure variations near Portsmouth, N. H. for May; 4*b* Pressure variation is due to acoustic excitation of a 500 cycle/sec audible tone.

sure changes occur at a rate of 15 to 20,000 cycles per second can such changes be sensed as an audio signal. If, indeed, atmospheric changes occurred very rapidly, we would not only hear those changes, but the sounds would be very loud.

Based upon the discussion above we can adopt a practical definition for a sound wave in air: *a sound wave is a modulation of the atmospheric pressure that is discernible to the human ear.*

A purely physical definition would be more general: *sound is an alternation in pressure, stress, particle displacement, or particle velocity, which is propagated in an elastic material, or the superposition of such propagated vibrations.*[1] Based upon this latter definition signals outside the range of human hearing would be included. The definition is also rather broad in specifying the type of physical disturbance that is being propagated. The sound may constitute a single pulse or transient, an irregular though repetitive modulation, or a simple periodic change.

FREQUENCY AND WAVELENGTH

The number of complete oscillations that a sound source undergoes per second is termed the frequency, f. The sound wave emitted from the source will propagate with this same frequency. For example, the vibration rate of an "A" string of a violin is 440 cycles/sec; this is the same frequency that will be detected via the transmitted sound wave by a listener. The common nomenclature for frequency is in Hertz* units (1 Hertz = 1 cycle/sec).

For any source executing repetitive motion, the emitted sound wave is also characterized by a wavelength, λ. Figure 5 indicates the wavelength for sound that is being radiated from a point source.

The wavelength λ can be defined as the distance that sound travels during each complete vibration or cycle of the sound source. We could also say that the wavelength is simply the distance between successive *compressions* or *rarefactions*. A compression (or condensation) is the region in which the molecular density and pressure is slightly greater than the ambient; a rarefaction is the region characterized by a reduction in molecular density and pressure relative to normal atmospheric pressure.

The frequency and wavelength of a sound wave are related by the simple formula:

$$c = f\lambda \tag{1}$$

where

$$c = \text{velocity of propagation, in cm/sec}$$
$$f = \text{frequency, in Hz (cycles/sec)}$$
$$\lambda = \text{wavelength, in centimeters}$$

The units of velocity and wavelength in this equation can also be expressed in meters/sec and meters respectively, or in other consistent units.

In air at 20°C the speed of sound is approximately 344 m/sec (1130 ft/sec). Thus, for the range of audio frequencies from 20 to 20,000 Hz,

---

* This unit of frequency was adopted in 1965 to honor the German physicist, Heinrich Hertz (1857-1894). One cycle per second is abbreviated as 1 Hz; 1 kilocycle/sec = 1 kHz.

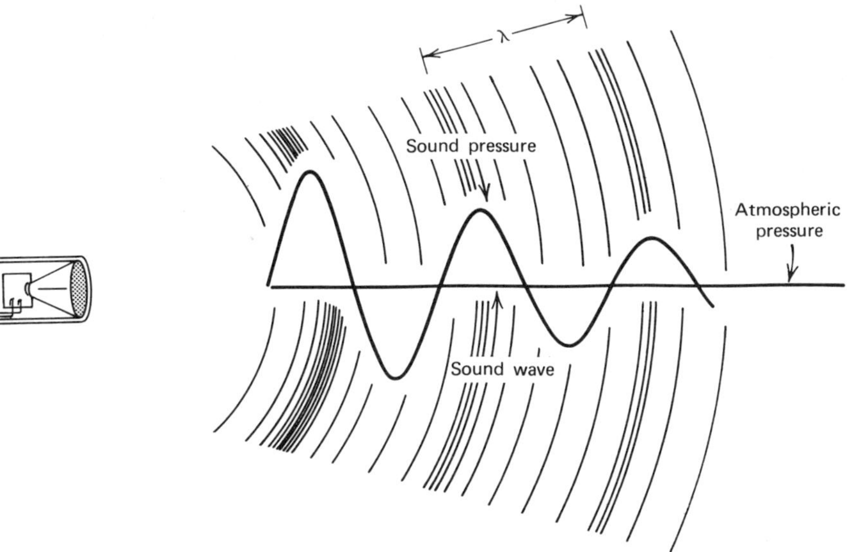

Figure 5　Pattern of compressions and rarefactions surrounding a point source. The distance between successive compressions and rarefactions characterizes the wavelength, λ.

the associated wavelengths will extend from about 17 m (56 ft) to 1.7 cm (0.7 in.). A short wavelength limit in air (in the ultrasonic) is determined by the mean free path between air molecules ($\sim 10^{-5}$ cm); there is no clearly defined long wavelength limit for infrasound.

In principle, it is possible to measure either the frequency or wavelength of sound. In practice, it is substantially more convenient to detect frequency by electronic apparatus. A direct measurement of the wavelength of sound is possible only in a laboratory situation.

Accurately known reference frequencies are often important for acoustics and other fields. Accordingly, audio frequencies of 60, 440, and 1000 Hz are broadcast from the National Bureau of Standards station WWV in Washington, D. C., at hourly intervals.

VELOCITY OF SOUND

The velocity of sound in a given medium is independent of frequency, as implied by equation 1. We are accustomed to taking this frequency independence of velocity for granted. Imagine the musical chaos that would exist if we listened to a symphony orchestra and the bass and treble tones arrived at our ears at different time intervals. We would also be in some difficulty if we did not hear simultaneously all frequencies of speech from a single voice.

However, sound is propagated in different media at different speeds, and a small temperature dependence is one of the prime causes for the bending of sound waves in the atmosphere. The theoretical expression for the speed of sound, c, in an ideal gas is

$$c = \sqrt{\frac{\gamma P}{\rho}} \tag{2}$$

where P is the ambient pressure, ρ is the gas density, and γ is the ratio of the specific heat of the gas at constant pressure to its specific heat at constant volume.

The thermodynamic basis for the relationship is the perfect gas law, which states that $PV = $ constant for an *isothermal* expansion or compression, and $PV^\gamma = $ constant for the *adiabatic* process, accompanying a sound wave. By an adiabatic process we mean that the condensations and rarefactions in the atmosphere take place so fast that the gas molecules actually experience a small change in kinetic energy or temperature. The dimensionless parameter γ (ratio of specific heats) is dependent upon the so-called degrees of freedom of the gaseous molecule. The number of degrees of freedom depends upon the complexity of the molecule. For monatomic molecules γ is 1.66, for diatomic molecules it is 1.40, and for triatomic molecules it is 1.29. Because air is composed primarily of diatomic molecules, the velocity of sound in air is

$$c = \sqrt{\frac{1.4P}{\rho}} \tag{3}$$

If we substitute the values of P and ρ for air under standard conditions at 0°C, we have

$$c = \sqrt{\frac{(1.4)\,(1.013 \times 10^6 \text{ dynes/cm}^2)}{1.293 \times 10^{-3} \text{ g/cm}^3}\left(\frac{1 \text{ g cm/sec}^2}{\text{dyne}}\right)} = 3.3 \times 10^4 \text{ cm/sec}$$

a value that is consistent with experimental determinations.

In view of the equation of state of air of an ideal gas ($PV = RT$) and the definition of density ρ, equation 2 may be written as

$$c = \sqrt{\frac{\gamma R T}{M}} \tag{4}$$

where R is the gas constant, T is the absolute temperature, and M is the molecular weight of the gas. From this equation the speed of sound in an *ideal* gas depends only on the type of gas and the temperature, and it is independent of changes in pressure. Thus the speed of sound at the top of a mountain should be the same as that at the foot of the mountain if the temperature is the same at both elevations. In practice this is almost true, for in equation 2, P and ρ both decrease with increasing altitude above sea level, and the net effect is that atmospheric pressure has only a slight effect upon sound velocity.

The accepted value[2] of the speed of sound in air is

$$c = 33,145 \pm 5 \text{ cm/sec}$$
$$c = 1087.42 \pm 0.16 \text{ ft/sec} \tag{5}$$

under the conditions of (a) audible frequency range (b) temperature = $0°C$, (c) 1 atmosphere pressure, (d) 0.03 mole % of CO_2 and (e) 0% water content.

To compute the speed of sound, c, at ordinary temperatures, the following relationships may be used

$$c = 33,145 + 60.7t \text{ cm/sec}$$
$$c = 1087.42 + 1.99t \text{ ft/sec}$$
$$c = 1052.03 + 1.106F \text{ ft/sec} \tag{6}$$

where t is the temperature in degrees centigrade and F is the temperature in degrees Fahrenheit. For temperatures much different than $20°C$, a more accurate calculation can be made using

$$c = 33,145 \sqrt{\frac{T}{273}} \text{ cm/sec} \tag{7}$$

where T is the temperature in degrees Kelvin. The speed of sound is seen to increase as the square root of the absolute temperature.

The velocity of sound in several other gases at $0°C$ is given in Table 1.

MOLECULAR VELOCITY AND DISPLACEMENT

In the preceding section we considered the velocity of sound waves, that is, the macroscopic propagation of sound. What connection is there between this wave velocity and the velocities of the individual air molecules undergoing compression and rarefaction?

Table 1   Velocity of Sound in Common Gases[3]

Gas	Formula	Velocity	
		m/sec	ft/sec
Air (dry, 0°C)		331.45	1087.42
Argon	A	319	1047
Carbon monoxide	CO	338	1109
Carbon dioxide	CO_2	259	850
Helium	He	965	3166
Hydrogen	H_2	1284	4213
Methane	CH_4	430	1411
Neon	Ne	435	1427
Nitrogen	N_2	334	1096
Oxygen	O_2	316	1037
Steam (134°C)	H_2O	494	1621

First, we should recall from kinetic theory that all air molecules are undergoing a random motion and the "average" velocity, u, associated with this random motion is given by

$$u = \sqrt{\frac{3kT}{M}} \tag{8}$$

where k is the Boltzmann constant, T is the temperature in degrees Kelvin, and M is the mass of the molecule. A simple calculation shows that at $T = 300°$, an oxygen molecule will have a velocity of 4.83×10^4 cm/sec.

Under usual conditions most molecules that constitute air will have velocities of the order of 10^4 to 10^5 cm/sec. In other words, the random velocities of molecules in the absence of sound are comparable to the velocity of a sound wave.

Now consider an ensemble of such molecules in front of a radio loud-speaker. For fairly loud sounds, the speaker cone may be oscillating with a mean velocity of about 1 cm/sec. Considering only one degree of motion, we must assume that additional momentum is imparted to the ensemble of molecules in front of the speaker, but that this incremental momentum will be small compared to the momentum possessed by the air molecules before the speaker is turned on.

This statement is at least consistent with the fact that the speed of sound propagation is essentially independent of the intensity of the sound wave. What parameters, then, affect the individual molecular velocities, that is, the additional velocity component in the direction of sound propagation?

The parameters can be identified by considering a "pure tone" or single frequency propagated in the x direction only, with the speaker cone undergoing simple harmonic motion

$$x = A \sin 2\pi f t \tag{9}$$

where A = maximum displacement or amplitude, f = frequency in Hz, t = time in seconds from equilibrium position, ($t = 0$ when $x = 0$). We assume that the air molecules undergo the same displacement as the speaker cone. Thus a molecule will attain an incremental velocity in the x direction of

$$\frac{dx}{dt} = 2\pi f A \cos 2\pi f t \tag{10}$$

From equation 10 we see that the particle velocity will be dependent upon both the frequency and the maximum amplitude of oscillation of the sound source (and thus the intensity of the sound or the driving force of the speaker cone). If the amplitude is constant, it follows that the particle velocity will be directly related to frequency. If the frequency is constant, the molecular or particle velocity will depend upon the power output of the speaker. The instantaneous velocity, of course, will depend upon the phase of the periodic vibration. At maximum displacement, the kinetic energy is minimum and the potential energy of the particle is a maximum. The maximum velocity will correspond to values when $\cos 2\pi f t$ is unity, and the potential energy of the particle is zero.

For the frequency range and sound intensities involved in speech and music, the maximum molecular velocities will usually be less than 1 cm/sec.

The relation between particle velocity u and the *rms** sound pressure (see subsequent section) is based upon the conservation of momentum principle that expresses the motion of a particle in terms of the forces that act upon it. For a free-traveling plane wave it can be shown that the *rms* particle velocity is[4]

$$u = \frac{p}{\rho c} \tag{11}$$

where p is the effective or *rms* sound pressure above the ambient, ρ is the density of air, and c the velocity of sound in air. We see that a con-

---

* Root mean square.

stant ratio exists between sound pressure and particle velocity. At normal temperature and pressure this particle velocity is approximately[5]

$$u = \frac{p}{40.6} \text{ cm/sec} \tag{12}$$

when p is expressed in microbars.

The exceedingly low particle velocity, compared to the velocity of sound can be estimated by substituting values corresponding to very intense sounds (200 μb or 120 dB) and the very low pressure corresponding to the threshold of human hearing (0.0002 μb or *zero* dB).* The former pressure yields a particle velocity of 4.9 cm/sec; the latter yields a particle velocity of 4.9×10^{-6} cm/sec.

The *rms* amplitude or *displacement* of an air molecule can also be calculated in terms of the *rms* pressure generated by a sound wave. The displacement, D, (0.707 of the maximum amplitude) is given by[6]

$$D = \frac{p}{2\pi f \rho c} = \frac{3.9 \times 10^{-3} p}{f} \text{ cm} \tag{13}$$

where p is the *rms* sound pressure expressed in microbars, f is the frequency (Hz), ρ is the air density, and c is the velocity of sound in air. For a 20 Hz tone at 2000 μb (140 dB), this displacement will be

$$D = \frac{3.9 \times 10^{-3} \times 2000}{20} = 0.39 \text{ cm}$$

The molecular displacement corresponding to the sound pressure at the threshold of human hearing (about 2×10^{-4} microbar at roughly 3000 Hz) is

$$D = \frac{3.9 \times 10^{-3} \times 2 \times 10^{-4}}{3000} = 2.6 \times 10^{-10} \text{ cm}$$

This displacement is substantially less than the "radius" of a hydrogen molecule. In normal conversational speech at a distance of about 10 feet from a speaker, the molecular displacement is of the order of one-millionth of an inch.

Qualitatively, the small displacement undergone by molecules at intermediate intensities can be verified by simply placing a finger on the vibrating cone of a loudspeaker. The power output of the speaker can be decreased until there is no tactile sensation of vibration (at high frequencies), yet the sound will be clearly audible.

---

* The decibel, dB, is formally defined in a subsequent section.

VELOCITY OF SOUND IN COMMON MATERIALS

The velocity of sound waves in common materials ranges from 3000 to 6000 m/sec, but in especially hard metals the velocity can be as high as 12,000 m/sec.

The formula for the velocity of sound in solids is similar to equation 2:

$$c = \sqrt{\frac{E}{\rho}} \tag{14}$$

where E, the elasticity of the medium replaces γP. In solids, very dense materials often have much higher moduli of elasticity than less dense media. For this reason, the velocity of sound propagation can be much greater than in air. Table 2 includes the sound velocity of some common materials. Elasticity, E, is given in dynes/cm^2; density, ρ is in g/cm^3, and c is in m/sec.

Table 2   Velocity of Sound in Various Solids[a,7]

Solid	$E (\times 10^{11})$ (dynes/cm^2)	ρ (g/cm^3)	$c = \sqrt{E/\rho}$ (m/sec)
Aluminum	7.00	2.70	5100
Copper	12.45	8.90	3740
Iron (cast)	9.1	7.65	3440
Lead (rolled)	1.6	11.35	1200
Steel	20.0	7.83	5060
Wood:			
Fir	0.88	0.45	4430
Oak	1.3	0.80	4050
Brick	2.67	2.0	3650
Glass (average)	—	2.4 to 5.9	5500

[a] In the form of bars, rods, or columns.

SOUND INTENSITY

Sound intensity is defined as the average rate of flow of sound energy per unit area normal to the direction of propagation. For a one-dimensional or *plane wave,* the intensity in ergs/sec per square centimeter is[8]

$$I = \frac{p^2}{\rho c} = pu = u^2 \rho c \tag{15}$$

where p is the root-mean-square pressure (defined in the next section) in dynes/cm^2, u is the molecular velocity in cm/sec, c is the velocity of wave propagation in cm/sec, and ρ is the density of the medium in g/cm^3. The product, ρc is termed the characteristic *acoustic impedance* of the medium. For a spherical wave*

$$I \text{ (at radius } r) = \frac{p^2}{\rho c} \text{ (at } r) = \frac{W}{4\pi r^2} \tag{15a}$$

where W is the total acoustic power radiated by the source of sound.†

From equation 15 we see that intensity is the product of the sound pressure and molecular velocity measured in the direction of the wave. We should also note that the equation has the same form as the equation for electrical power $(W = Ei = E^2/R)$ with pressure replacing electrical potential, E, and ρc corresponding to electrical resistance, R.

The sound intensity or the flow of acoustic power from ordinary sound sources is exceedingly low. For example, it would take about 100 years to heat a cup of coffee from the energy of conversational speech, assuming perfect insulation, no heat losses, and providing the cup was held in close proximity to the speaker's mouth. Furthermore, the direct measurement of such low intensities or energy flow is impractical. Our choice is to measure molecular velocity or pressure, and the latter is the only parameter that can be determined with simple instrumentation.

SOUND PRESSURE

Because sound consists of alternate positive and negative pressures (compressions and rarefactions) with respect to the steady state, the mean value of a sound-pressure disturbance will be zero. At least this is true in most practical situations. Thus it is useful to adopt a system of pressure measurement that permits the negative pressures to be added to, rather than subtracted from the positive pressure values. The convention that is employed is identical to that used in measuring alternating currents in electrical circuits. A *root-mean-square (rms)* sound pressure is obtained by a specific procedure. First, the values of the sound pressure, at each

---

* A spherical wave will be generated by a balloonlike surface, pulsating uniformly about some equilibrium radius.

† In the *mks* system intensity is expressed in watts/m^2. The diverse fields of applied acoustics use *mks, cgs,* and *English* units. Throughout the book units will usually be used that are consistent with the references cited.

instant of time, are squared. The squared values are then added and averaged over the time that the sound pressure is monitored. Finally, the square root is taken of this average.

In this procedure the squaring operation converts all negative pressures to positive squared magnitudes, so that all sound disturbances have nonzero values. The *rms* value is also termed the *effective* sound pressure.

For the specific case of a pure sinusoidal traveling sound wave,

$$p_{rms} = \frac{p_{max}}{\sqrt{2}} \tag{16}$$

Hence, the *rms* value of a sinusoidal wave equals $1/\sqrt{2}$ or 0.707 of its amplitude. In complex waves generated from either single or multiple sources, the maximum or peak value may vary considerably, depending on phase relations, but the *rms* pressure is relatively stable. Essentially *rms* measurements can accurately reflect changes in sound pressure whether the sound is sinusoidal and from a single source, complex and from many sources as in music, or random as in noise. This *rms* pressure or effective pressure is the basis for all microphone calibrations and sound level meter measurements.

THE DECIBEL SCALE

Because both the human ear and our acoustical instruments respond to changes in pressure, it is desirable to have a scale that will measure these changes in convenient notation. Since the ear can detect changes in pressure amplitude of 10^6 to 1, the scale should encompass a large dynamic range. Furthermore, it would be desirable if the physically measurable scale of pressure increments corresponded to our auditory perception of increasing "loudness." The decibel scale satisfies both of these requirements, being formulated from observations and experiments in both physiology and electrical engineering.

The relationship between a physical stimulus and a corresponding physiological sensation was first studied by E. H. Weber (1825) in connection with human estimates of the weights of various objects.[9] A person was blindfolded and asked to respond when various weights were placed in his hand. It was found that the person could perceive differences in very small weights, but if his hand already contained an appreciable load, a large weight had to be added if the additional weight was to be detected.

In this and subsequent studies (regardless of the particular physical stimulus and sensation) it was found that *the increase in stimulus necessary to produce a given increase in sensation is proportional to the pre-existing stimulus*. This can be expressed mathematically as

$$ds = k\, \frac{d\omega}{\omega} \tag{17}$$

where ds is the minimum perceptible increase in sensation, $d\omega$ is the weight producing it, and ω is the total weight originally present. An integration of equation 17 is now known as the Weber-Fechner law:

$$s = k \log \omega \tag{18}$$

The above two expressions suggest why the ear is able to respond to both small and exceedingly large pressure differentials, and they justify the use of a rather arbitrary logarithmic scale.

The *decibel* is used in the following sense. The difference in *decibels* between two quantities P_2 and P_1 is *defined* to be

$$10 \log_{10} \frac{P_2}{P_1}$$

$$= 10 \log_{10} P_2 - 10 \log_{10} P_1 \tag{19}$$

where P_2 and P_1 are "powerlike" quantities (e.g., watts). It is clear that we are simply expressing two quantities as a ratio (P_2/P_1), taking the logarithm of this ratio to the base 10, and multiplying the number resulting from this operation by a factor of 10. The definition is quite an arbitrary one, and the result of the operation is a dimensionless number. Historically, the nomenclature, bel $= \log_{10}(P_2/P_1)$ was coined to express power ratios, the term bel being adopted in honor of Alexander Graham Bell, a pioneer in telephony. This larger unit was found to be less convenient than 0.1 bel $= 1$ decibel, which was adopted by acousticians and which is closer to sound pressure ratios that can be distinguished by the human ear.

If we wish to extend the decibel notation to express the magnitude of a *power level* on an absolute scale, it is essential that one power in the ratio serve a reference, that is,

$$\text{magnitude in dB} = 10 \log_{10} \frac{P}{P_{\text{ref}}}$$

$$= 10 \log_{10} P - 10 \log_{10} P_{\text{ref}} \tag{20}$$

where P_{ref} is a clearly defined reference value.

Most of us are more familiar with electrical power ratings (light bulbs, devices, and appliances) than with magnitudes relating to acoustics, so we will consider an electrical example.

The same number of decibels represents the ratio between 10 W compared to 5 W, as 20 to 10, or 100 to 50. Suppose now our reference power, P_{ref}, is equal to 1 W.

A decibel rating for a 100 watt light bulb (with a 1 watt reference) would be

$$\text{dB} = 10\left(\log_{10}\frac{100}{1}\right) = 10 \times (2) = +20$$

It is also possible, of course, to construct a light source whose power consumption is less than our 1 W reference power. Let us procure a non-commercial light bulb with a 0.1 W rating. The "dB" rating of this device would be

$$\text{dB} = 10\left(\log_{10}\frac{0.1}{1}\right) = 10(-1) = -10$$

Thus we can have negative dB ratings, based upon our rather arbitrary definition and choice of a reference value, in the same sense that we have positive and negative values for our Fahrenheit and Centigrade temperature scales.

We now proceed to consider quantities that are relevant to acoustics. The mean square value of a waveform, and the *root-mean-square* (rms) value are important in sound because they are related to the power, or amount of energy per unit time, contained in the waveform. It can also be shown that the instantaneous power associated with a sound wave is proportional to the *square of the instantaneous pressure* associated with an acoustic wavefront, or to its mean square value value, p^2. Thus, if we are comparing two power quantities in acoustics, we will obtain

$$\text{dB} = 10\log_{10}\frac{(p_2^2)}{(p_1^2)} = 10\log_{10}\left(\frac{p_2}{p_1}\right)^2$$

$$= 20\log_{10}\frac{p_2}{p_1} \tag{21}$$

where p_1 and p_2 are the *rms* values of pressures. Thus, if we are comparing *sound pressures* in acoustics, we must employ a factor of 20 rather than 10 in order to be consistent with our original definition.

Sound Pressure Level

The common reference pressure for airborne acoustic measurements is 2×10^{-4} μb $= 2 \times 10^{-5}$ Newton/meter^2 (20 μN/m^2) $= 20$ micropascal (20 μPa)* This *rms* reference pressure ($\sim 2.9 \times 10^{-9}$ lb/in.^{2}) is somewhat arbitrary, but it is convenient. First, it is the pressure that approximates the threshold of human hearing, and the acoustic power at this threshold is 10^{-16} W/cm^2 (10^{-12} W/m^2). Second, this 2×10^{-4} μb reference permits us to establish a positive set of "dB" values (from 0 to 120) that encompass most aural environments. Note that at 120 dB, the sound pressure will be a million times that of threshold, that is,

$$L_p = 20 \log_{10} \frac{200 \ \mu b}{2 \times 10^{-4} \ \mu b} = 20 \log_{10} 10^6 = 120 \text{ dB}$$

A list of pressures and the corresponding *sound pressure levels*, L_p, in decibels, using 2×10^{-4} μb as a reference, is presented in Table 3.

At a sound pressure level, $L_p = 120$ dB, a person with normal hearing sensitivity will experience discomfort, in the frequency range of 200 to 10,000 Hz. Within this same frequency band, pain is experienced at about 140 dB. With respect to our 0.0002 μb reference, 1 lb/ft^2 is approximately equal to 148 dB; 1 lb/in.^{2} $\simeq$ 171 dB; and a sonic signal of 1 atmosphere (above the 1 atmosphere ambient) is approximately 194 dB.

Features of the Decibel Scale

The decibel scale has several advantages over linear scales some of which have already been mentioned. First, provided that we establish clearly a *reference* (pressure, intensity, power) we can indicate the magnitude of these parameters. With a reference for pressure being 2×10^{-4} μb (20μPa), the reference for power being 10^{-12} W,† and the reference for intensity being 10^{-16} W/cm^2 we can talk about the *sound pressure level* (L_p) the sound *intensity level* (I_L) and power level (L_W). A *sound level meter* is calibrated to measure the sound pressure level directly (see Chapter 7).

Second, it permits us to express changes in "loudness" by means of small numbers, that is, a 10 dB change is close to the hearing sensation

* The unit of pressure, called the Pascal is equal to 1 N/m^2. It is named after the noted French mathematician and physicist, Blaise Pascal (1623-1662), who measured the decrease of pressure with increasing altitude.

† 1 W $= 10^7$ ergs/sec.

Table 3 Sound Pressure Levels

Sound Pressure		Sound Pressure Level[b]	Environmental Condition
(μb)	(N/m² or Pa)		
0.0002	0.000020	0	Threshold of hearing
0.00063	0.000063	10	Rustle of leaves
0.002	0.0002	20	Broadcasting studio
0.0063	0.00063	30	Bedroom at night
0.02	0.002	40	Library
0.063	0.0063	50	Quiet office
0.2	0.02	60	Conversational speech (at 1 m)
0.63	0.063	70	Average radio
[a]1.0	0.1	74	Light traffic noise
2.0	0.2	80	Typical factory
6.3	0.63	90	Subway train
20	2.0	100	Symphony orchestra (fortissimo)
63	6.3	110	Rock band
200	20.	120	Aircraft takeoff
2000	200.	140	Threshold of pain

[a] 1 μb = 1 dyne/cm² = 74 dB is a common reference pressure for instrument calibration. It is also a reference pressure for measurements in underwater acoustics.
[b] Reference is 2×10^{-4} μb = 2×10^{-5} N/m² = 2×10^{-5} Pa.

of twice or one-half the "loudness" (see Chapter 6). Table 4 indicates approximate "dB" values corresponding to perceived changes in loudness. Furthermore, in computational work, because decibels are logarithmic, data multiplications are reduced to additions, and data divisions are reduced to subtractions. This feature is also useful in power amplifier system analysis because the gain or attenuation of each interlinked component can simply be added, if the dB ratings are known.

For adding two power ratios, expressed in the decibel notation, we must be sure that the same reference value applies. Consider the case where we have two machines in a factory. With a microphone in a fixed position we measure the ambient noise, first with one machine on, then with the second one on. How do we add the two decibel readings to predict the noise level with both machines on? We could convert to antilogarithms, add, and convert back to the logarithmic decibel scale. In practice, we resort to simple tables.

Table 4

Change in Sound Level	Change in Perceived "Loudness"
3 dB	Just perceptible
5 dB	Noticeable difference
10 dB	Twice (or $\frac{1}{2}$) as loud
15 dB	Large change
20 dB	Four times (or $\frac{1}{4}$) as loud

Table 5 permits the addition of decibel readings. Thus, suppose the first machine produced a reading of 90 dB, and the second machine yielded a reading of 85 dB. What would be the reading with both machines running simultaneously? From Table 5 we note that for a difference between the two of 5 dB we should add 1.2 to 90 dB, giving us a resultant of 91.2 dB. If we had three sound sources in a common enclosure, we could use the table to obtain the sum of the first two, and then use again the table for getting the resultant of the first two plus the third.

Table 5   Addition of Decibel Readings for Two Sound Sources

Difference between Two dB Readings	Add to Larger dB Value
0.0 dB	3.0 dB
0.5	2.8
1.0	2.6
1.5	2.2
2.0	2.1
2.5	2.0
3.0	1.8
4.0	1.5
5.0	1.2
6.0	1.0
7.0	0.8
8.0	0.6
10.0	0.4
13.0	0.2
15.0	—

This type of tabular addition applies in those practical instances where the several sources of sound are not phase-coherent, as is usually the case in auditoria, offices and factories, and most conventional closed spaces.

Power Level

Instrumentation is not generally available to measure directly the acoustic power output of a source or the resultant intensity of sound in the surrounding medium. However, it is often desirable to calculate these values from measurements of sound pressure.

Sound power level is often abbreviated to L_W and is defined as

$$L_W = 10 \log_{10} \frac{W}{W_{ref}} \text{ dB} \tag{22}$$

where W is the acoustic power of the source and W_{ref} is the reference acoustic power ($W_{ref} = 10^{-12}$ W). For example, to determine the sound power level of a source radiating 100 milliwatts (0.1 watt):

$$L_W = 10 \log_{10} \frac{10^{-1} \text{ W}}{10^{-12} \text{ W}}$$

$$= 10 \log_{10} 10^{11}$$

$$= 110 \text{ dB } re \ 10^{-12} \text{ W}$$

Some typical power levels are given in Figure 6.

The sound power in normal conversational speech is only about 20 μW (2×10^{-5} W) whereas the acoustic power generated by a Saturn rocket may approach 10^8 W. The acoustic power of a loudspeaker may be only 1 W even though the electrical power rating may be 40 W, because of the inefficient process of converting electrical to acoustic energy.

Intensity Level

Sound intensity level is sometimes abbreviated L_I and is defined as

$$L_I = 10 \log_{10} \frac{I}{I_{ref}} \tag{23}$$

where I is the sound intensity (power passing in a specified direction through a unit area) and I_{ref} is the reference intensity ($I_{ref} = 10^{-12}$ W/ m^2 = 10^{-16} W/cm^2). For example, to determine the sound intensity level

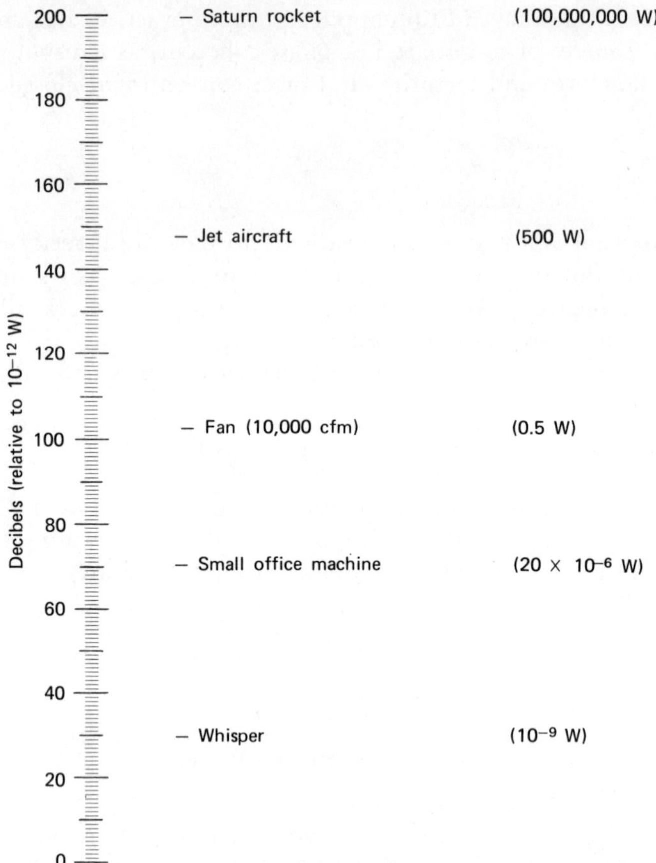

Figure 6   Typical sound power levels for various sound sources. (Courtesy of M. J. Crocker.)

generated by the source of our previous example (0.1 W), radiating uniformly in free space and at a distance of 10 meters:

$$I = \frac{W}{S} = \frac{0.1}{4\pi r^2} = \frac{0.1}{400\,\pi} = 7.96 \times 10^{-5} \text{ W/m}^2$$

$$L_I = 10 \log \frac{7.96 \times 10^{-5}}{10^{-12}}$$

$$= 10 \log 7.96 - 10 \log 10^5 + 10 \log 10^{12}$$

$$= 9 - 50 + 120$$

$$= 79 \text{ dB } re \; 10^{-12} \text{ W/m}^2$$

SOUND WAVE INTERACTION PHENOMENA

Dissipation of Energy

In equation 11 we showed that a relationship existed between sound pressure, air density, velocity of the sound wave, and the velocity of the molecules undergoing oscillatory motion. This relationship can be written as

$$p = \rho c u \tag{24}$$

where we have previously identified ρc as the characteristic impedance, and u is the *rms* velocity of the molecules in a plane wave for an *rms* pressure, p.

It is instructive to see the quantitative relationship between the sound pressure levels on our decibel scale and the *rms* molecular velocities and peak molecular amplitude displacements at various frequencies. Approximate values are given in Table 6.

Table 6   Relationship Between Sound Pressure, Molecular Velocity, and Amplitude at Several Frequencies[10]

Pressure Level (dB)	Molecular Velocity rms (cm/sec)	Peak Amplitude (cm)		
		100 Hz	1 kHz	10 kHz
0	5×10^{-6}	1×10^{-8}	1×10^{-9}	1×10^{-10}
60	5×10^{-3}	1×10^{-5}	1×10^{-6}	1×10^{-7}
120	5	1×10^{-2}	1×10^{-3}	1×10^{-4}

It can be seen that at 1000 Hz an amplitude of vibration of only 10^{-6} cm corresponds to an easily discernible sound ($\sim$60 dB). Table 6 also provides some indication of the problem of attenuating noise from radiating surfaces; displacements of the radiator must be constrained to exceedingly small values if there is to be no detectable sound.

At great distances from a sound source, of course, there will be a substantial attenuation of sound energy and the molecular displacements will decrease to a value where the sound is inaudible. In the open air and with spherical sound divergence the sound pressure level will decrease about 6 dB for each doubling of the distance from a uniformly radiating source because of the simple inverse square law (Chapter 3).

There is also a small decrease in molecular kinetic energy because of air absorption, in which sound energy is transformed to heat.

Reflection

Reflection is one of the most important characteristics of sound, and many of the defects (as well as desirable qualities) in room acoustics are due to this phenomenon. If a sound wave in air is incident upon a surface whose characteristic impedance, ρc, is much greater than for air, most of the sound energy will be reflected. Furthermore, if the surface is very far from the sound source, the radius of the sound wave fronts will be large and the problem of reflection can be treated as one involving plane waves. Then we can consider a wave front as comprising a beam of nearly parallel rays, as in optics.

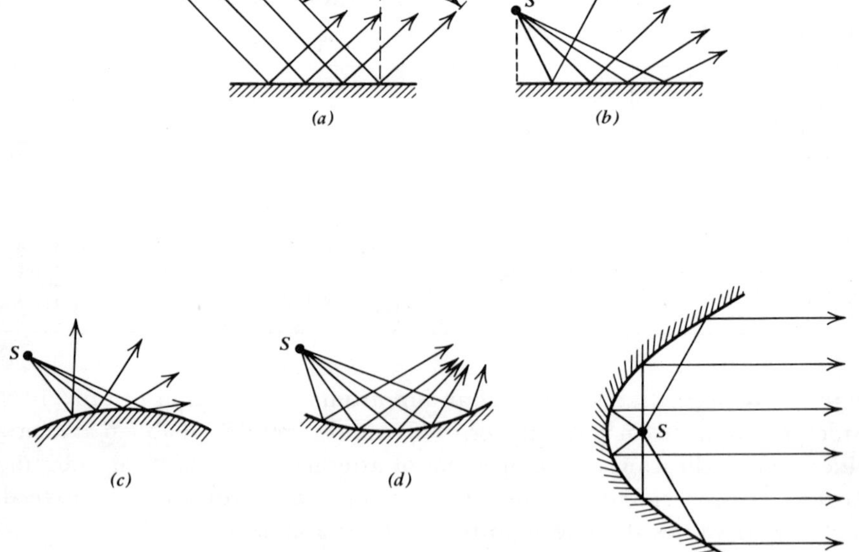

Figure 7 Examples of the geometric reflection of sound from various surfaces: (a) distant source reflecting from a plane surface, (b) point source from plane surface, (c) point source near convex surface, (d) point source near concave surface, (e) point source with parabolic reflecting surface.

The *law of reflection* then applies, that is, the angle of incidence of the wavefront is equal to the angle of reflection. Thus, in Fig. 7 where a wavefront is incident upon a plane surface $<i = <r$. When the surface is concave or convex, the nature of the reflection will be analogous to the optical case provided that the wavelength is small compared to the dimensions of the reflecting surface. Of course, no surface is a perfect reflector. Furthermore, any practical surface will absorb and reflect sound in a frequency-dependent manner, so that the spectral composition of broad-band sound after even a single reflection will be altered.

Diffraction

Pure reflection takes place only when the dimension of the reflecting surface is large in comparison with the wavelength of the incident radiation. Thus pure reflection rarely occurs in acoustics because the wavelengths of sound are in the range of 2 cm to 20 meters; such dimensions are comparable to the size of typical objects in rooms, door openings, windows, and hallways. Thus, we perceive a substantial amount of sound energy by diffraction, that is, the bending or spreading out of a sound wave after it intersects an aperture, a solid object, a recess, or a surface protrusion. The primary factor in diffraction, of course, is not the dimension of the wavelength alone or that of the intercepting object or aperture, but the ratio of wavelength to object. For light, all wavelengths are less than 10^{-4} cm, a dimension that is very small compared to virtually all light sources, mirrors, lenses, or obstacles. Hence, diffraction is of no consequence and sharp shadows are perceived. If light is diffuse it is via multiple reflections and scattering. With sound, geometric acoustics or ray tracing is not applicable for frequencies below 250 Hz, at least in closed spaces of normal dimensions (Chapter 4).

Figure 8*a* shows the diffraction of a sound wave through an aperture, where the wavelength is much longer than the aperture, and Fig. 8*b* shows the diffraction of a sound wave past a partition. For a given sized obstacle or aperture, the longer the wavelength, the greater will be the bending into the "shadow zone." One can detect this bending qualitatively by listening to music being emitted through a doorway (as in Fig. 8*b*). It is easy to hear low pitched sounds (long wavelengths) in the "shadow zone," but on approaching the doorway the sound becomes more brilliant as the higher frequencies and shorter wavelengths are detected at a greater intensity.

The phenomenon of diffraction is used extensively in the acoustical design of auditoria where it is often desirable to break up the regularity

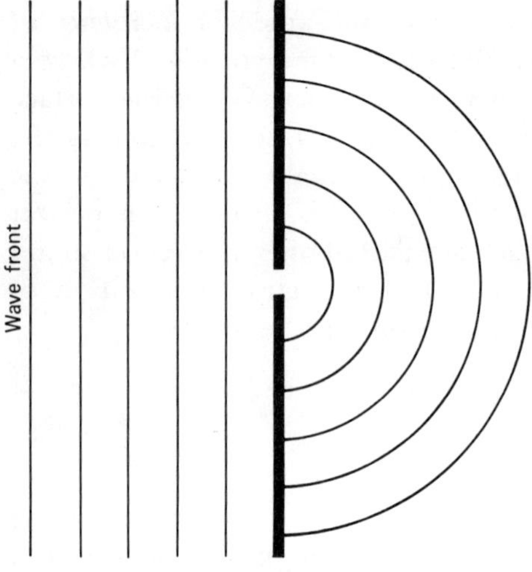

Figure 8*a* Diffraction of a sound wave passing through a small aperture.

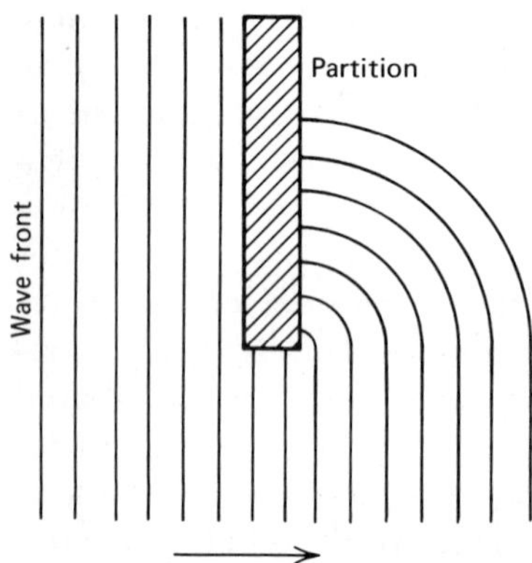

Figure 8*b* Diffraction of a sound wave past a partition or barrier. (From V. O. Knudsen and C. M. Harris, *Acoustical Designing in Architecture*. Wiley, 1950.)

of reflections from large walls. For example, if a reflecting wall is lined with strips of absorbing material, alternating with reflecting areas or panels, it is possible to minimize the effect of regular reflections. Diffraction can thus be produced by selective absorption as well as by the transmission of a sound wave through an opening. In other words, not only a geometric shape but the "boundary impedance" of a surface are important in the diffusion process.[11]

Refraction

When a wave front passes from a medium in which the velocity of propagation is c_1 to another medium in which the velocity of propagation is

c_2, the direction of the wave front will be changed unless the angle of incidence is perpendicular (Fig. 9). If the angle of incidence is θ_1 and the angle of refraction is θ_2, the angles are related by the expression

$$\frac{\sin \theta_1}{\sin \theta_2} = \frac{c_1}{c_2} \tag{25}$$

The experiment indicated in Fig. 10 is a common one to show how an enclosed dense gas can refract sound. A balloon is filled with carbon dioxide whose refractive index for sound (ratio of the velocity of sound in CO_2 to velocity in air) at 0°C is 1.28. If the ear is positioned on one side of the balloon and a watch on the other, a point can be determined where the ticks are the loudest. The balloon thus acts as an acoustic lens in a manner analogous to the refraction of light waves. As we shall see in the next chapter, refraction also explains why sounds can sometimes be heard at very great distances.

Superposition or Interference

It is a general characteristic of many mechanical systems that a resultant total displacement (when two motions are taking place simultaneously) may be described as the sum of the two displacements considered inde-

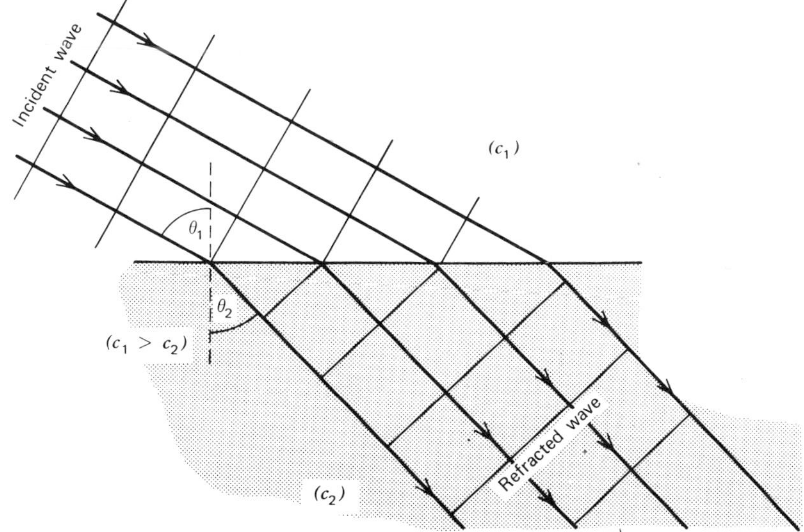

Figure 9   Refraction of a sound wave front at the boundary of two media in which the velocity of sound is different. In this illustration, $c_1 > c_2$.

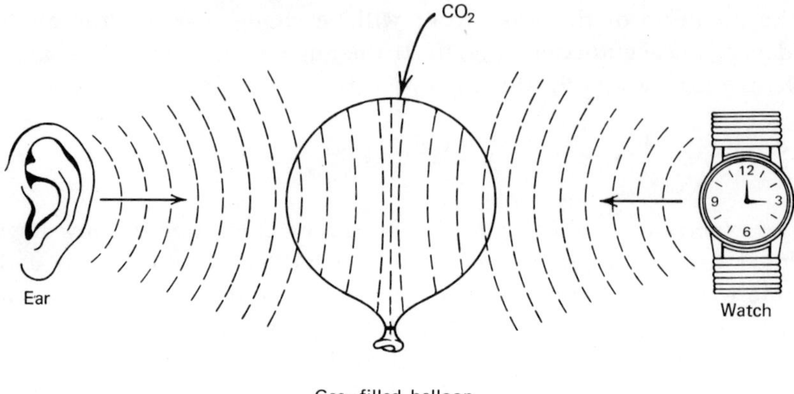

Figure 10   Refraction of sound in a gas-filled balloon, where the balloon acts as an acoustic lens.

pendently. This is one example of the *"superposition principle,"* which states that whenever a system is "linear" (amenable to description by a linear differential equation), certain parameters are simply additive: for example, the composition of force vectors, potentials in electrical circuits, or the displacements of air molecules accompanying a sound wave. Thus we can determine the net effect of two interacting sound waves by the simple linear addition of their independent compressions or rarefactions.

When the result of the superposition or linear addition of two sound pressure waves yields a spatial distribution of energy that is markedly different from what results from the sound waves being considered separately, the superposition is called *"interference."* Thus two sound waves can combine to give regions of greater compression or rarefaction than in either sound wave. Superposition or interference may result from two or more sound sources; however, superposition or interference may also result from a single sound source whose wave fronts subsequently interact because of multiple reflection or diffraction, within a closed space.

We shall consider the single example of two sound sources, S_1 and S_2 emitting waves of the same frequency, and of the same phase, as shown in Figure 11. The solid concentric circles surrounding S_1 and S_2 represent surfaces of maximum compression, and the dotted circles indicate surfaces of maximum rarefaction. Maxima of pressure thus occur where

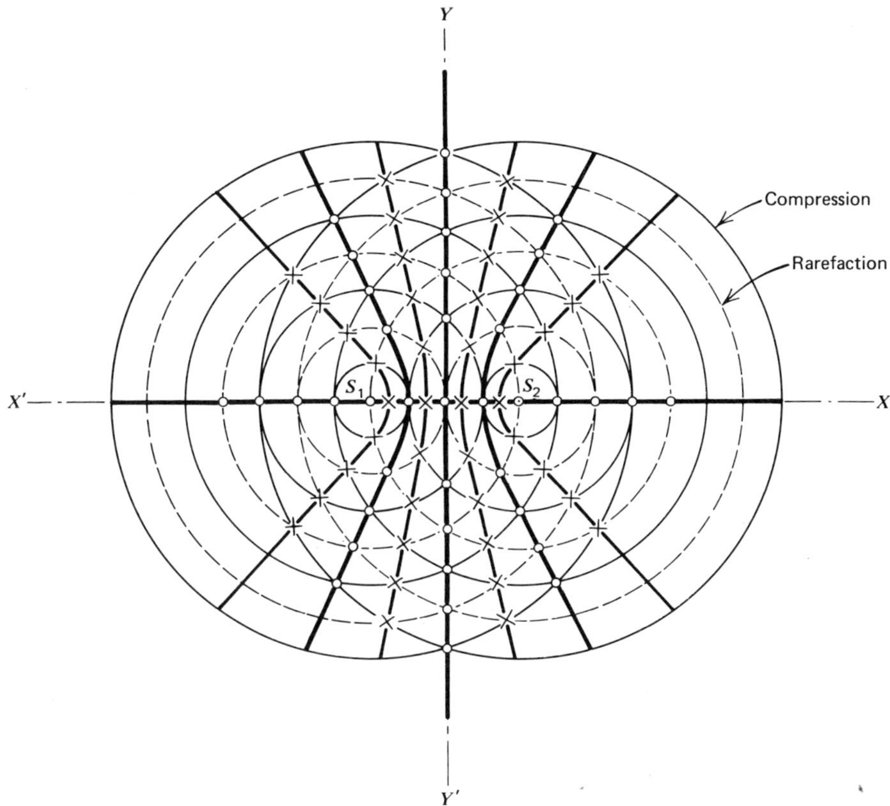

Figure 11   The interference of waves from two sources that radiate sound of the same frequency. Maxima of pressure occur where a compression from S_1 meets a compression from S_2.

a compression due to S_1 meets a compression due to S_2; minima occur where a rarefaction meets a rarefaction; both are indicated by the open dots. Regions where a compression is matched by rarefaction, resulting in a null, are indicated by crosses.

Thus, along the hyperbolic curves marked by dots and along $X - X'$ and $Y - Y'$ we have a maximum flow of energy, except between s_1 and s_2 where opposing wavefronts produce a stationary wave. We shall later consider this phenomenon in connection with "standing waves" produced in an enclosure (Chapter 4).

SOUND SPECTRA

Line Spectra

By analogy with light, sound that consists only of a few discrete frequencies is said to constitute a *line spectrum*. An oscillator may emit a single frequency or it may emit a fundamental frequency plus multiples of this frequency. The resultant sound spectrum, however, will consist of a limited number of pure tones. Musical instruments also generate line spectra of harmonically related frequencies, and power transformers generate frequencies that are multiples of the frequency of the input current.

Continuous Spectra

A *continuous sound spectrum* is built up by a very large number of frequencies so as to form a continuum. Many sounds in nature are characterized by continuous spectra that include a sizable portion of the audio frequency range. Noise may consist of either line spectra, continuous spectra, or a combination of each. Figure 12 is a sound pressure level vs. frequency graph for (a) line, (b) continuous, and (c) a complex spectrum—line and continuous.

Bandwidth

We introduce the concept of bandwidth here because we shall refer to it in the next section and in many subsequent chapters. In a general sense, every source inherently has a bandwidth; so also has every receiver.

In optics it is possible to produce a light source that emits line spectra or a broad band of frequencies or wavelengths. The eye as a receiver can detect wavelengths from about 3500 to 7000 Å (1 Å $= 10^{-8}$ cm), and photographic film is available that will respond to most of this range. For this reason we could consider the human eye and photographic film as broad band detectors. Optical filters can be inserted between sources and detectors; some of these can be designed so as to transmit a wavelength band of less than 10 Å, and then they are termed narrow band filters.

In acoustics we have a similar situation. Sound sources range from oscillators or tuning forks that emit but a single frequency to broad band sounds that encompass the entire audio range. Conventionally, we refer

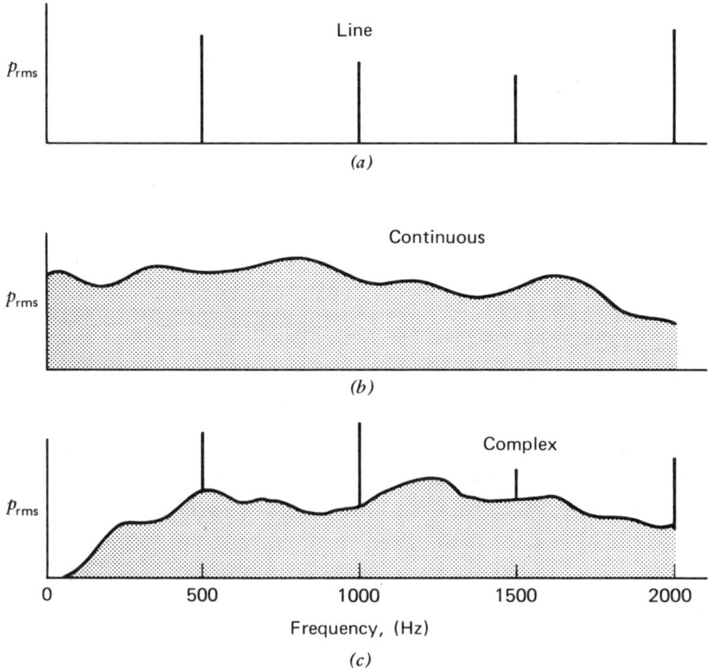

Figure 12   Line, continuous, and complex sound spectra.

qualitatively to bandwidth in comparison to the range of frequences that can be sensed by the ear (15 to 20,000 Hz). A microphone is a broad bandwidth receiver as it responds to most of this frequency range.

The term *transmission bandwidth* refers to an acoustical filter, that is, to the range of frequencies that it will pass without appreciable attenuation. (The term passband is synonomous with transmission bandwidth.) The bandwidth of filters may be a single frequency, narrow band ($\Delta f/f \ll 1$), octave band, or "weighted" to simulate the frequency response of the ear.

Bandwidth is thus a very general term, referring to the spectral range of sound sources, filters, or detectors.

NOISE

Noise is often referred to as any *unwanted* sound. This definition emphasizes the highly subjective aspect of hearing. Sounds that might be enjoyable to some people will be intolerable to others, depending upon a person's interests, activity, and mood. Some forms of music, for example, even though they may be damaging to the hearing mechanism, are wanted by some people, unwanted by others. There is always the question of degree. Some type of sound sensation is usually wanted, assuming the sound intensity is at a low level. Indeed, few persons are able to function normally for any extended period in the total absence of sound.

Another common concept about noise is that it is comprised of a broad band of frequencies as opposed to a pure tone. Actually, noise consists of virtually all kinds of sounds: periodic, aperiodic, and impulse sounds. Rotating machinery noise will usually include dominant frequencies plus a broad band of frequencies of varying intensity. Noise is associated with every type of human activity, and it is emitted from almost all apparatus having moving components. Machinery, office equipment, fans, ducts, gears and bearings, plumbing, traffic, aircraft— all generate a noise spectrum. Electrical equipment is also a significant noise source: for example, fluorescent lamp ballasts, transformers, chokes, and high voltage transmission lines.

It is the function of acoustic instrumentation to analyze the frequency spectrum or impulse nature of noise, because only with such equipment can the spectral composition of noise be made quantitative. Broad band noise implies the inclusion of both high and low frequencies. Jet noise, for example, includes frequencies from infrasonic to ultrasonic. Narrow band noise implies a restriction to a fraction of the audible frequency range. *White noise* and *pink noise* are technical terms.[12] "White noise" is defined as noise having a uniform distribution of energy as a function of frequency, over the entire frequency range, that is, constant energy-per-Hertz bandwidth. Such noise will show an increasing sound pressure level of 3 dB per octave, when measured with "constant-percentage" bandwidth filters such as octave or 1/3 octave analyzers. (See Chapter 7.) "Pink noise" is defined as having constant energy per octave bandwidth.

General categories of noise include (1) ambient or background noise, (2) steady-state noise, (3) fluctuating or intermittent noise, and (4) impulsive noise. Ambient or background noise is the noise in an environment from both near and far sources. Usually it refers to minimum levels when no strong sources of noise or sound are turned on. Steady-state noise often refers to machinery or apparatus where sound levels are reasonably constant during the period of measurement. Fluctuating or intermittent noise is exemplified by traffic where the sound may vary in level but is "on" for times longer than the integration time of the ear[13] (about 200 msec). An impulsive noise is of very short duration for its peak pressure level. It has a very short rise time, sometimes in the microsecond range, and special instrumentation is required to detect the transient nature of the pressure wave. Impulsive sound is sometimes defined as a change in *rms* air pressure at greater than 40 dB/0.5 sec.[14]

Also of importance is the concept of a "signal-to-noise" ratio. This term is used as a figure of merit in electrical engineering, optics, acoustics, and other fields. It is a measure of the magnitude of the desired signal in comparison to the undesired noise from any or all sources. This ratio, in part, also determines the environmental quality of speech and music.

FOURIER REPRESENTATION OF A SOUND WAVE

Because sounds originate from vibrating solids, liquids, and gases, it is reasonable to presume that many sounds (including noise) might be represented by a mathematical formulation composed of many frequencies. This statement can be made somewhat plausible from the graphs of Fig. 13. If we add, graphically, a sequence of frequency components, each of which is an integral multiple of some fundamental frequency, we will obtain a complex curve. This curve will be periodic, and the period repeats according to the smallest component frequency. Thus in Fig. 13, let a second harmonic (twice the fundamental frequency) of smaller amplitude be added to the fundamental frequency. Curve C is the result of the summation. Addition of the third harmonic will yield curve E, and the addition of the fourth harmonic will yield curve G.

It was the French mathematician J. B. J. Fourier (1822), however, who showed that any repetitive wave form can be represented as the sum of an infinite sequence of harmonics, the amplitude and phase of which are both permitted to be variables, and that the fundamental frequency is determined by the wave-form repetition rate. Thus the analytical expression that is universally used to represent a periodic sound wave is called a Fourier series. The basic premise of the Fourier series formulation is that any function* $f(x)$ defined in the interval from $x = -\pi$ to $x = +\pi$ can be expanded in a series of trigonometric functions (sines and cosines).

A general discussion of Fourier Series will be found in standard mathematical texts and the Appendix contains a single example. For-

---

* Subject only to exceptions that usually do not apply in engineering.

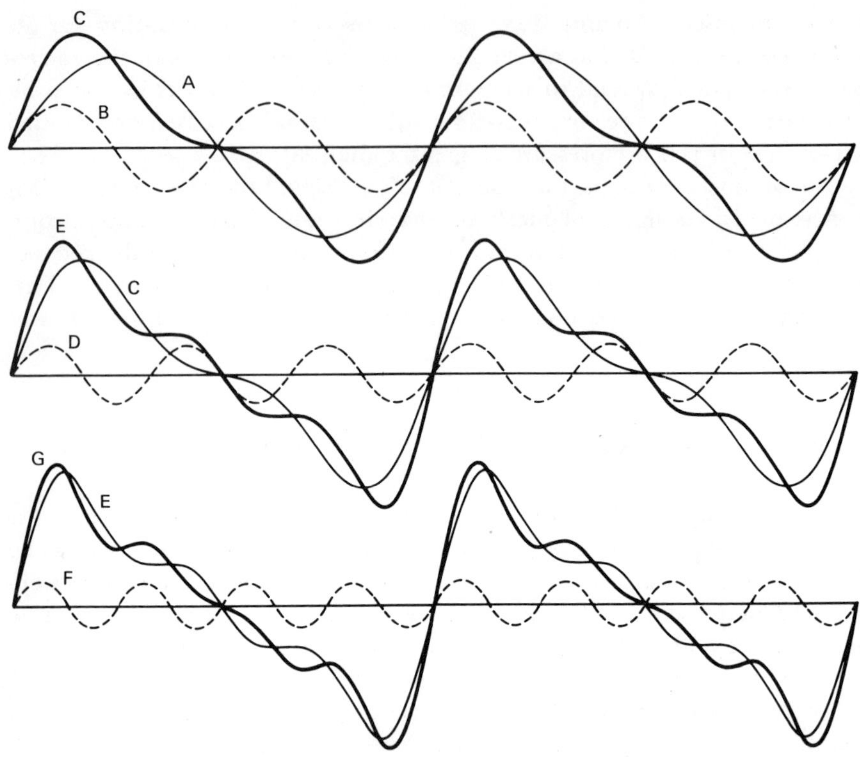

Figure 13 Pure tone frequencies added to produce a complex wave. A = fundamental, B = second harmonic, C = fundamental plus second harmonic, D = third harmonic, E = fundamental plus second and third harmonics, F = fourth harmonic, G = fundamental plus second, third and fourth harmonic.

tunately, with the aid of special instrumentation and computers, this analysis of any periodic acoustic signal can now be easily achieved without laborious mathematics. Even very complex wave shapes can be "Fourier analyzed," so that the many frequencies and their respective amplitudes, which contribute to a resultant complex sound wave, can be determined automatically.

REFERENCES

1. A definition by the American National Standards Institute (formerly American Standards Association) 1430 Broadway, New York, N. Y. 10018.

2. H. C. Hardy, D. Telfair, and W. H. Pielemeier, *J. Acoust. Soc. Am.*, **13**, 226 (1942).

3. *Am. Inst. of Physics Handbook*, McGraw-Hill, New York, 1972, Vol. 3, p. 74.

4. L. L. Beranek, *Noise and Vibration Control*, McGraw-Hill, New York, 1971, p. 12.

5. M. Rettinger, *Acoustic Design and Noise Control*, Chemical, New York, 1973, p. 11.

6. Rettinger, *op. cit.*, p. 10.

7. M. Y. Colby, *Sound Waves and Acoustics*, Henry Holt, New York, 1934, p. 196.

8. H. F. Olson, *Acoustical Engineering*, Van Nostrand, Princeton, 1957, p. 15.

9. C. A. Taylor, *The Physics of Musical Sound*, Elsevier, New York, 1965, p. 149.

10. A. J. King, *The Measurement and Suppression of Noise*, Chapman and Hall, London, 1965, p. 71.

11. E. Meyer and E. G. Neumann, *Physical and Applied Acoustics*, Academic Press, New York, 1972, p. 51.

12. A. P. G. Peterson and E. E. Gross Jr., *Handbook of Noise Measurement*, General Radio, Concord, Mass., 1972, p. 132.

13. Beranek, *op. cit.*, p. 93.

14. K. D. Kryter, *The Effects of Noise on Man*, Academic Press, New York, 1970, p. 3.

3

SOUND PROPAGATION IN THE ATMOSPHERE

PROPERTIES OF THE ATMOSPHERE

Because we have defined airborne sound as an audio frequency pressure variation in the atmosphere, it is appropriate to examine this sound medium and its general characteristics. Over short distances the atmosphere is a very homogeneous fluid. Furthermore, in small enclosures it is even legitimate to neglect any absorption effects in the air. But in outdoor environments, air absorption is not negligible, and for sound propagated over large distances, inhomogeneities in the medium will affect the direction of sound transmission.

First, let us recall that the total mass of the earth's atmosphere is 5×10^{18} kg. About half of this total mass is confined to the 6 km shell or layer in which most of the clouds are formed, and about 99% of the earth's total atmosphere is confined to an altitude of <30 km. At sea level there is about one kg of air above each cm^2, the molecular density is about 2.5×10^{19} molecules per cm^3, and the molecular mean free path is about 6.6×10^{-6} cm. The particle density is very high, and sound energy can be transmitted between neighboring molecules. However, all acoustic signals vanish when the atmosphere is rarefied to the extent that the distance between molecules is comparable to the wavelength of sound. Thus space silence begins at about 130 km and the darkness of space also begins at roughly this same altitude.[1]

The primary constituents of the standard "dry" atmosphere, by volume percent, are: nitrogen, 78.09; oxygen, 20.95; argon, 0.93 and carbon dioxide, 0.3. Trace quantities of neon, krypton, helium, ozone, xenon, and hydrogen are also present. Water vapor also constitutes an appreciable percentage of the total atmosphere, but this percentage depends chiefly upon the temperature. For a given pressure, the higher the temperature the greater the possible, and usually the actual, percentage of water vapor that is present. Volume percentages may range from about 3 to 0% (annual average values are 2.63 at the equator and 0.22 at a latitude of 70°N).[2] We shall observe subsequently that the relative humidity has a measurable effect on the absorption of sound over large distances. The average total amount of water vapor in the atmosphere is an amount just sufficient to cover the entire surface of the earth with a layer 2.54 cm (1 in.) deep.[3]

The temperature profile of the atmosphere from sea level to 100 km is shown in Fig. 1. The lowest atmospheric region, characterized by decreasing temperature with height, is called the *troposphere*. For acoustical considerations, we shall only be concerned with this zone and, more important, the local variations and temperature "inversions" that occur within it. The vertical extent varies with season and latitude. In tropical

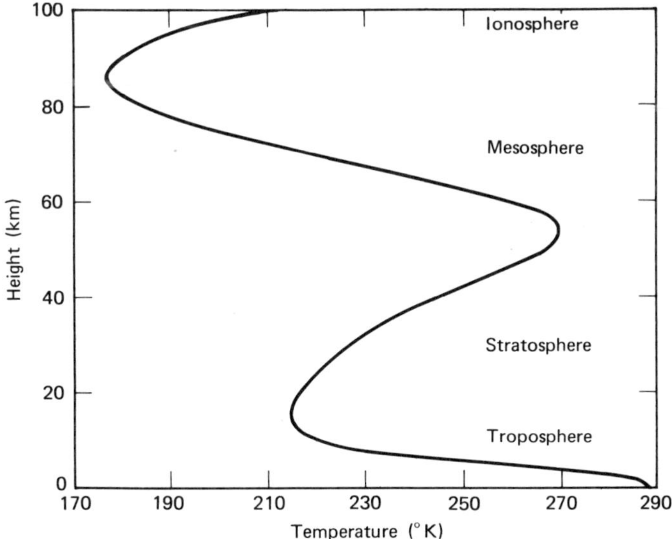

Figure 1   Average temperature vs. altitude profile of the International Reference Atmosphere.

regions it varies between 16 and 18 km. Over the polar regions it is 8 to 10 km in summer; in winter seasons the temperature profile will change markedly. In this zone the pressure decreases with altitude by nearly 10% per 3300 ft (1 km) up to 10,000 ft (~3 km) where the atmospheric pressure is approximately 0.7 atmosphere.[4] Contained within the troposphere is also the *jet stream,* in which a high velocity core of air moves rapidly from west to east. At an altitude of roughly 40,000 ft (12 km) winds attain speeds of 150 mph, carrying air around one-fifth of the earth in one day at latitudes of 40 to 50°.[5] Westerly wind speeds increase from the ground to these jet altitudes; thus, sound propagation is indirectly affected at lower levels by this phenomenon.

SOUND PROPAGATION IN THE OPEN AIR

Before discussing specific atmospheric effects on sound transmission, we will consider idealized point and line sources and determine the intensity of sound at a distance from these sources. The point source is the most common one, and it produces a spherical sound field. A line source is exemplified by a busy highway in which sound is being radiated from many vehicles over an extended distance. Sound from such a line source will radiate with a cylindrical pattern.

The Inverse Square Law

Figure 2 represents an idealized point source and the resultant sound distribution pattern. Let S be the sound source and R_1 and R_2 be the radii of two concentric spheres. We know that equal amounts of acoustic energy flow through the two spheres. The surface area subtended by any arbitrary solid angle is proportional to the square of the distance from the source. The sound intensity will therefore be inversely proportional to the square of this distance. Thus,

$$\frac{I_{R_1}}{I_{R_2}} = \frac{R_2{}^2}{R_1{}^2} \tag{1}$$

where I_{R_1} is the intensity at R_1, and I_{R_2} is the intensity at R_2. How does this relate to a relative sound pressure level reading at R_1 and R_2? The difference in sound pressure level is

$$\Delta L_p = 10 \log_{10} \frac{I_{R_1}}{I_{R_2}} \tag{2}$$

$$\Delta L_p \;=\; 10 \log_{10} \frac{R_2{}^2}{R_1{}^2} \;=\; 20 \log_{10} \frac{R_2}{R_1} \tag{3}$$

From the above expression it is clear that *the sound pressure level will decrease 6 dB for each doubling of the distance,* that is,
Let

$$R_2 = 2R_1$$

$$\Delta L_p \;=\; 20 \log_{10} \frac{2R_1}{R_1}$$

$$=\; 20 \times 0.301 \;\approx\; 6 \text{ dB} \tag{4}$$

We should also note the relationship of the *rms* pressure, p, as a function of the distance, r, from a point source. In Chapter 2 we observed that the intensity of a sound wave was proportional to the square of the sound pressure. If this intensity is also inversely proportional to r^2, it then follows that

$$p \;=\; k \,\frac{1}{r} \text{ where } k \text{ is a constant.} \tag{5}$$

This relationship is called the *inverse-distance law,* which governs sound propagation in the "far" field.*

Intensity from Line Sources

For sound emerging as a cylindrical wave front from a line, the sound intensity at some distance R will be

$$I \;=\; \frac{W_l}{2\pi R} \tag{6}$$

where W_l is the sound power per unit length. The change in intensity and sound pressure level as a function of distance can be readily computed by a consideration of Fig. 3. The surface areas of the two concentric cylinders will be proportional to their radii. Thus, if sound is being radiated uniformly from the axis, the intensity will decrease in direct proportion to an increase in radius. Hence

$$\frac{I_{R_1}}{I_{R_2}} \;=\; \frac{R_2}{R_1} \tag{7}$$

* The "far" field is defined as the sound field sufficiently distant from the source so that the particle velocity is primarily in the direction of the sound wave and the acoustic intensity is proportional to the square of the *rms* sound pressure.

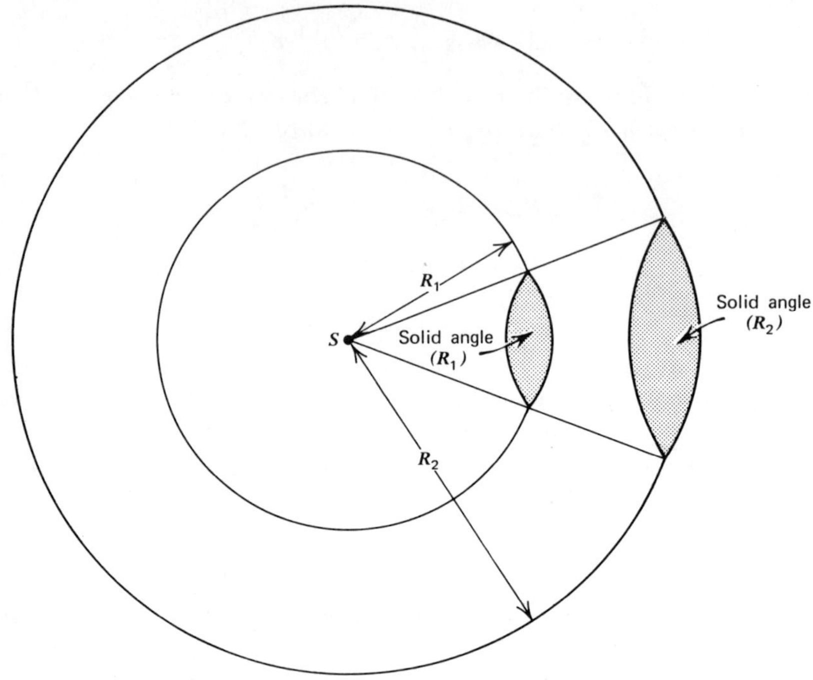

Figure 2   Sound attenuation from a point source. The sound intensity will decrease in proportion to the square of the distance from the source.

Therefore

$$\Delta L_p = 10 \log_{10}(R_2/R_1) \tag{8}$$

Then if

$$R_2 = 2R_1,$$

$$\Delta L_p = 10 \log_{10} 2$$

$$= 10 \times 0.301$$

$$\approx 3 \text{ dB} \tag{9}$$

Thus for every doubling of the distance from a "line" source the sound pressure level will decrease by 3 dB.

Situations in which sound calculations can be roughly estimated by assuming a cylindrical sound field include dense automobile traffic on multilane highways, long railroad trains, and noise from high voltage transmission lines.

The calculated decrease in sound pressure levels vs. the ratio of distances (R_2/R_1) from point and line sources is graphed in Fig. 4. In prac-

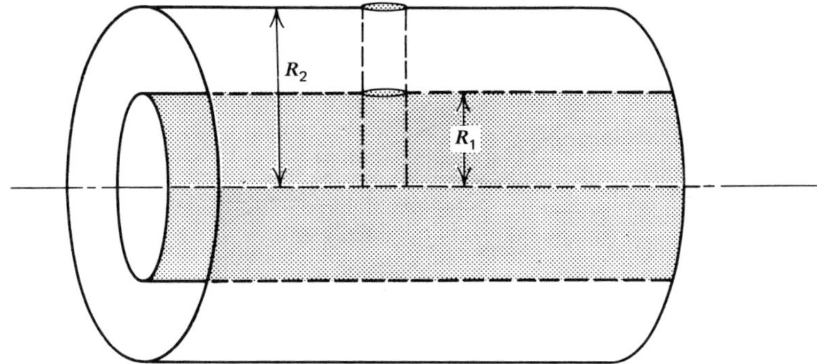

Figure 3   Because the surface of a cylinder is proportional to its radius, the sound intensity decreases directly with the radial distance from a line source.

tice, nearly all sound sources are somewhat directional, and some absorption and reflection will occur.

AIR ABSORPTION

Because air is not a perfectly elastic medium, during successive compressions and rarefactions several complex irreversible processes occur, all of which are dependent on frequency. As a result, some sound attenuation must be attributed to absorption as well as to divergence. Thus in a spherical wave, a decrease in sound pressure with distance, r, can be expressed as

$$p = k\left(\frac{1}{r}\right)e^{-\alpha r} \tag{10}$$

where α is a frequency-dependent parameter.

This attenuation constant, which is expressed in dB/100 ft, is comprised of two parts:[6]

$$\alpha = \alpha_1 + \alpha_2 = 4.34 \times 10^{-9}f^2 + \alpha_2 \tag{11}$$

where f is the frequency in Hz. The first part, α_1, is due to the combined effect of viscosity, heat conductivity of the air, and a dissipation of energy due to rotational energy states of air molecules. This is the so-called "classical absorption"; it is independent of the humidity of the air, and it can be neglected except at very high frequencies.

The second term α_2 is a result of a complex "molecular relaxation absorption," and it is frequency, temperature, and humidity dependent.[7]

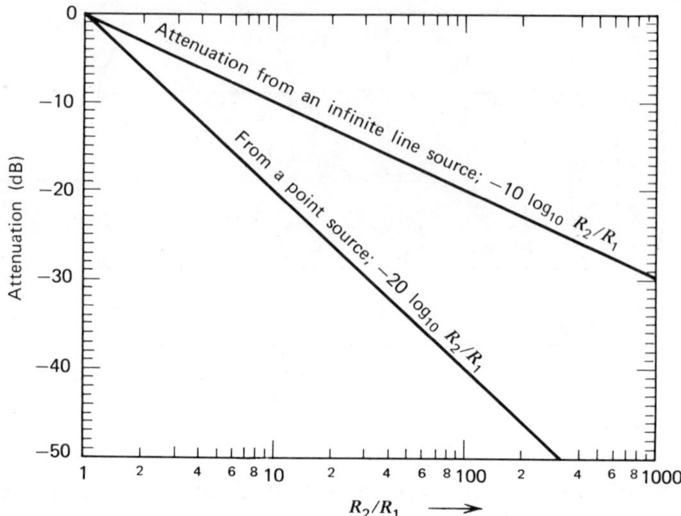

Figure 4   Decrease in the sound pressure levels vs. the ratio of distances R_2 and R_1 from a point and line source.

Essentially, a polyatomic gas can absorb energy by means of its several degrees of vibrational and rotational modes, at the expense of translational energy.[8] For a sound wave of 4000 Hz, at 40% relative humidity and a temperature of 70°F, this molecular absorption causes an attenuation of greater than 10 dB/1000 ft.[9]

Recently, it has been the practice to combine both absorption terms (primarily because α_2 is dominant) and refer to a total attenuation coefficient m for the case of a plane traveling wave. This nomenclature is used widely in room acoustics. For a plane wave traversing a distance x in a homogeneous medium, where P_0 is the *pressure amplitude* of the wave at position $x = 0$, and m is the total attenuation coefficient, the pressure P at a distance x is:[10]

$$P = P_0 e^{-mx/2} \tag{12}$$

Values of the total attenuation coefficient, m, vs. relative humidity for air at 20°C and normal atmospheric pressure for frequencies between 2 and 12.5 kHz are shown in Fig. 5.[11] The data show the dramatic effect of molecular resonance absorption as a function of relative humidity, which is defined as the ratio of the actual vapor pressure to the saturation vapor pressure at a given temperature. Humidity attenuation curves will also change with temperature.

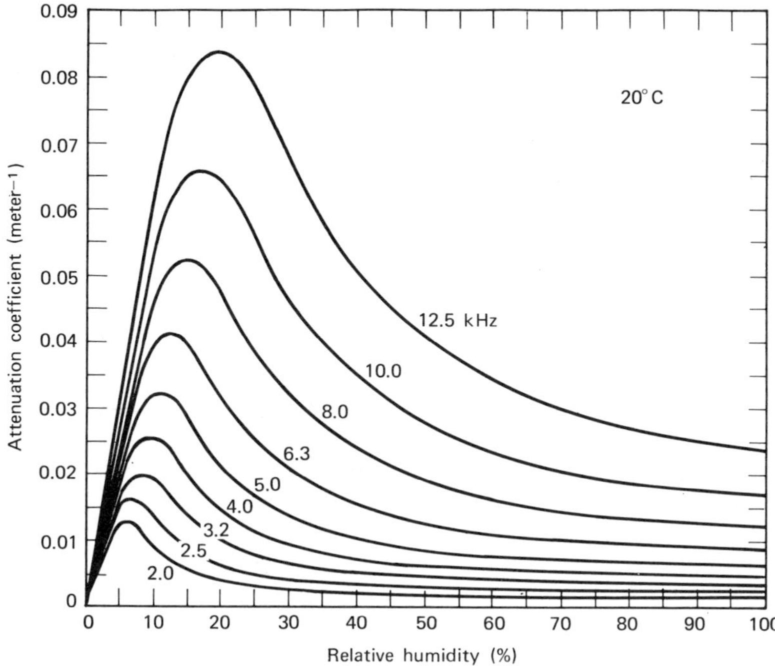

Figure 5 Values of the total attenuation coefficient m vs. percent relative humidity for air at 20°C for frequencies of 2 kHz to 12.5 kHz. [From C. M. Harris, *J. Acoust. Soc. Am.*, **40**, 148 (1966).]

There are some immediate practical consequences to the above. In an open-air theater of large dimensions we must expect some losses other than from pure divergence. Also, a given musical performance may even lose some "brilliance" because of meteorological conditions. Even for large closed spaces, the air absorption is of some consideration, and in large auditoriums air absorption can decrease the reverberation time (Chapter 4), although the relative humidity in a closed auditorium can usually be controlled.

EFFECT OF WIND AND TEMPERATURE

Both the wind and temperature variations in the atmosphere affect substantially the energy distribution of sound, by deflecting the sound rays from their normal paths.

From experience we know that it is easier to hear a distant sound if the sound is traveling with the wind direction (i.e., we are downwind), than if the wind is opposing the sound wave front that is coming to us. Consider the diagram shown in Fig. 6.

First, we note that a vertical gradient normally exists for winds; the wind velocity increases with the height above the ground. For this situation a larger velocity vector must be added to sound waves at higher altitudes than at a lower altitude. The net effect is to have sound "bent" toward the ground for sound traveling with the wind, and to have sound rays deflected up when the wind is in opposition to the sound wave front. Over long distances, even a slight upward bending is sufficient to reduce greatly the amount of acoustic energy that will be received by an observer upwind, and a "shadow zone" results. Also, if the sound source is at a reasonably high elevation (e.g., a 200 ft tower) the downward bending of the sound wave front with the wind causes sound to be detected over abnormally long distances.

The average wind velocity will depend upon latitude, season, diurnal variations, and local disturbances. Figure 7 shows average wind velocities, measured from 0 to 3 km, at Drexel, Nebraska. Seasonal winds will be at a maximum whenever the temperature differential between the air of higher and lower latitudes is the greatest. In the northern hemisphere, this occurs in the summer at a latitude of about 45°. In general, the belt of maximum winds will shift north with the coming of summer and toward the equator with onset of winter.[12] Over level land the velocity of surface wind also undergoes a well-defined daily cycle, being least about sunrise and maximum from 1 to 2 p.m. This change is more pronounced on clear than on cloudy days.[13] The phenomenon is related to the presence or absence of thermal convection, and the mixing of surface and upper air layers. On mountain tops, this diurnal change of wind velocity is reversed, being a maximum at night and a minimum by day.

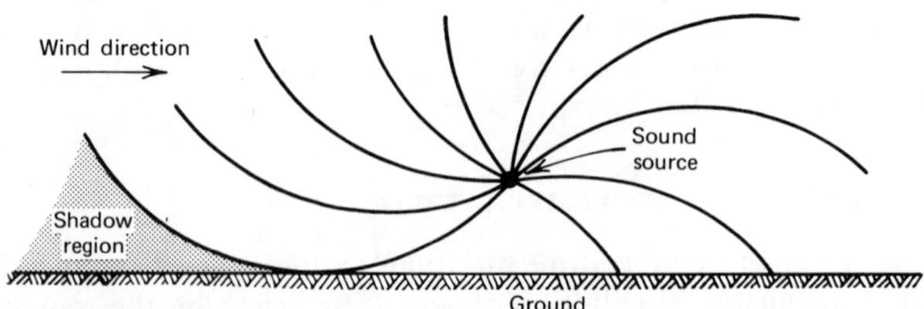

Figure 6   Effect of a wind gradient on sound propagation. The sound waves are refracted by a vertical wind shear, and a "shadow region" can exist where little sound is detected.

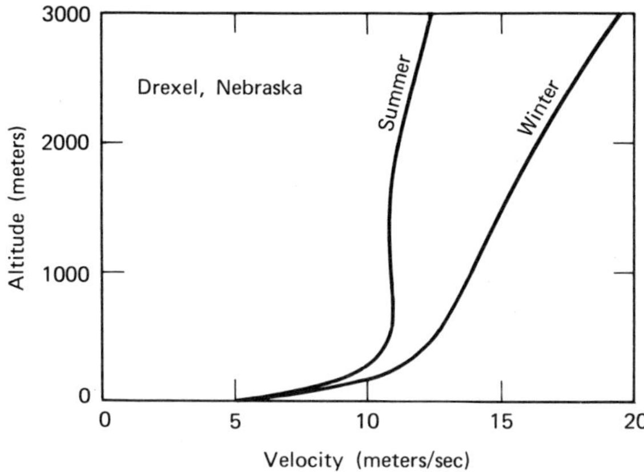

Figure 7   Average wind velocity as a function of altitude over Drexel, Nebraska —based on observations of the U.S. Weather Bureau taken over a period of several years.

Figure 8 shows qualitatively how "shadow" or "skip zones" can be created when there is a sudden change in wind velocity with altitude.[14] Sound rays that are tangent to the two-layered interface represent the limiting rays for the first audible region. Rays with a larger angle of elevation will enter the second layer and be bent downwards to form a second audible region.

The temperature gradient also has a pronounced effect on sound propagation. The "average" altitude-temperature profiles over Europe for both summer and winter, up to 10 km, are given in Table 1.

Table 1 represents a *normal* temperature gradient in which the temperature of the air decreases with height, about 0.5°C/100 meters at low altitudes. Since the velocity of sound decreases with a decrease in temperature, waves leaving a source will be bent upward as shown in Fig. 9*a*. The opposite effect takes place with a "temperature inversion" in which cool air is close to the ground and the air temperature increases with altitude. As shown in Fig. 9*b*, with such a gradient the top of the wavefronts will travel faster than the bottom of the wavefronts with the result that sound is effectively bent toward the ground.

Consequences of such an inversion are often dramatic. In winter, when such inversions often occur, explosions and loud sounds can be heard over very great distances. Some temperature inversions frequently occur over land surfaces just after sundown. The effect causes marked changes in the sound pattern from airports, trains, highways, and so on.

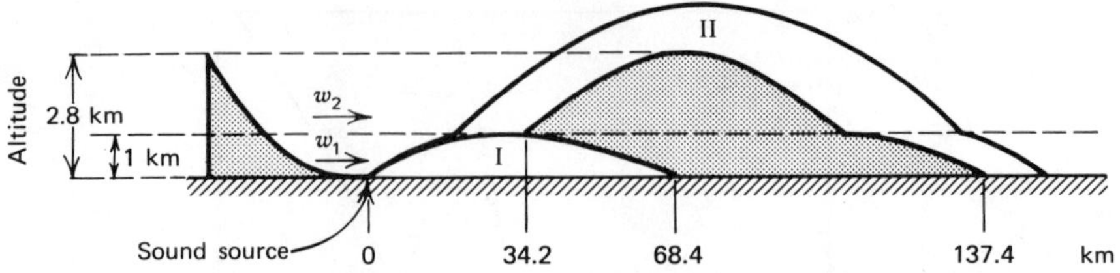

Figure 8   Both "shadow" and "skip" zones can be created if the wind velocity changes markedly at different altitudes. In the illustration $W_2 > W_1$ the sound ray that is tangent to the top of the first layer represents the limiting ray for the first audible zone. Rays with larger angles of elevation will enter the second and will subsequently be deflected downward to form a second audible zone. [From U. Ingard, *J. Acoust. Soc. Am.*, **25**, 405 (1953).]

When the sound is bent toward the ground, multiple reflections can occur. The multiple reflections are very effective in transmitting sound, especially if the surface is a frozen lake. Occasionally, with an inversion over a lake and on a quiet day, it is possible to understand speech at distances up to a half a mile or more. Also, the sound of an aircraft rising through such an inversion may drop suddenly, as noted by an

Table 1   Summer and Winter Temperatures vs. Altitude[15]

	Temperature (°C)	
Altitude (km)	Summer	Winter
10.0	−45.5	−54.5
9.0	−37.8	−49.5
8.0	−29.7	−43.0
7.0	−22.1	−35.4
6.0	−15.1	−28.1
5.0	−8.9	−21.2
4.0	−3.0	−15.0
3.0	+2.4	−9.3
2.5	+5.0	−6.7
2.0	+7.5	−4.7
1.5	+10.0	−3.0
1.0	+12.0	−1.3
0.5	+14.5	0.0
0.0	+15.7	+0.7

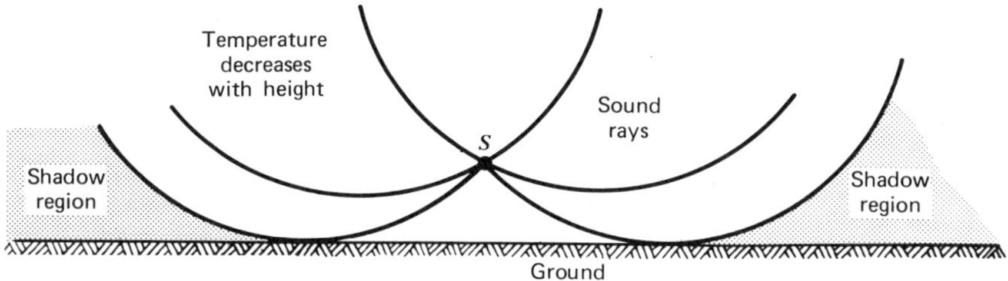

Figure 9*a* Upward bending of sound rays with normal temperature gradient.

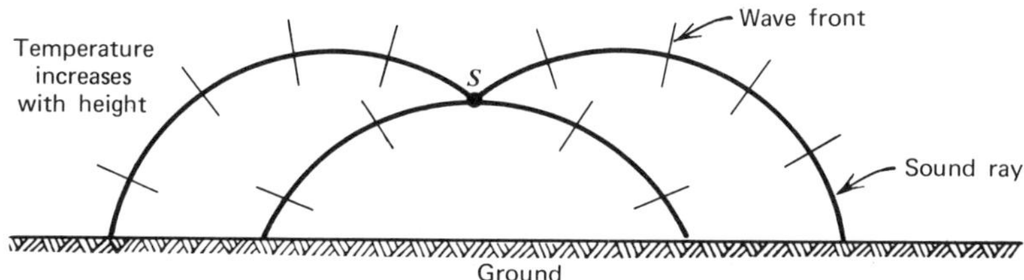

Figure 9*b* Downward bending of sound rays because of a temperature inversion.

observer on the ground, because of the highly inhomogeneous sound radiation pattern.

The combined effects of wind and temperature near the ground may be very large, and the resultant sound velocity gradient as a function of altitude may be steeply negative or positive. Figure 10 shows the substantial variation of sound velocity with altitude where c is the velocity due to temperature differentials, and $(c + w)$ indicates the sum of both temperature and wind effects.[16] These and other measurements have now fully explained the anomalous propagation of intense sounds over large distances, in which sound originating from a ground source to a height of some 25 miles is returned to earth beyond a "zone of silence" at distances of more than 100 miles.

Figure 11 is presented to show the relative contribution to sound attenuation (apart from divergence) by the several phenomena that have been discussed: classical absorption, humidity or molecular absorption, and irregular winds.[17] The bottom curve is due to classical absorption, and the next curve shows the more important effect of molecular absorption of air at 40% relative humidity. The top curve represents measurements made for irregular winds in the range of 6 to 12 m/sec, and the intermediate curve represents the attenuation from rather low velocity winds. Such data partially explains why intense sounds are not normally

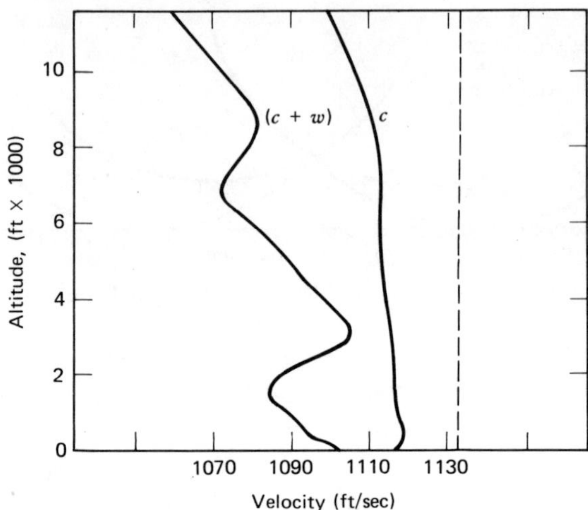

Figure 10 Sound velocity profile as a function of altitude showing the combined effect of wind and temperature [From **P.** Rothwell, *J. Acoust. Soc. Am.,* **28,** 656 (1956).]

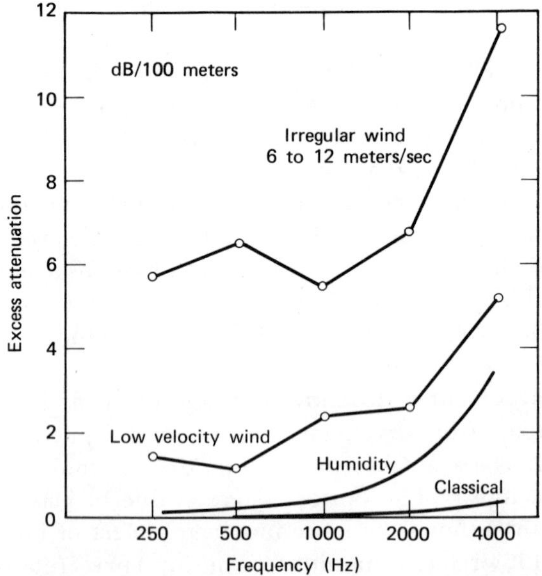

Figure 11 Diagram showing the relative attenuation per 100 meters because of clasical absorption, humidity, and moderate wind turbulence. [From U. Ingard, *J. Acoust. Soc. Am.,* **25,** 405 (1953).]

heard over very great distances. For example, if there were no other attenuation than that caused by the inverse square law, we would be able to detect the sound of an aircraft 100 miles distant.

EFFECTS OF CLOUDS, FOG, AND SMOKE

If sound waves in a clear atmosphere impinge normally on a fog bank in which the density is less ($\sim 1\%$) than the average atmospheric value, there will be little acoustic energy reflected back. In fact, reflected sound is negligible until an angle is reached corresponding to "total reflection" angle in optics. Several conditions must be fulfilled for "total reflection." First, the size of the cloud must be greater than the wavelength of the sound. Second, the boundary must be reasonably uniform. Third, the angle for total reflection will be approximated by the expression[18]

$$\frac{\sin i}{\sin r} = \left(\frac{\rho'}{\rho}\right)^{1/2} \tag{13}$$

or

$$i = \sin^{-1}\left(\frac{\rho'}{\rho}\right)^{1/2} \tag{14}$$

where i is the angle of incidence in the air whose density is ρ, and r is the angle of refraction in the cloud whose density is ρ'. If the cloud has a *relative* density, $\rho' = 99$, compared to $\rho = 100$ for a clear atmosphere

$$i = \sin^{-1}\left(\frac{99}{100}\right)^{1/2} \cong 84°$$

Thus, it is only for glancing angles that total reflection will occur, and the physical conditions are not likely to be ideal. Clouds are highly irregular in shape, the boundaries are not sharply defined and, hence, a rather small amount of sound is actually reflected from clouds or cloud banks. Conceiveably, multiple reflections could substantially alter the direction of wave propagation, but this situation is unlikely. If there is a very gradual density change, through a distance of many wavelengths, reflection is practically absent. For example, there can be no reflection from the gradually rarefied layers of the upper atmosphere.

There is a small attenuation due to fog and smoke; most of it is attributable to the molecular absorption mechanism mentioned above.[19] However, if the particle size is very minute, then at low frequencies the particles can move with a velocity that approaches that of air molecules.

The net effect is to cause a very slight additional absorption. The precise degree of additional attenuation depends on particle size, species of smoke particulate, and frequency. Tables 2 and 3 give attenuation values, in decibels per meter, of "thick fog" and artificial smoke (NH_4Cl) at several frequencies.

Table 2   Attenuation Due to Thick Fog; Visibility Limited to 50 Meters[20]

Frequency (Hz)	Added Attenuation (dB/m)
500	0.014
1000	0.02
4000	0.03
8000	0.04

Table 3   Attenuation Due to Thick Artificial Smoke (NH_4-Cl)[20]

Frequency (Hz)	Added Attenuation (dB/m)
500	0.003
1000	0.009
2000	0.035
4000	0.10
6000	0.17

EFFECT OF GROUND COVER

Sound traveling close to the ground will be attenuated to a limited extent by shrubs, bushes, and trees, and there will be some absorption by the soil itself.[21] Leaves provide some absorption, and tree trunks and limbs cause multiple scattering. An estimate of the effect of excess noise reduction, resulting from dense woods comprised of a mixture of deciduous and evergreen trees and heavy ground cover,[22] is shown in Fig. 12. It is apparent that a 100 ft wide strip of foliage will reduce traffic noise only about 2 dB at low frequencies, although there is a substantial reduction at high frequencies. Other specific measurements have been reported for the transmission of random noise through dense corn, a hemlock plantation, an open pine stand, dense hardwood brush, and over-cultivated soil.[23] Pine and brush reduced noise by only about 5 dB/100 ft at 4000 Hz, but a very dense corn planting provided an excess attenuation of about 15 dB/100 ft at 1000 Hz and 30 dB/100 ft at 4000 Hz. In general, however, ground cover must be very dense and of considerable depth to act as a practical barrier to highway or other ground-level noise.

ECHOES AND THUNDER

Echoes are distinct, separate reflections of a sound that can be discerned in open environments and in large rooms. The echoes that are perceived

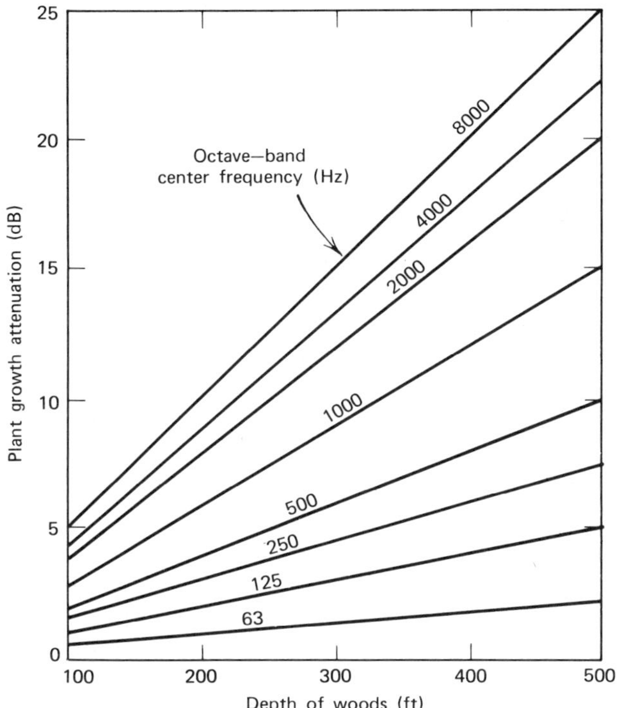

Figure 12   Approximate attenuation of sound because of dense woods having a visibility penetration of 70 to 100 ft. (Courtesy of the American Petroleum Institute, ref. 22).

in outdoor situations are often dramatic because of the large time intervals between direct and reflected sound. It has been determined experimentally that if reflected sound of sufficient intensity is received later than 50 msec after the arrival of the direct sound, an echo may be perceived. The distance between the direct and reflected sound must be at least 50 ft. In chasms or deep gorges, of course, the dimensions may be hundreds or thousands of feet, so that multiple reflections can be obtained with long time intervals between successive reflections.

Furthermore, in order to clearly discern an echo, the sound signal must be a short impulse sound, that is, syllables of speech, a whistle, a short call, or clap. Sustained tones will simply blend in with the reflections, so that reinforcement of sound may take place without evidence of discrete echoes. Reflected sounds that are heard within 50 msec are not resolved by the ear but are simply perceived as a single sound of somewhat greater loudness.

The phenomenon of thunder results from the violent electrical breakdown between the ionosphere (80–300 km) and the earth. Both the ionosphere and the earth are excellent electrical conductors, so that charges are readily conducted within each region. Normally, however, there is little vertical electrical current flow from the ionosphere to the earth (about 4×10^{-16} ampere/cm^2 or 2×10^3 amperes over the entire earth).[24] On the occasion of an electrical storm, however, a discharge current may approach a million amperes with intense heating and molecular dissociation accompanying the lightning. The result is a rapid gaseous expansion simulating an explosion that produces a compressional sound wave of large amplitude in the surrounding atmosphere.

The sound of thunder is seldom heard at distances greater than 20 miles, which is less than the distance over which cannon are sometimes heard. A partial explanation is that under the usual atmospheric conditions during a thunderstorm the sound is refracted in a manner to make it inaudible in too remote locations. Figure 13 shows the general refraction phenomenon. If the thunder occurs at a height z in kilometers, a mean temperature of $T°K$, at a lapse rate γ (temperature decrease as a function of height, in °C/km), the "critical" ray that strikes the earth at grazing incidence is determined by the equation of a parabola:[25]

$$x - x_0 \approx 2\left(\frac{T_z}{\gamma}\right)^{1/2} \tag{15}$$

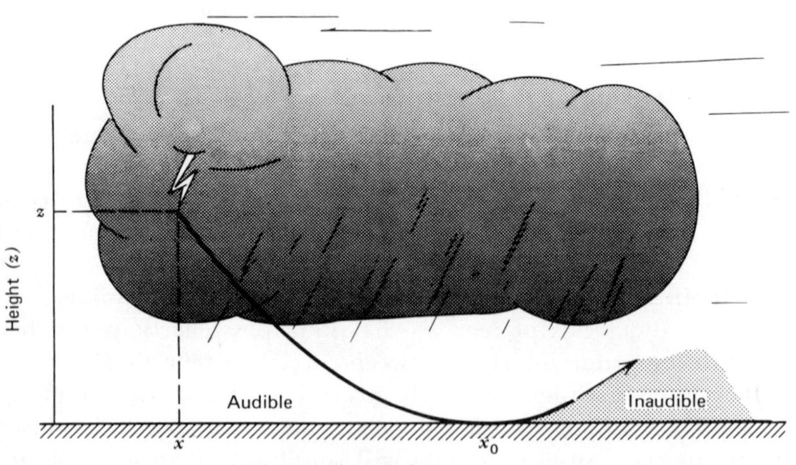

Figure 13   Refraction of a sound ray produced by a lightning flash. The path of a "critical ray" depends upon the temperature gradient and an inaudible zone may exist beyond x_0 for sound produced at an altitude, z.

where $x - x_0$ is the range of audibility. When lightning is visible at distances greater than the range of audibility, it is called *heat* or *sheet* lightning.

The differences in the patterns of sound from thunder can be explained generally on the basis of (1) the intrinsic nature of the discharge (2) the lightning-observer distance, and (3) the complexity of the lightning discharge path. The primary discharge will usually emit a very wide band of frequencies. Indeed, a substantial amount of acoustic energy may be in the infrasonic. If this is the case, we observe low frequency vibrations in windows and building structures, for example.

For lightning discharges that are more or less vertical (close to earth) the length of the lightning path may be approximately a mile.

In Fig. 14*a* the distances from all points of the discharge to the observer are approximately equal, and all sound generated by the lightning will reach the observer about the same time. If the lightning-observer distance is small (~1/2 mile), there will be substantially less absorption of the high frequencies than at extended distances, and the thunder will be short and intense.

Consider now a cloud-to-cloud discharge extending over many miles over a complex path (Fig. 14*b*). In this case the thunder will be substantially prolonged. If the total path is 20 miles (*A* to *F*), the sound from *F* will arrive at the observer long after the sound generated at *A*. The more distant sources of sound will also have much of their high frequency components filtered out by air absorption, the intensity will be less, and the net effect of the complex path is to generate a "rumbling" whose intensity and frequency vary throughout the entire thunder period. In most cases this "rumbling" is about the same over such natural areas as oceans, prairies, and mountains. In general, the sound pattern is not due to echoes or "complex reflections from the clouds." Rather, it is explained on the basis of observer-discharge distance, and the crookedness of the electrical discharge path.

THE DOPPLER EFFECT

In the first chapter historical reference was made to the discovery of the Doppler effect. The change of frequency as a result of relative motion was also observed by Doppler with respect to light. If a source of light such as a star is moving rapidly toward the earth, the wavelength of light observed on the earth is shortened, and if the star is receding the wavelength is increased. Astronomers thus see characteristic spectral lines shift

Figure 14*a*   Path of lightning in a vertical cloud-to-earth discharge.

to the blue or shorter wavelengths of the optical spectrum when a star is approaching the earth, and when the star is receding, they see spectral lines from the star shift toward the red or longer wavelengths.

In an analogous fashion, a change in frequency can be observed for sound if there is relative motion between source and observer. If an ambulance equipped with a siren is approaching, one can observe a decrease in frequency to a lower pitch just as the vehicle passes. In general, the pitch of a tone is raised when observer and sound source are approaching, and the pitch is lowered when observer and source are becoming relatively more distant.

For sound waves there are actually three velocities to be considered in determining an overall frequency shift: (1) the velocity of the source (2) the velocity of the observer, and (3) the velocity of the air—the sound medium. Figure 15*a* suggests the manner in which the sound wavelength

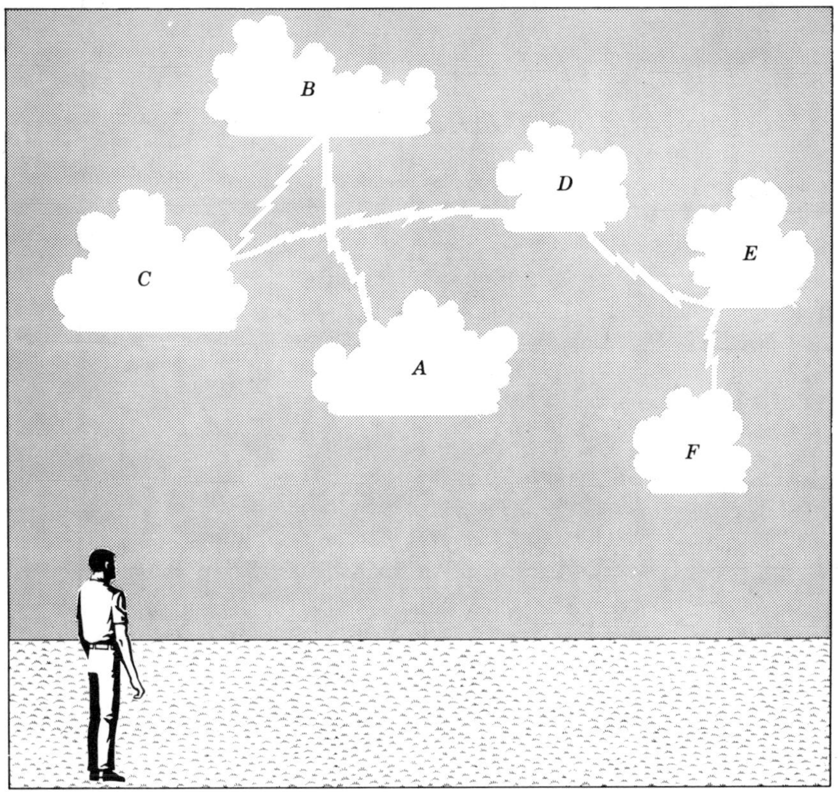

Figure 14*b* Path of lightning in a cloud-to-cloud discharge over an extended path.

is shortened (and frequency increases) for a source moving toward a stationary observer. Let us consider first the emission of sound waves from the stationary source S, for the interval of a single second. In this time the wave front will get as far as the point P, that is, a distance of SP. The wavelength will be determined by

$$\lambda = \frac{SP}{f} = \frac{c}{f} \tag{16}$$

where f is the frequency and c is the veloctiy of sound in air. In Fig. 15*b* the source S is moving to the right toward the observer with a velocity v_s. The first emitted sound wave will arrive at P in one second as in Fig. 15a, but the source itself will have reached S' where $SS' = v_s$ (for a one

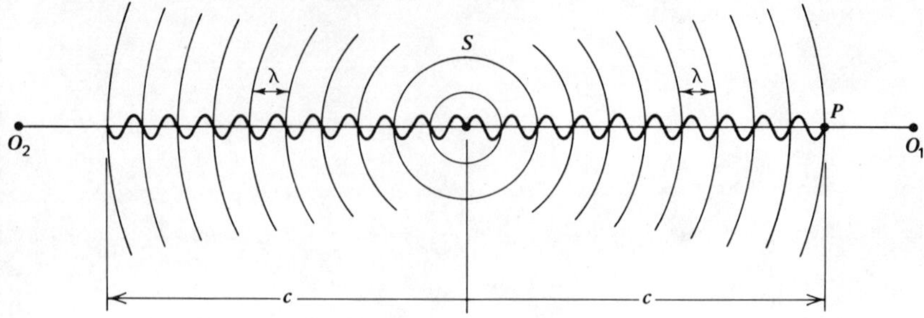

Figure 15*a*   Sound wave fronts when source and observer are stationary.

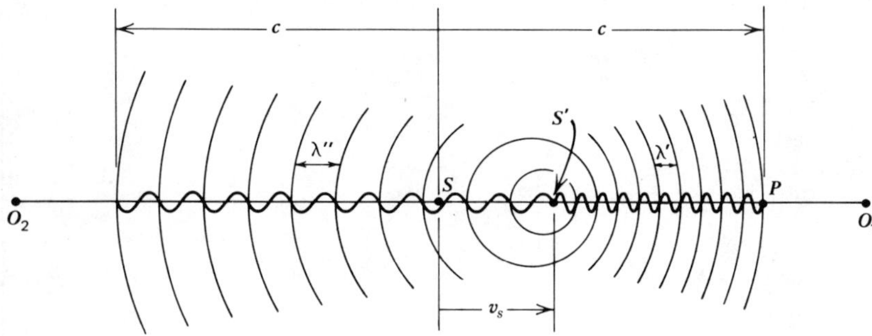

Figure 15*b.*   Sound wave fronts when source is moving and observer is stationary.

second interval). Thus, the f waves emitted by the source in this one second period are compressed into the distance $S'P$ and the wavelength λ' is

$$\lambda' = \frac{S'P}{f} = \frac{c - v_s}{f} \tag{17}$$

We thus note that the net effect of the motion of the source toward a fixed observer is to shorten the wavelength, and because the velocity of sound is independent of wavelength, the observer at 0_1 in Fig. 15*b* will detect a frequency

$$f' = \frac{c}{\lambda'} = \frac{c}{c - v_s} f \tag{18}$$

In a similar manner it can be shown that an observer at 0_2, Fig. 15*b*, hearing the source that is receding would detect a frequency

$$f'' = \frac{c}{c + v_s} f \tag{19}$$

It is interesting to note the difference resulting from equations 18 and 19 when the source is moving (1) toward or (2) away from the observer, at the velocity of sound, c. In the first case the frequency heard by the observer approaches infinity; in the second case (equation 19) the observer hears a frequency of $f/2$. Of course, when the velocity of the source is small compared to that of sound, the frequency change (approaching and receding) will be approximately the same. For fast-moving objects, however, the increase in frequency for an approaching object is substantially greater than the decrease in frequency when an object recedes.

For example, let us consider a jet plane that is flying at a low altitude over an airport toward an observer at 500 mph (733 ft/sec). Assume that the jet is emitting a 1000 Hz sound. The approach frequency will be

$$f' = \frac{1100}{(1100 - 733)} \times 1000 = 2997$$

that is, the frequency increment is approximately 2 kHz.

When the plane is receding from the observer the frequency would be

$$f'' = \frac{1100}{(1100 + 733)} \times 1000 = 600 \text{ Hz}$$

that is, the frequency decrement is 400 Hz.

If the observer rather than the source is in motion, it can be shown that the observed frequency will be

$$f' = \frac{c + v_0}{c} f \tag{20}$$

where v_0 is the velocity with which the observer is directly approaching the source. If the observer is leaving the source with velocity v_0, the observed frequency is

$$f'' = \frac{c - v_0}{c} f \tag{21}$$

A careful comparison of all the above equations indicates that the precise frequency change for a given relative velocity of source and observer depends upon which is in motion and which is at rest.

General Formula for the Doppler Effect

A general expression for the Doppler shift must also take into account the velocity of the medium. Wind velocities are generally small compared to the velocity of sound, and in most cases it is impossible to actually hear a frequency shift due to wind effects. The calculation, however, can be made easily. We can simply replace the sound velocity c, by a velocity $c + w$, where w is a steady wind velocity in the direction in which the sound is traveling. Hence, a relationship that takes into account the motion of observer, source, and wind is

$$f' = \frac{c + w - v_0}{c + w + v_s} f \qquad (22)$$

Equation 22 shows that the wind does not produce any frequency change unless there is some relative motion of the sound source and the observer.

Doppler Effect Phenomena

Some Doppler shifts are observable; others obviously are not. As we shall see in Chapter 6 the human ear can detect a very small frequency shift. A shift of 6% is easily discernible, since this is the approximate increment between successive notes of our musical scale. We have all heard the whine of heavy truck tires on a highway and the apparent decrease in pitch as a vehicle passes, if we are stationary. Generally, we do not detect an increase in frequency upon approach, unless the sound source is accelerating toward us. What we usually hear is the drop in frequency (Δf, approach minus the receding frequency) just at the time the object passes. There is a subjective effect of pitch relating to sound intensity (i.e., the apparent pitch change of a very loud siren), but this should not be confused with the Doppler effect.

What happens when the approaching sound source is traveling with the velocity of sound? In this case all wavelengths are compressed to zero, that is, there are no discrete waves, but a common wave front exists. For this condition there is no defined frequency increment; there is just an overpressure that gives rise to a sonic boom, an interaction that we will treat in Chapter 10.

We should also ask what effect, if any, can be attributed to the Doppler phenomenon, with respect to musical sounds? If a performer moves a musical instrument, such as a trumpet or violin, while generating sustained tones, would the audience hear a narrow "frequency band" rather than a single frequency? It is true that some performers choose to oscillate their instruments while playing and there is sometimes a noticeable effect. The effect, however, is primarily in changing the ratio of direct to indirect sound to the listener—which is substantially different than a

Doppler effect. On the other hand, some electronic organs have been built with rotating baffles in front of large output speakers. If the velocity of rotation is sufficiently high, such baffles can "broaden" a sustained tone in frequency, as well as change the entire pattern of the direct and diffuse sound field.

An especially interesting application of the Doppler effect relates to the motion picture industry. We have all seen movies in which some person falls from a great height into a chasm or abyss. The sound effects for this type of scene are very important from the standpoint of emotional impact. There are actually three effects that the sound engineer can apply to add realism to the action. First, the sound intensity (voice of the falling person) decreases during the descent. Second, the engineer can simulate the reverberation or echoes of the deep chasm. Finally, the engineer can change the frequency of the voice of the victim in a manner that either matches or overemphasizes the frequency shift detected by a person at the top of the chasm or gorge. In a free gravitation fall we recall that $v = gt$, when $g = 32$ ft/sec^2. Thus the velocity of fall is 32 ft/sec after the first second, 64 ft/sec after two seconds, and 96 ft/sec after three seconds. The latter velocity, compared to the velocity of sound, is large enough so that a decrease in frequency can be easily detected. Other "engineered" sound effects, utilizing the Doppler effect, include acoustic-simulated scenes of approaching trains and jet aircraft, for example.

SPEED OF EXCEEDINGLY LOUD SOUNDS

We should also note that some sonic phenomena are not amenable to simple analysis, nor do they conform to usual sound waves. If the sound is very intense, the displacement of the individual air molecules will not conform to our concept of adiabatic expansion. The density and compressibility of the air are no longer constant, and our equations do not apply. Explosions, for example, generate shock wave fronts of great steepness of slope that contain many frequencies and propagate with supersonic velocities.

In other instances, as a consequence of the moderate velocity of sound, the perception of classical sound is the last phenomenon detected. For example, consider a high velocity shell. An observer might hear, in sequence, (1) the whine of the shell, (2) its explosion, and (3) the actual firing of the cannon. The whine is noted first because the shell is continually producing sound far in advance of the initial wave front from the cannon—because of its higher velocity.

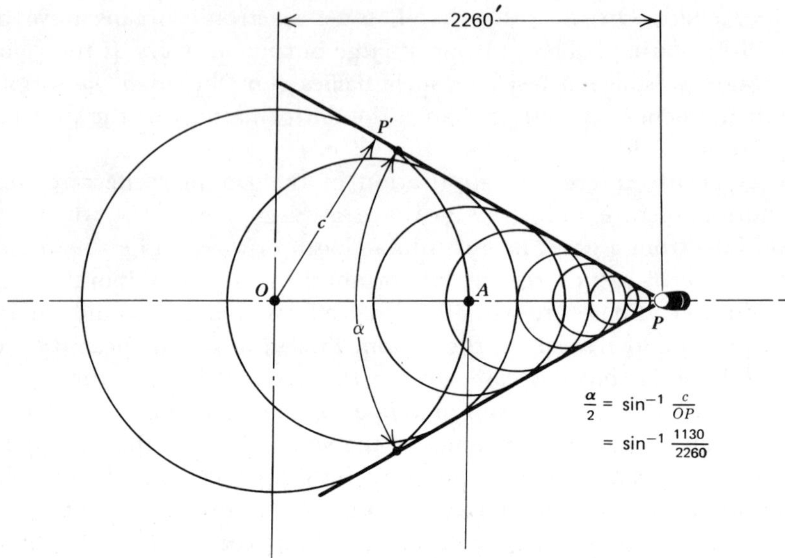

Figure 16   Angular divergence of a shock wave for a high velocity projectile.

We might also note in Fig. 16 that if a projectile moves faster than sound, the wave-front envelope is a cone whose vortex is at the source, and whose angular divergence, α, is given by the equation

$$\sin \frac{\alpha}{2} = \frac{c}{s} \tag{23}$$

where c is the velocity of sound and s the velocity of the projectile. For example, consider an object moving with twice the velocity of sound, $2c$, or approximately 2260 ft/sec. In one second the object will move from 0 to P (2260 ft), but the sound radiating from 0 will travel only to P' (1130 ft). The half-angle of divergence $\alpha/2 = \sin^{-1} c/2c = 30°$, and $\alpha = 60°$.

REFERENCES

1. F. A. White, *Mass Spectrometry in Science and Technology*, Wiley, New York, 1968, p. 233.
2. W. J. Humphreys, *Physics of the Air*, Dover, New York, 1964, p. 68.
3. *Ibid*, p. 81.

4. H. Riehl, *Introduction to the Atmosphere*, McGraw-Hill, New York, 1965, p. 19.

5. *Ibid.*, p. 12.

6. C. M. Harris (ed.) *Handbook of Noise Control*, McGraw-Hill, New York, 1957, Chapt. 3, p. 3.

7. L. B. Evans, H. E. Bass, and L. C. Sutherland, *J. Acoust. Soc. Am.* **51**, 1565 (1972).

8. E. Meyer and E. Neumann, *Physical and Applied Acoustics*, Academic Press, New York, 1972, p. 98.

9. L. L. Beranek, *Noise Reduction*, McGraw-Hill, New York, 1960, p. 189.

10. V. O. Knudsen and C. M. Harris, *Acoustical Designing in Architecture*, Wiley, New York, 1950, p 158.

11. C. M. Harris, *J. Acoust. Soc. Am.*, **40**, 148 (1966).

12. Humphreys, *op. cit.*, p. 145.

13. *Ibid.*

14. U. Ingard, *J. Acoust. Soc. Am.*, **25**, 405, (1953).

15. L. L. Beranek, *Acoustic Measurements*, Wiley, New York, 1949, p. 41.

16. P. Rothwell, *J. Acoust. Soc. Am.*, **28**, 656 (1956).

17. Ingard, *op. cit.*, p. 406.

18. Humphreys, *op. cit.*, p. 427.

19. F. M. Wiener, *J. Acoust. Soc. Am.*, **33**, 1200 (1961).

20. W. Furrer, *Room and Building Acoustics and Noise Abatement*, Butterworths, Washington, D. C., 1964, p. 9.

21. D. I. Cook and D. F. Van Haverbeke, "Trees and Shrubs for Noise Abatement," *Res. Bull. No. 246* (1971), The Forest Service, U. S. Department of Agriculture.

22. Research Report EA 7301, Committee on Medicine and Environmental Health, American Petroleum Institute, 1973.

23. D. Aylor, *J. Acoust. Soc. Am.*, **51**, 197 (1972).

24. R. G. Fleagle and J. A. Businger, *An Introduction to Atmospheric Physics*, Academic Press, New York, 1963, p. 123.

25. *Ibid.*, p. 305.

4

SOUND IN ENCLOSED SPACES

SOUND FIELDS

The distribution of sound energy from a single or multiple sound source in an enclosure will depend on the room size and geometry, and the combined effects of reflection, diffraction, and absorption. When the number of superposed direct, reflected, and diffracted sound waves is large, one can no longer talk about individual wave fronts, and reference is made to a sound *field*.

A *free field* is a region surrounding the source where the sound pattern is comparable to that in an open space. For a point source sound-wave fronts will be spherical, and the intensity will approximate the inverse square law. This *free field* in most rooms will be of limited extent because of the interaction of sound at the room boundaries and with objects within the room. But if one is close to a sound source in a large room having reasonably absorbent surfaces, the sound energy will be detected predominantly from the primary sound source and not from multiple reflections. A free field, however, will extend throughout an entire enclosure only in an anechoic (echoless) chamber of the type shown in Fig. 1. Such a chamber is typically lined with long wedges of highly absorbent material, and the "floor" consists of wire mesh suspended from the side walls. In such chambers it is possible to conduct highly controlled experiments on sound sources and directivity patterns of sound propagation.

82

Figure 1   An anechoic chamber. (Courtesy of Bell Laboratories.)

A *diffuse field* is one in which a large number of reflected or diffracted waves combine to make the sound energy reasonably uniform throughout a space. Figure 2 indicates how diffusion is achieved by multiple reflections. The degree of diffuseness will be improved if the room surfaces are nonparallel so there is no preferred direction for sound propagation. Concave surfaces having radii comparable to sound wavelengths must be avoided to prevent focusing, but convex surfaces aid in the diffusion process. Diffusion is also achieved in auditoria by the use of multiple speakers in amplifying systems, and by employing specially designed baffles that are hung from ceilings to deflect sound. In general, the spatial diffusion of sound is measured during the steady-state emission of sound from a source.

A *reverberant field* is created by reflected sounds and is time-dependent. Even after the cessation of sound source, a sound field persists for a finite interval (due to multiple reflections and the low velocity of sound

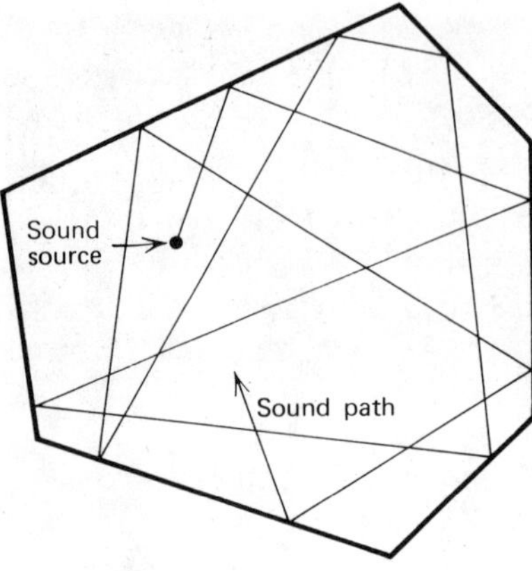

Figure 2   Diffusion of sound by multiple reflections.

propagation) and we call this residual acoustical energy the reverberant field. A somewhat broader definition than the above is simply to classify sound in an enclosure as comprising direct (free field) and indirect (reverberant) sound. A listener sitting in an average auditorium, will thus receive both primary or direct sound waves and indirect or reverberant sound as indicated in Fig. 3. The amount of sound energy arriving at the listener's ear by any single reflected path will be less than that of the direct sound for two reasons. First, since the reflected path is longer than the direct source-listener distance, there is a greater loss due to divergence; second, all reflected sound results in an energy decrease. However, the indirect or reverberant sound that the auditor hears may come from a very large number of reflection paths. As a result, the con-

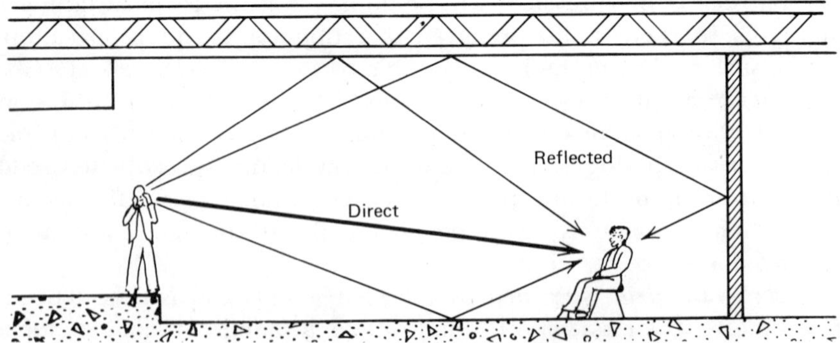

Figure 3   Acoustic energy is received by a listener through both direct and indirect sound.

tribution of reflected sound to the total intensity at the listener's ear can be larger than the contribution of direct sound if the room surfaces are highly reflecting.

Both the growth and decay of sound can be explained with the aid of Figs. 3 and 4. Assume that a short pulse of sound is emitted at $t = 0$ and that the direct sound wave reaches the auditor at t_1. The listener will detect the direct sound wave plus signals coming from a number of reflected rays at times t_2, t_3, t_4 and so on. If the arrival times between successive sound waves at the listener's ear are sufficiently short, the ear will not sense discrete signals, and the graph shown in Fig. 4 suggests the integrated sound growth curve. Furthermore, if the sound source is a steady-state tone rather than a pulse, the growth curve will continue to an equilibrium value where the total acoustic absorption rate is equal to the source power. When the source is turned off, the auditor will first be deprived of direct sound and subsequently from the sound of the many reflected wave fronts. The decay will continue until virtually all sound has been absorbed by the surfaces of the room and by the air.

If a speaker is addressing an audience in a large room (with no amplification system), it is clear that the listener will receive varying amounts of both direct and reverberant sound. At some distance from the speaker, however, the energy or sound pressure level of the direct sound field will decrease (due to spherical divergence) to a level roughly equal to that generated by the reverberant sound field. This distance is called the *room radius*, and it can be estimated by the expression[1]

$$r = \sqrt{\frac{A}{16\pi}} \tag{1}$$

where A is the total absorption in the room. This total absorption A* is computed by summing the products of all surface areas times an appropriate absorption coefficient, that is, $A = S_1\alpha_1 + S_2\alpha_2 + S_3\alpha_3 + \ldots$ where α_1 is the absorption coefficient of area S_1 ft^2, and so on. An open window will have an absorption coefficient of 1.0 because no sound is reflected. A highly reflecting surface will have an absorption coefficient approaching zero.

Clearly, a major reduction of the sound pressure level at a distance *less* than the room radius from a source of sound cannot be realized by an acoustic treatment of the walls, floor, or ceiling, since the direct sound

---

* The unit of absorption is called the *sabin*, it is the equivalent of 1 ft^2 of a perfectly absorptive surface. A surface of area S, ft^2, having an absorptive coefficient α has a total absorption of $S\alpha$ *sabins*.

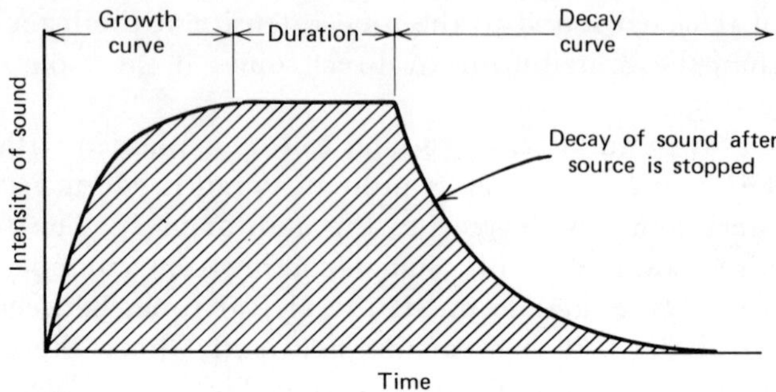

Figure 4   Buildup and decay of sound.

field predominates. Only the indirect or reverberant sound field can be attenuated by the use of absorbent materials.

GROWTH OF A SOUND FIELD

We have seen qualitatively that the sound pressure within an enclosure will increase after a source is turned on if the enclosure is "live" or reverberant. If the walls could be made perfectly reflecting and there was no air absorption, the sound level would tend to increase indefinitely because no sound energy would ever be dissipated. In practice there will be both boundary and air absorption, but the increase in sound level can be very large at a remote point in a room, compared to an "open" space. For a rectangular room that is quite reverberant and where all directions of a propagation of sound are equally probable, the growth in sound pressure at any point is approximated by the equation[2]

$$P = 1300\left(\frac{W}{A}\right)^{1/2}(1 - e^{-cAt/4V})^{1/2} \qquad (2)*$$

where P is the *rms* pressure in dynes/cm², W is the acoustical power of the steady-state source in watts, V is the volume of the enclosure in ft³, c is the velocity of sound in ft/sec, t is the time in sec, and A is the total absorption in *sabins*. (i.e., $A = S_1\alpha_1 + S_2\alpha_2 + S_3\alpha_3 + \ldots$)

* This equation (and equations 1, 3, and 4) inherently assume that the average absorption coefficient of the six surfaces of the room is not too high (of the order of 0.2 or less). Hence, equations 1 to 4 do not apply to anechoic rooms where α is typically 0.99 or more.

When t becomes large in comparison with $4V/cA$, the exponential term can be neglected and a steady-state or maximum value is

$$P = 1300\left(\frac{W}{A}\right)^{1/2} \tag{3}$$

The corresponding sound pressure level is:[3]

$$L_p = 10 \log_{10} \frac{W}{A} + 136 \text{ dB} \tag{4}$$

From equation 4 we note that for a reverberant room the steady-state sound level is independent of the volume and shape of the room, and that it depends only upon the acoustic power of the source and the total absorption. We have assumed that we are beyond the *room radius* and tacitly assumed that the buildup of sound is frequency independent.

As an example of the steady-state pressure level that is generated by a sound source, consider an enclosure having dimensions 30 by 50 by 100 ft. Assume an average absorption coefficient of 0.05, and let the acoustic power of the source be 0.005 W. The total absorption, A will be the product of the surface area (19,000 ft²) and absorption coefficient (0.05) or 950 ft² units (sabins). Hence, according to equation 4

$$L_p = 10 \log_{10} \frac{0.005}{950} + 136 \simeq 83 \text{ dB}$$

If the same enclosure had an average absorption coefficient of 0.20 instead of 0.05, L_p would only be about 77 dB.

REVERBERATION TIME

We shall now consider, quantitatively, the decay of sound after a sound source has ceased. Since energy is no longer being supplied by the source, sound waves will lose energy at each successive reflection and eventually the sound will become inaudible. In considering this decay, let us define an average absorption coefficient, α, for an enclosure by the expression

$$\alpha = \frac{\alpha_1 S_1 + \alpha_2 S_2 + \alpha_s S_3 + \dots}{S_1 + S_2 + S_3 + \dots} = \frac{A}{S} \tag{5}$$

where α_1, α_2, α_3 are the absorption coefficients of the areas S_1, S_2, S_3, and so on. Each time a sound wave intercepts a room boundary, a fraction, α, of the incident energy is absorbed, and a fraction $(1 - \alpha)$ is reflected. As indicated in Fig. 5, if the initial energy of a sound wave is E_0, the

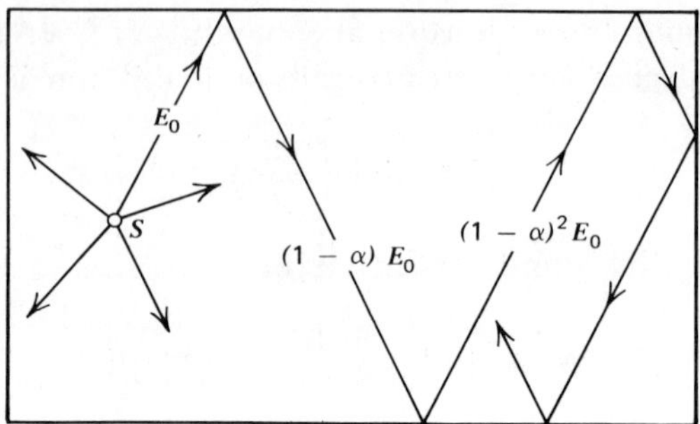

Figure 5   Decay of sound by successive reflections.

energy will decrease to $(1 - \alpha)\, E_0$ after one reflection, $(1 - \alpha)^2 E_0$ after two reflections, and so on until after n reflections, the residual energy, E_n, is

$$E_n = (1 - \alpha)^n E_0 \tag{6}$$

Assume that after n reflections the sound energy is reduced to one-millionth of its original value, that is,

$$E_n = 0.000001 E_0 \tag{7}$$

If we know the average absorption coefficient α of the enclosure, we can then compute the number of reflections, n, required to attenuate the sound. For example, if $\alpha = 0.05$, then $(1 - \alpha)$ is 0.95 and $E_n = (0.95)^n E_0 = 0.000001 E_0$. The number of reflections, n, is then found to be 269. Thus we see that in any "live" room the number of reflections that will take place before sound becomes inaudible is very large. Corresponding to this 10^6 to 1 decrease in sound intensity will be a thousand-to-one decrement in sound pressure and a $\Delta L_p = 20 \log_{10} (10^3/1) = 60$ dB.

Reverberation time, T_{60}, is now defined as the time, in seconds, that is required for sound to decay to one-thousandth of its initial sound pressure level[4] or $\Delta L_p = 60$ dB. The definition is somewhat arbitrary but it is convenient, and it corresponds roughly to the time that a loud sound in a room decays to a level of inaudibility.

Let us now determine the reverberation time of an enclosure in terms of practical room parameters, instead of in terms of the number of discrete reflections.[5] From equation 6 we have $E_0/E = (1 - \alpha)^{-n}$. Taking natural logarithms and solving for n, we have

$$n = \frac{\log_e (E_0/E)}{[-\log_e (1 - \alpha)]} = \frac{2.3 \log_{10} (E_0/E)}{[-\log_e (1 - \alpha)]} \tag{8}$$

and we have defined the reverberation time, T_{60}, as the time required for E_0 to decrease to one-millionth of its value.

$$n = \frac{2.3 \times 6}{[-\log_e (1 - \alpha)]} \tag{8a}$$

Additionally, it can be shown that the average distance between reflecting surfaces in a room, that is, *the mean free path, l,* is equal to $4V/S$,[6] where V is the room volume and S is the surface area.

The number of reflections occurring in a time T_{60} is then equal to

$$n = \frac{cT_{60}}{4V/S} = \frac{2.3 \times 6}{-\log_e (1 - \alpha)} \tag{8b}*$$

where c is the velocity of sound (1130 ft/sec). Solving for T_{60} we get

$$T_{60} = \frac{4 \times 2.3 \times 6}{1130} \frac{V}{S[-\log_e (1 - \alpha)]} = \frac{0.05V}{S[-\log_e (1 - \alpha)]} \tag{9}$$

Thus, equation 9 allows us to calculate the reverberation time of an enclosure when its volume, V, and surface area S, are expressed in cubic feet and square feet, respectively.

If V and S are expressed in cubic meters and square meters, respectively, we have

$$T_{60} = \frac{0.16V}{S[-\log_e (1 - \alpha)]} \tag{9a}$$

For the special case where α is small compared to unity ($\alpha < 0.1$), $[-\log_e (1 - \alpha)] = \alpha$ approximately, and

$$T_{60} = \frac{0.05V}{S\alpha} \qquad \text{(English units)} \tag{10}$$

$$T_{60} = \frac{0.16V}{S\alpha} \qquad \text{(metric units)} \tag{11}$$

A more exact relationship, including the effect of air absorption, is given in Chapter 16.

Reverberation time is the most important single acoustical property of a room designed for speech and music. Figure 6 is a diagram showing

---

* It can be seen that the general equation for the decay of sound in a room has the form $E = E_0\, e^{-bt}$, where

$$b = -\frac{cS \log_e (1 - \alpha)}{4V}$$

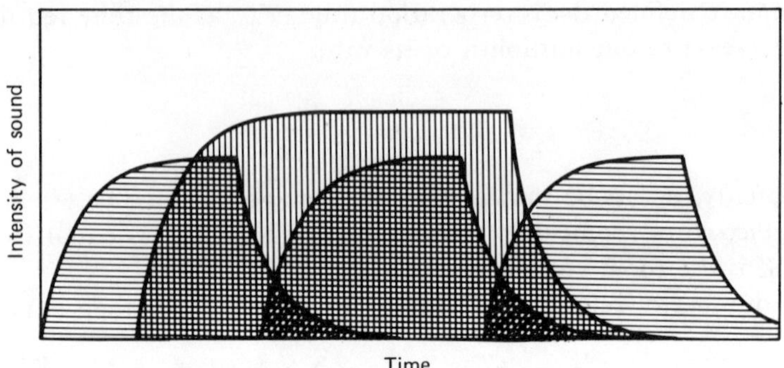

Figure 6   Overlapping sound signals in a reverberant room resulting from a sequence of musical notes.

the sound signals that overlap because of a sequence of musical notes. A very long reverberation time will lead to blurring in music and a loss of distinctness in speech. Also, a highly reverberant industrial space, in which there are many noise sources, will have a high ambient noise level.

STANDING WAVES

Almost everyone on some occasion has detected a resonance associated with certain sustained tones in rooms or auditoria. Loud sustained sounds may even have caused windows to vibrate. In many such instances vibrations can be explained by *standing waves*. Consider an enclosure consisting of parallel walls. When the wavelength of sound is equal to one of the room dimensions, a standing wave can be generated in this dimension. The word *standing* or *stationary* wave is appropriate because it implies that the *pattern* of compressions and rarefactions is fixed in space.

Consider one dimension in the room only (Fig. 7) with sound energy impinging perpendicular to a wall. Sound will be reflected back along the same path as the incident wave. If the reflected wave form exactly matches that of the incident wave, there will be no random interference and the sound waves reinforce themselves. Furthermore, if the sound source continues to emit a single frequency and the room dimension is an integral multiple of the wavelength, reinforcement of sound will continue to occur as a function of time. This type of sound reinforcement can, for highly reflecting parallel surfaces, generate increments in sound pressure level as high as 20 dB. Thus, when walking across a room in the

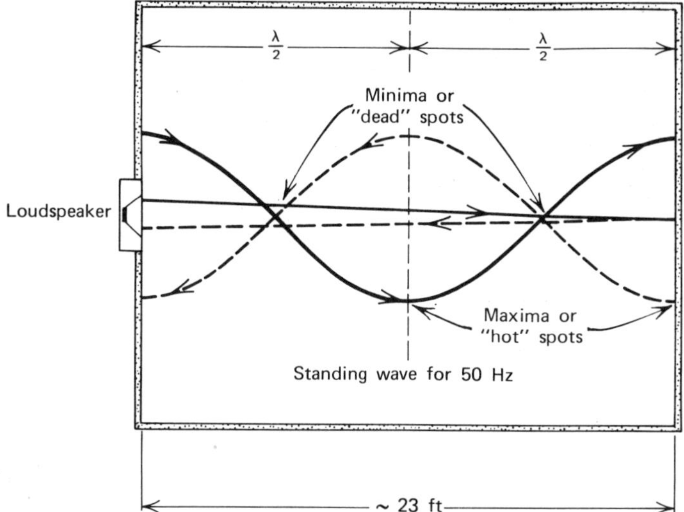

Figure 7   Standing waves in a room; a stationary pattern of compression and rarefactions can exist between parallel walls for certain frequencies.

direction of a standing wave pattern, one can detect an increase and decrease of sound intensity, because one has passed through the maxima and minima of the wave pattern. For rooms having shapes of rectangular parallelipipeds, there can be several frequencies that appear to have a preferred acoustical intensity. Also, if the sound is complex, certain frequencies in such a sound will tend to be accentuated and they may appear to decay more slowly than other frequencies. These frequencies, associated with standing waves, are sometimes called *eigentones* and they can be calculated from the room dimensions. If l_x, l_y, and l_z are the length, width, and height, respectively, of a rectangular room, these eigentones can be calculated from the expression,[7]

$$f = \frac{c}{2} \sqrt{\left(\frac{n_x}{l_x}\right)^2 + \left(\frac{n_y}{l_y}\right)^2 + \left(\frac{n_z}{l_z}\right)^2} \tag{12}$$

where c is the velocity of sound, and n_x, n_y, and n_z are integers or zero. The integral values for n_x, n_y, and n_z represent specific *modes* of vibration, that is, whether one, two, or three wavelengths correspond to a specific room dimension. The possible eigentones that may be generated in a room are then obtained by considering all the permutations of n_x, n_y, and n_z and inserting the appropriate dimensions for l_x, l_y, and l_z.

In a rectangular room, the possible eigentones can be classified as:

a. "Axial" eigenfrequencies in which two $n's = 0$; that is, waves are parallel to a room axis and only two walls are intercepted.

b. "Tangential" frequencies in which one $n = 0$; that is, waves touch four room boundaries but are parallel to the two other boundaries.

c. "Oblique" eigenfrequencies in which no $n = 0$; that is, waves progress so as to intercept all six room boundaries.

For case (a) where only an axial standing wave is generated, the lowest eigentone frequency will be

$$f = \frac{c}{2l_x} \tag{13}$$

Standing waves are an important consideration in musical acoustics where every attempt is made to maintain balance and fidelity. Architects will design auditoria so that room dimensions are not simple multiples of each other, and parallel walls are usually avoided. In pipe organ installations, builders will also decrease the loudness of the low frequency pipes whose notes coincide with the resonant frequencies of the space. This lessens the effect of standing waves. On some occasions, however, professional singers will actually take advantage of standing wave phenomena in a concert hall in order to attain a greater dynamic range for certain notes.

Standing waves have also been a contributing factor to high noise levels within industrial plants, where rotating machinery emits strong pure tone components. Standing waves can even be excited in buildings very distant from the sound. In one instance, some high-powered industrial exhaust fans were found to be disturbing people inside houses as far as a mile distant.[8] An acoustician discovered that the residents affected were primarily those with bedrooms having a 12 ft dimension, and that the high noise level was due to the radiation of a tone that generated standing waves in these rooms.

ABSORPTION IN ACOUSTIC MATERIALS

All materials comprising the boundaries of an enclosure will both absorb and reflect sound. The *reflection coefficient, r,* is defined as the ratio of the amplitude of the reflected wave to the amplitude of the incident wave (Fig. 8). The fraction of the energy that is absorbed when a sound is incident upon a material is called *sound absorption coefficient, α,* of the

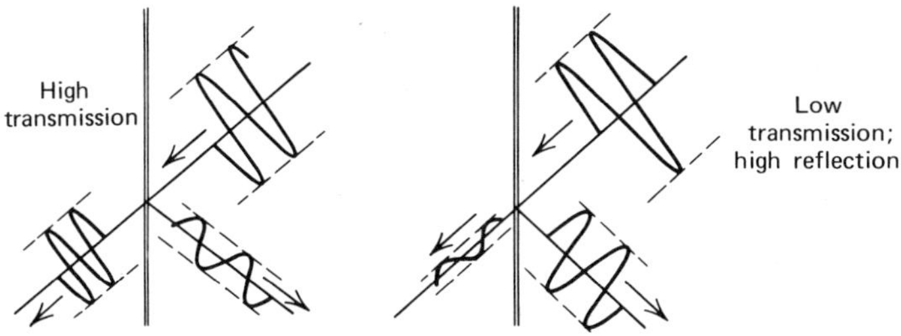

Figure 8   Schematic diagram of a sound wave being transmitted, absorbed, and reflected at a boundary.

material. Because the energy in a sound wave is proportional to the square of the amplitude, these two coefficients are related by

$$\alpha = 1 - r^2 \tag{14}$$

The actual fraction of absorbed energy will vary with the angle of incidence and the frequency. Materials that are used for room surfaces, however, are usually exposed to sound waves that impinge on them from many different angles. This is especially true of multiple reflections. Hence, published coefficients of most materials are for "random incidence" as distinguished from "normal (perpendicular) incidence." Actually, there is no good angle-absorption correlation for surfaces and acoustical materials. At high frequencies, the absorption coefficient in some materials is roughly the same at all angles; at low frequencies, random-incidence absorption tends to be greater than for normal incidence. However, the coefficient of absorption, α, for materials varies substantially as a function of frequency, and it is common practice to measure coefficients at six standard frequencies, 125, 250, 500, 1000, 2000, and 4000 Hz.

The actual absorption process can be explained qualitatively. In order that a material absorb sound energy, three conditions must be fulfilled. First, the *surface* must be relatively transparent to sound. In a real material this *acoustical transparency* may be found in a highly porous material, a perforated board that covers a porous backing material, or a thin but flexible membrane that covers a fiber-type material. Second, the material must possess sufficient depth and an internal structure so that within the material the vibratory energy of the sound wave is transformed, by friction, into heat. A third parameter affecting sound absorption is the structure to which the acoustical material is bonded. If the material is mounted directly to a rigid wall, there will be reflections back through the material. If, however, acoustic tile is suspended a foot or more below a structural ceiling, the volume of air between the ceiling and tile acts as an additional enclosure to trap sound waves. The mounting of acoustic material is especially important at low frequencies.*

Absorption Coefficients

A list of absorption coefficients for a selected group of building materials and furnishings is presented in Table 1. Note that some materials are quite frequency dependent. An average or effective absorption coefficient is determined by an arithmetical average of the values at the six frequencies mentioned above. Table 2 lists only a few of the large number of "acoustic tile" that are employed for sound conditioning. The absorption coefficients of these materials at frequencies of 125 and 250 Hz can be increased to 0.5 or greater if they are suspended from ceilings with an appreciable air space (~16 in.) between the tile and the structural floor above. A general theoretical analysis of the acoustical properties of fibrous absorbent materials is reported elsewhere.[9]

Noise Reduction Coefficients (NRC)

Commercial acoustical materials such as those listed in Table 2 are used widely to achieve noise reduction (Chapter 11). Because noise is often broad band, it has been customary to utilize a single figure of merit for a material, which provides an index of its absorption and noise reducing efficiency. Such an index is called a *noise reduction coefficient* (NRC). It is obtained by averaging arithmetically, the absorption coefficients at the four frequencies 250, 500, 1000 and 2000 Hz. The NRC of a material is then expressed to the nearest multiple of 0.05. This figure is admittedly somewhat empirical, but it is useful in estimating the amount of absorbent material that will be required to reduce noise levels or shorten the reverberation time. When the NRC coefficient is given for commercial products, it is necessary to include the particular type of mounting to which the rating applies.

GENERAL NOISE REDUCTION IN ROOMS

The use of appropriate acoustical materials can be effective in (1) limiting the growth of a sound field (2) decreasing the reverberation time, and

---

* Low frequency absorption can sometimes be increased if the acoustic material is mounted one-quarter wavelength out from the structural surface.

Table 1 Sound Absorption Coefficients of General Building Materials and Furnishings[a]

Materials	Coefficients					
	125 Hz	250 Hz	500 Hz	1 kHz	2 kHz	4 kHz
Brick, unglazed	.03	.03	.03	.04	.05	.07
Brick, unglazed, painted	.01	.01	.02	.02	.02	.03
Carpet, heavy, on concrete	.02	.06	.14	.37	.60	.65
Same, on 40 oz hairfelt or foam rubber	.08	.24	.57	.69	.71	.73
Same, with impermeable latex backing on 40 oz hairfelt or foam rubber	.08	.27	.39	.34	.48	.63
Concrete block, coarse	.36	.44	.31	.29	.39	.25
Concrete block, painted	.10	.05	.06	.07	.09	.08
Fabrics						
Light velour, 10 oz/yd², hung straight, in contact with wall	.03	.04	.11	.17	.24	.35
Medium velour, 10 oz/yd² draped to half area	.07	.31	.49	.75	.70	.60
Heavy velour, 18 oz/yd², draped to half area	.14	.35	.55	.72	.70	.65
Floors						
Concrete or terrazzo	.01	.01	.015	.12	.02	.02
Linoleum, asphalt, rubber or cork tile on concrete	.02	.03	.03	.03	.03	.02
Wood	.15	.11	.10	.07	.06	.07
Wood parquet in asphalt on concrete	.04	.04	.07	.06	.06	.07
Glass						
Large panes of heavy plate glass	.18	.06	.04	.03	.02	.02
Ordinary window glass	.35	.25	.18	.12	.07	.04
Gypsum board, ½ in., nailed to 2 × 4's 16 in. on center	.29	.10	.05	.05	.07	.09
Marble or glazed tile	.01	.01	.01	.01	.02	.02
Plaster, gypsum or lime, smooth finish on tile or brick	.013	.015	.02	.03	.04	.05
Plaster, gypsum or lime, rough finish on lath	.02	.03	:04	.05	.04	.03
Same, with smooth finish	.02	.02	.03	.04	.04	.03
Plywood paneling, ⅜ in. thick	.28	.22	.17	.09	.10	.11
Water surface, as in swimming pool	.008	.008	.013	.015	.020	.025

[a] Data furnished by the Acoustical and Insulating Materials Association.

Table 2[a]

Manufacturer	Tile[b]		Absorption Coefficient				
		125Hz	250Hz	500Hz	1kHz	2kHz	4kHz
Armstrong Cork	Minatone	.12	.25	.83	.87	.64	.52
Baldwin-Ehret-Hill	Hansonite	.06	.17	.61	.99	.78	.47
Celotex	Safetone[c]	.09	.30	.82	.95	.76	.61
Conwed	Lo-Tone	.05	.16	.48	.99	.86	.58
Johns-Manville	Spintone	.18	.28	.68	.95	.84	.66
Kaiser Gypsum	Mineral Fiber	.16	.16	.64	.99	.95	.65
National Gypsum	Solitude	.07	.28	.69	.99	.78	.53
Owens-Corning	Fiberglass-Vinyl	.03	.14	.52	.99	.91	.51
Simpson Timber	Classé	.06	.17	.61	.99	.78	.47
U.S. Gypsum	Auratone	.04	.16	.65	.97	.70	.50

[a] Acoustic Tile of $\frac{5}{8}$ in. thickness; mounting cemented to plaster board with $\frac{1}{8}$ in. air space. Performance data on many types of acoustical materials is available through the Acoustical and Insulating Materials Association, 335 East 45th Street, New York, New York 10017.

[b] Trade name.

[c] $\frac{3}{4}$ in. thickness.

(3) eliminating the presence of standing waves. However, no materials can affect the direct sound field (unless they are used as baffles), and there are limits to the amount of sound-absorbing material that is really useful in attenuating the overall noise level. The improvement will depend, substantially, upon the amount of sound absorption that already exists.

Consider a room wherein the energy of *reflected* sound is essentially uniform throughout. For a given sound source (or sources) the reflected sound energy will be inversely proportional to the total room absorption, A. If E_1 and E_2 represent the reflected sound energy before and after sound conditioning and A_1 and A_2 are the corresponding total amounts of room absorption, then

$$\frac{E_1}{E_2} = \frac{A_2}{A_1} = \frac{\bar{\alpha}_2}{\bar{\alpha}_1} \tag{15}$$

where $\bar{\alpha}_2$ and $\bar{\alpha}_1$ are the effective absorption coefficients corresponding to A_2 and A_1, respectively.

The noise reduction or sound pressure level decrease will then be

$$\Delta L_p = 10 \log_{10} \frac{E_1}{E_2}$$

$$= 10 \log_{10} \frac{A_2}{A_1} = 10 \log \frac{\bar{\alpha}_2}{\bar{\alpha}_1} \tag{16}$$

The graph of Fig. 9 shows the calculated sound pressure level change in terms of the absorption ratio $\bar{\alpha}_2/\bar{\alpha}_1$.

TRANSMISSION OF SOUND FROM EXTERIOR SOURCES

Sound may enter an enclosure by one or all of the following means:

1. Transmission of airborne sound through openings.
2. Transmission of sound through interior or exterior walls, floors, and ceilings.
3. Transmission of sound energy through structures that are set in motion by impact or mechanical vibration.

Even the best sound-insulating walls can be ineffective if there are small apertures around a door, and even a keyhole will transmit substantial sound. Interconnecting ventilating ducts also function as effective transmission paths.

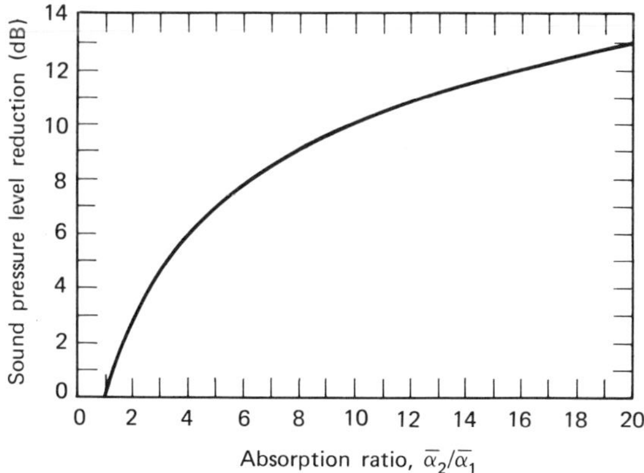

Figure 9   Decibel reduction of reverberant sound in a room as a function of the absorption ratio $\bar{\alpha}_2/\bar{\alpha}_1$; $\bar{\alpha}_2$ and $\bar{\alpha}_1$ are effective absorption coefficients of the room after and prior to acoustic treatment, respectively.

Transmission Loss

Consider a homogeneous partition separating rooms 1 and 2 (Fig. 10) and assume the partition is the only leakage path. The partition will have a characteristic *transmission coefficient* τ, analogous to an absorption coefficient. It is defined as the ratio of transmitted to incident sound energy, and the source of sound is assumed to be a diffuse sound field (in room 1). Since the transmission coefficient of most building walls or partitions is quite small (0.0001 to 0.01), it is convenient to use an equivalent nomenclature termed the *transmission loss* (TL) expressed in decibels. Transmission loss is defined by the relation

$$TL = 10 \log_{10} \frac{1}{\tau} \; \text{dB} \tag{17}$$

Using this definition a transmission coefficient of 0.0001 would correspond to a 40 dB transmission loss.

Note that for a given sound pressure level in room 1, the total sound power transmitted to room 2 will depend upon the partition area, S. Also the sound pressure level in room 2 will increase with increasing partition area and decrease with its total room absorption, A_2. The noise reduction or difference in sound pressure levels between the two rooms can then be estimated by the expression:[10]

$$L_{p_1} - L_{p_2} = TL - 10 \log_{10} \frac{S}{A_2} \; \text{dB} \tag{18}$$

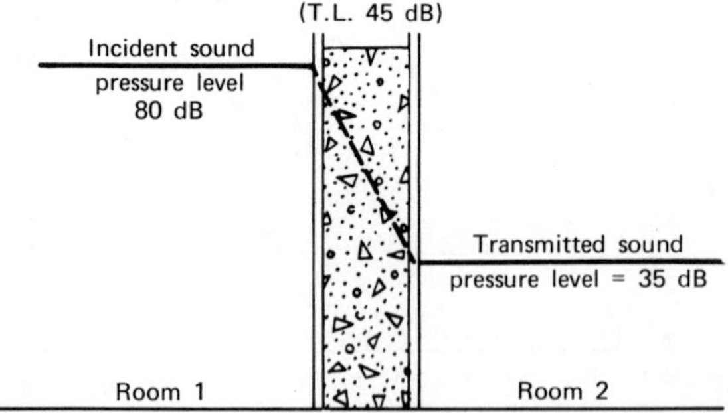

Figure 10   Transmission loss of 45 dB through a homogeneous partition. Persons in room 2 will not be disturbed by sound from room 1 unless the ambient noise level is less than 35 dB.

where

$$TL = \text{transmission loss of partition, in dB}$$
$$S = \text{area of partition in ft}^2$$
$$A_2 = \text{absorption in ``receiving'' room in sabins}$$

This formula is valid only when the sound pressure level is reasonably uniform throughout the enclosure.*

The Mass Law

Basically, a wall attenuates the transmission of sound by the inertia of its mass. If the wall is very massive, the transmitted wave will, of course, be of the same frequency as the incident wave, but the amplitude of the transmitted wave will be severely "damped." We can obtain an approximate relation for the transmission loss through a wall as a function of its mass by equating inertial forces per unit area of the wall to the difference of the sound pressures on its two sides.[11] Let p_i, p_t, and p_r be the acoustic pressure of the incident, transmitted, and reflected waves, respectively; $m = $ the mass of the wall per unit area; $v = $ the velocity amplitude of the vibrating wall; $\omega = $ the radian frequency of vibration (i.e., $2\pi f$); $i = \sqrt{-1}.$ A *sinusoidal* pressure-time dependence is assumed. A balance of inertial forces gives:

$$p_i + p_r - p_t = im\omega v \tag{19}$$

However, the transmitted energy will always be small ($p_i \approx p_r >> p_t$), so that

$$p_i \approx \frac{im\omega v}{2} \tag{20}$$

On the side of the transmitted sound wave, the displacement velocity of the wall will be the same as molecular velocity of the air, and the pressure of the transmitted wave will be

$$p_t \approx \rho c v \tag{21}$$

where ρ and c represent the mass density and speed of sound for air, respectively. Also, since the energy transmitted in a plane wave is propor-

---

* For low TL (i.e., < 10 dB) a better relation is $L_{p_1} - L_{p_2} = 10 \log_{10} (1 + A_2/S\tau)$. This takes into account the transfer of sound energy back and forth between rooms 1 and 2.

tional to the square of the sound pressure, the sound reduction or trans-mission loss, in dB, is given by

$$TL \approx 10 \log_{10} \left(\frac{p_t}{p_i}\right)^2 \tag{22}$$

so that

$$TL \approx 20 \log_{10} \frac{m\omega^*}{2\rho c} \tag{23}$$

This is the so-called *mass law*. It states that sound reduction increases logarithmically with (1) the mass of the wall per unit area and (2) the frequency of the sound wave.

Doubling the wall thickness (for a homogeneous material) should thus increase the sound reduction by $20 \log 2 = 6$ dB, for a given frequency. Likewise, a 6 dB decrement is predicted for each doubling of the sound frequency. In practice, the incident sound impinges upon a wall through-out a wide range of angles, and a more practical form of equation 23 is[12]

$$\begin{aligned}
TL &= 14.5 \log fm - 16 \\[6pt]
&= 14.5 \log m + 14.5 \log f - 16 \qquad (m \text{ in lb/ft}^2) \\[6pt]
&= 14.5 \log m' + 14.5 \log f - 26 \qquad (m' \text{ in kg/m}^2) \quad (24)
\end{aligned}$$

where TL is in decibels, f is the sound frequency, and m and m' represent the wall mass per unit area. A plot of this relationship for frequencies from 100 to 4000 Hz is given in Fig. 11.

There are definite frequency limits to the applicability of this mass law for practical materials. At low frequencies the wall or partition cannot be considered on the basis of its mass alone; its overall "stiffness" must be taken into account. The greatest displacement of a wall will occur at its center, and the least at its boundaries—if the boundaries are constrained by other forces. The net effect is that stiffness and possible resonant modes of vibration cause a practical wall to depart from a mass law response. At high frequencies the transmission of a wall also undergoes a "coincidence dip" at a *critical frequency* wherein the wavelength in air matches a "bending wave" of the structure. Essentially, there is always some angle and frequency at which a sound wave can match a bending wave in a solid, resulting in an additional transfer of sound energy and a lower effective insulation for the wall.[13] A qualitative representation of these two effects is indicated in Fig. 12.

---

* A slightly more accurate expression is $TL = 10 \log [1 + (m\omega/2\rho c)^2]$.

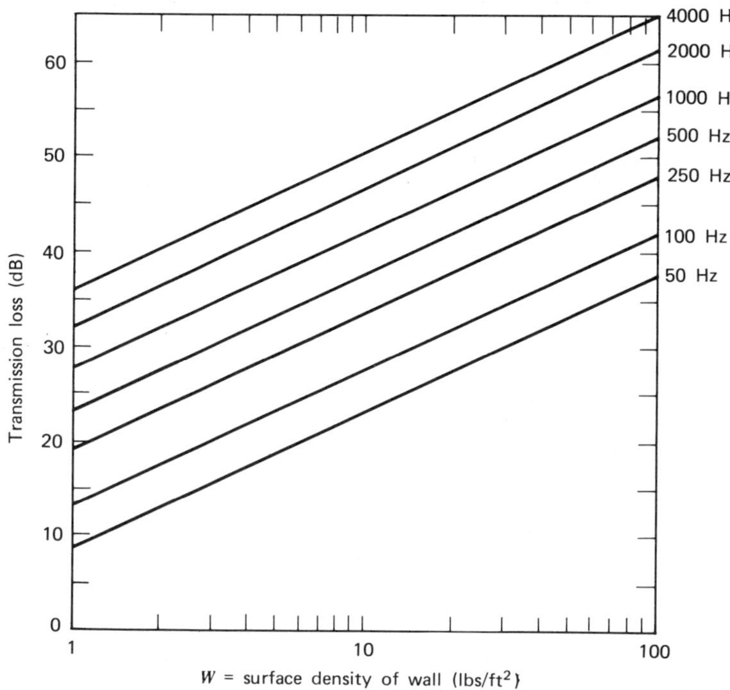

Figure 11   Transmission loss of a homogeneous wall as a function of surface density and frequency of sound. (From *Acoustic Design and Noise Control*, by Michael Rettinger, 1973, reprinted with permission of the Chemical Publ. Co., Inc.)

Common building materials of reasonable densities that are effective in providing acoustic insulation include concrete, plaster, brick, and glass (when used in sufficient thickness). Lead has also been used for building partitions, when adhesively bonded to plywood and gypsum board.[14] From the discussion above it is clear that ordinary floor carpeting will do little to attenuate the *transmission* of sound between floors, and that a door constructed of lightweight acoustic tile would be acoustically transparent.

There are some double walls, however, that are substantially more effective than the mass law predicts. For example, the typical aircraft cabin wall, having a surface density of 1.5 lb/ft², consists of two 0.050 in. thick aluminum alloy sheets spaced 4 in. apart with a glass fiber blanket between them. The transmission-loss properties of this double wall construction are indicated in Table 3. Such a double wall has sound-

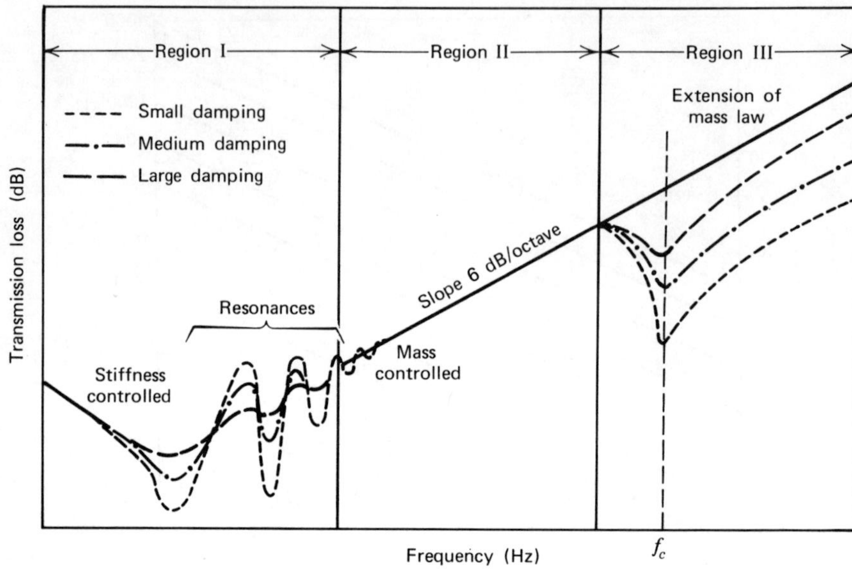

Figure 12   Over a limited frequency range (Region II) the sound insulation will be mass dependent. The critical frequency, f_c, will depend on the material and its thickness. (From E. M. Diehl, *Machinery Acoustics*, Wiley, 1973.)

insulating properties, at high frequencies, that are substantially better than the mass law predictions of Fig. 11.

Leaks and Flanking

In many instances the minimum sound pressure level that can be attained in a room is essentially determined by "leaks" and "flanking," rather than by the inherent transmission rating of one or more homo-

Table 3   Transmission Loss vs. Frequency for Aircraft Cabin Walls[15]

	Frequency (Hz)					
	125	250	500	1000	2000	4000
Transmission loss (dB)	15	21	31	44	60	65
Mass law estimate (dB)	17	21	26	30	34	39

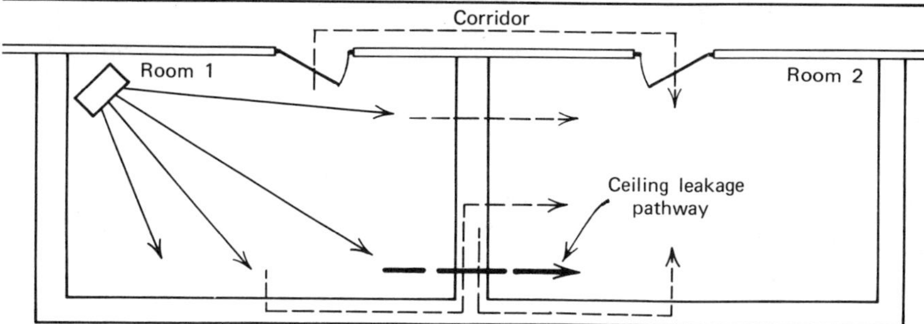

Figure 13   Flanking paths from room 1 to room 2 include floors, ceiling, and corridors.

geneous walls. Leaks usually refer to holes or openings that permit the direct transmission of airborne sound into a room. Common leaks include cracks around doors, windows, and apertures that accompany electrical and plumbing conduits into adjoining rooms through a common wall. Such leaks can degrade the acoustic properties of an otherwise acceptable enclosure. An aperture of only 1 in.^{2} through a 40 dB wall will transmit as much acoustic energy as nearly 100 ft^2 of the wall.[16] Also the better the acoustical quality of the wall, the more pronounced is the effect of a leak. Lightweight doors and windows are also sources of effective leaks; and on exterior walls, windows are usually easy transmission paths.

Flanking usually refers to the entry of sound into a room via indirect paths, either through structural vibrations or airborne paths. As indicated, in Fig. 13, flanking can occur via floors, ceilings, roof spaces, corridors, stairwells, and heating and ventilating ducts. In commercial buildings, flanking often occurs in the space between a "hung" acoustic ceiling and the structural ceiling, when room partitions do not extend to the full height of the structural ceiling.

REFERENCES

1. B. F. Day, R. D. Ford, and P. Lord, *Building Acoustics,* Elsevier, London, 1969, p. 96.

2. V. O. Knudsen and C. M. Harris, *Acoustical Designing in Architecture,* Wiley, New York, 1950, p. 145.

3. *Ibid.,* p. 146.

4. *Ibid.,* p. 152.

5. A. E. Bates and M. E. Pillow, *Proc. Phys. Soc.,* **59,** 535 (1947).

6. W. C. Sabine, *Collected Papers in Acoustics,* Dover, New York, 1964, p. 38.

7. W. Furrer, *Room and Building Acoustics and Noise Abatement,* Butterworths, London, 1964, p. 55.

8. J. Wargo, *New Eng.,* **1,** 42 (1972).

9. M. E. Delany and E. N. Bazley, *App. Acoust.,* **3,** 105 (1970).

10. L. L. Beranek, *Acoustic Measurements,* Wiley, New York, 1949, p. 870.

11. K. K. Schiller, *J. Sound Vib.* **6,** 283 (1967).

12. M. Rettinger, *Acoustic Design and Noise Control,* Chemical, New York, 1973, p. 130.

13. Day, *op. cit.,* p. 16.

14. P. B. Ostergaard, R. L. Cardinell, and L. S. Goodfriend, *J. Acoust. Soc. Am.,* **35,** 837 (1963).

15. Rettinger, *op. cit.,* p. 145.

16. L. F. Yerges, *Sound, Noise and Vibration Control,* Van Nostrand Reinhold, New York, 1969, p. 43.

5

AUDITORY PERCEPTION

ANATOMY OF THE EAR

The auditory apparatus of man is exceedingly complex. As a sensory system, the ear is comprised of three major components: the external ear, the middle ear, and the inner ear. Figure 1 shows the general anatomical features.

The External Ear

The visible portion of the auditory system is termed the *pinna*. It is a cartilaginous structure that extends beyond the head, and to some degree its shape aids in the reception of sound. Two external ears, of course, make possible a vastly improved sensitivity in the location of sounds in space compared to a single receiver. In certain animals the pinna is not immobile but, like an antenna, it can be voluntarily turned in the direction of a sound source. Sound energy is thus preferentially directed into the auditory canal, and some enhancement of hearing sensitivity and directional orientation is realized.

The connection between the pinna and the eardrum, or tympanic membrane, is called the external meatus. This meatus is the auditory canal transmitting the pressure fluctuations that cause the tympanic membrane to vibrate. Dimensions of the meatus are approximately 2.5 cm. in length and between 5 and 7 mm in diameter;[1] the longitudinal

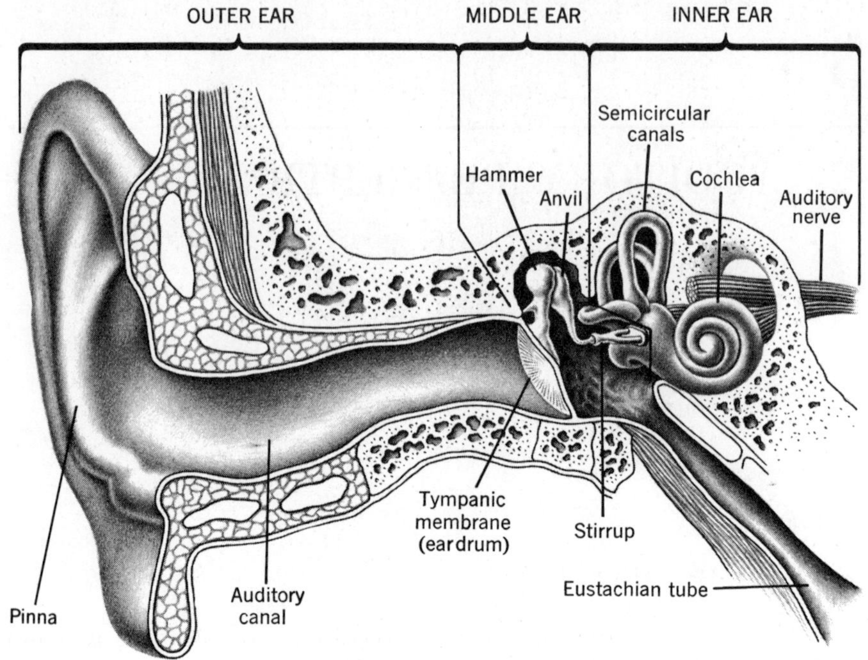

Figure 1   Sketch of the external, middle, and inner ear.

axis is roughly perpendicular to the plane of the pinna. The interior wall of this cavity is bony along two-thirds of its length, but the segment near the pinna is cartilaginous. Thus, the meatus cannot be characterized as a simple cylindrical resonator. It does, however, possess a natural resonant frequency close to 4000 Hz, and for this reason hearing sensitivity in this frequency range is increased by about 10 dB (Fig. 2).

The tympanic membrane or eardrum terminates the external meatus and separates the external and middle ear. It has the shape of a shallow cone, the apex points inward, and its axis is tilted with respect to that of the meatus. It is an exceedingly delicate membrane consisting of radial and circular fibrous tissues that are superposed. Despite its delicacy, the tympanic membrane can heal itself if not damaged too extensively. Attachment of the base of this conical membrane to the meatus is accomplished by means of a bony ring. As with a high fidelity loudspeaker, the tympanic membrane responds to pressure fluctuations over a wide frequency range, and at high frequencies the membrane is known to vibrate in many modes.

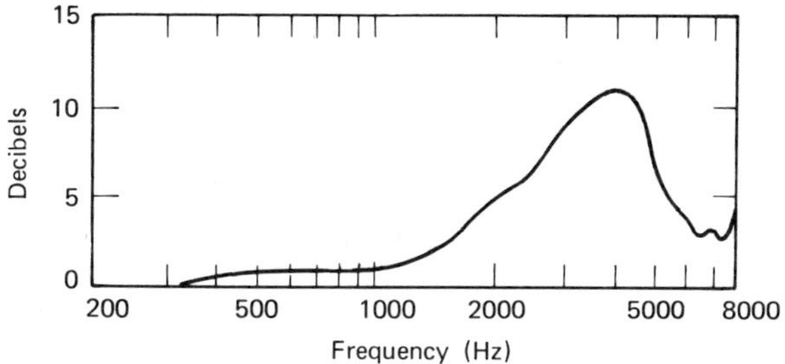

Figure 2   "Resonance" in the external ear. The ordinate shows the ratio in decibels between the sound pressure at the eardrum and the sound pressure at the entrance to the auditory canal. (From F. M. Wiener, and reported by K. D. Kryter in *The Effects of Noise on Man,* Academic Press, 1970.)

Middle Ear

The middle ear is the air-filled cavity of about 2 cm³ that contains the mechanism for transmitting the vibratory motion of the tympanic membrane to the inner ear. The air in this cavity is maintained at atmospheric pressure through the eustachian tube, for the static pressure on each side of the tympanic membrane must be equalized if the ear is to function normally. Swallowing will relieve any differential pressure that may build up in the middle ear, and the tympanic membrane can then respond to very small alternating pressures associated with incident sound waves. A chain of three small bones, termed the *ossicles,* provides a mechanical linkage between the tympanic membrane and a small membrane covering the entrance to the inner ear, called the *oval window.* This ossicular chain (Fig. 3) consists of (1) the *malleus,* or hammer attached to the tympanic membrane, (2) the *incus,* or anvil, which is an interconnecting lever, and (3) the *stapes* or stirrup that presses against the oval window covering the entrance to the *cochlea.*

This chain of tiny bones is suspended and constrained by ligaments, and by small muscles. Collectively, the ossicles, ligaments, and muscles function as a *mechanical transformer.* The ligaments and muscles provide a suspensory system for the mechanically levered bone chain. To a limited degree, the muscles also act to limit or damp out excessive mechanical motion induced by intensive sound levels. At the onset of sound waves of very large amplitude, some of these muscles contract (provided that the rise time of the pressure wave is not too short), stiffening the ossicular chain in order to attenuate sound energy transmitted to the inner ear. The ossicular chain, under intense sound, also tends to

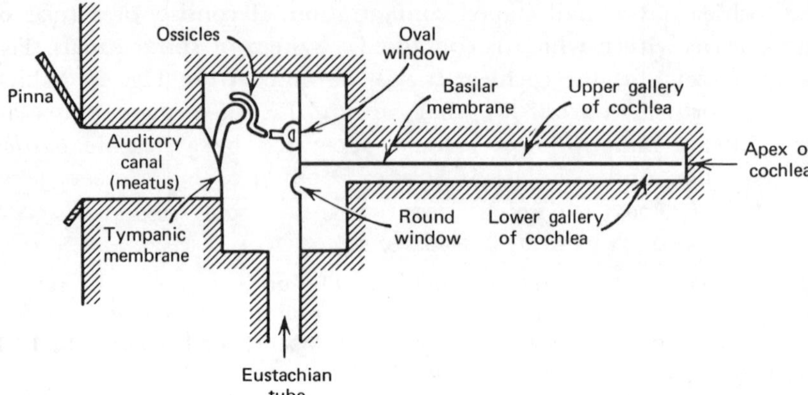

Figure 3   Schematic diagram of the ossicular chain, and its connection to the inner ear.

rotate from its normal axis to reduce the transmitted sound. In other words, the middle ear is a mechanically levered system with a built-in shock absorber or "volume control" to protect the inner ear against excessive overpressures associated with impact noise and other transients.

More important, of course, the ossicular chain serves to transfer the very small amount of energy impinging upon the eardrum to the *hydraulic* system of the cochlea. Normally, when sound waves in air impinge upon a liquid surface, most of the sonic energy is reflected because of the difference in the *acoustic impedance* of the different media. We can say that there is a poor impedance match because of the difference in the density and elasticity of the media. The middle ear thus provides us with a lever action that takes airborne energy from a large membrane (the tympanic) to move the small oval window covering the fluid system of the cochlea. The overall mechanical advantage of the ossicular chain and the membrane is in the range of 10 to 18.[2]

The Inner Ear

The inner ear is comprised of an entrance chamber that connects with the middle ear, some semicircular canals whose function is to furnish us with a sense of balance, and a *cochlea*. We shall focus our attention on the cochlea, for this organ contains the final receptor mechanism for hearing.

The cochlea has a snail-shaped configuration; it consists of a tube of about $2\frac{3}{4}$ turns within which is contained a system of spiral canals. Figure 4a is a sketch of the cochlea. It will be noted from Fig. 4b (which shows the cochlea "unrolled"), that within this tube there exists an "upper gallery" and a "lower gallery" separated by a flexible *basilar membrane*. The upper gallery is further subdivided by another membrane, called Reissner's membrane, so that there effectively is an upper, middle, and lower gallery. Each of these galleries or canals contain fluid and function as a hydrodynamic system. The upper and lower galleries are connected at the apex of the cochlea through a small opening called the *helicotrema*. An *oval* window and *round* window are also located at the base or *vestibule* of the cochlea.

The organ of Corti is the highly complex structure that rests in the basilar membrane, and it is within this organ of Corti that a large number ($\sim$23,500) of sensory hair cells are embedded.[3] They are spaced quite evenly along the basilar membrane, and they all have tiny cilia, or fila-

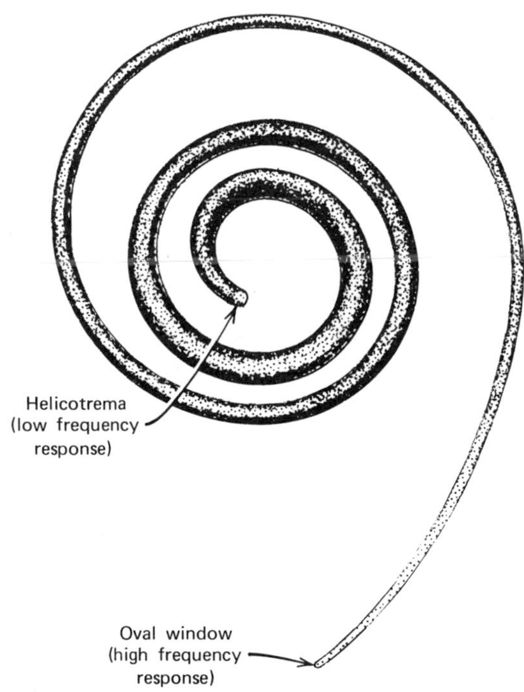

Figure 4a   Sketch of the cochlea consisting of $2\frac{3}{4}$ turns. [After E. G. Wever, *Ann. Otol., Rhinol. Laryngol.*, **47**, 37 (1938).]

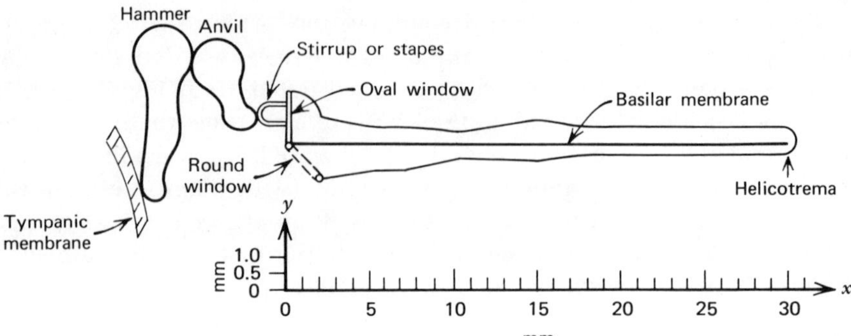

Figure 4*b*   Schematic diagram of the cochlear (unrolled), containing the basilar membrane.

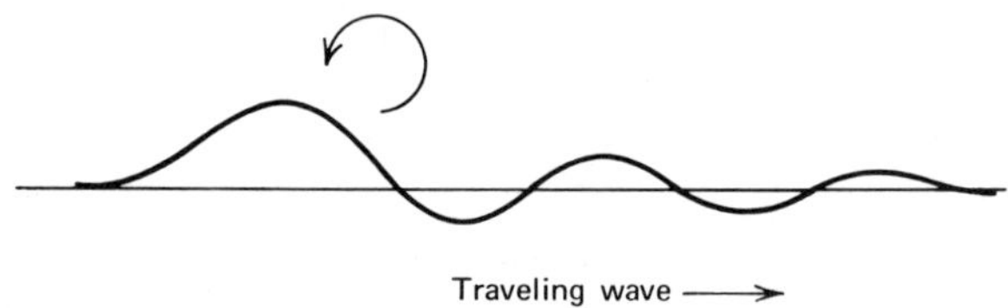

Figure 4*c*   Representation of a traveling wave on the basilar membrane, showing the displacement of the membrane at an instant in time. The curved arrow suggests eddy currents that are produced in the fluid of the inner ear.

ments, which extend into the fluid (endolymph) of the cochlear duct. These hair cells are connected to nerve fibers that are joined to the auditory nerve and thence to the brain.

Now consider the greatly simplified schematic diagrams of Figs. 3, 4*b*, and 4*c* that illustrate the action of the inner ear. The basilar membrane is flexible and is deformed when the stapes is pressed against the oval window. The fluid is quite incompressible and is relieved only by a small *round window*, that is, the fluid is forced around the apex end of the basilar membrane and the pressure causes the round window to bulge out. Basically, it is the fluid action initiated by the vibratory motion of the stapes that imparts a slight flexing motion to the basilar membrane.

A type of "traveling wave" is thus generated along the basilar membrane, accompanied by the generation of eddy currents in the surrounding fluid. Hair cells in the organ of Corti then undergo a type of shear motion or displacement, and the resulting motion generates small electrical potentials that stimulate the nerve endings.

Not all sound energy is transmitted to the inner ear only via the outer and middle ear. Sound reception by bone conduction is also possible. Experiments on bone conduction vs. air conduction in hearing one's own voice suggest that sound energy to the ear via air and bone are of the same magnitude. For this reason a person's own voice will sound different to himself than to others.

COCHLEAR POTENTIALS

The magnitudes of the cochlear potentials (CP) resulting from the motion of the hair cells have been measured by high gain amplifiers. It is now known that the inner ear functions much like a microphone, that is, a sinusoidal acoustic wave will generate a sinusoidal output voltage. The cochlear potential has no apparent threshold limit; the magnitude of the signal has been observed to extend down to about 0.0001 μV, the sensitivity limit of present experimental measurements.[4] Upper limit output voltages are close to 1 mV. The cochlear potentials observed in a cat as a function of acoustic stimulation are shown in Fig. 5.

The cochlea has a linear response over a wide dynamic range. At very high acoustic pressures, however, the output will be nonlinear, and the ear is said to be *overloaded*. For exceedingly high sound pressures, of course, the ear may suffer permanent damage as a result of an actual destruction of the hair cells. When these hair cells are destroyed, there is no means for restoring hearing.

The exact physiological process by which these cochlear potentials arise is not completely clear. Superficially, cochlear potentials seem to be generated in a manner similar to the action of a piezoelectric crystal, when a mechanical deformation of the crystal gives rise to an output voltage across it. A better analogy, perhaps, is the common water plant (*Nitella*), which when bent or pinched produces a sizeable electrical voltage.[5,6] In any event, a close correlation has been made between cochlear potentials and acoustic pressures for both humans and animals, and in congenitally deaf species where no cochlear potentials have been detected, the array of hair cells has been absent.

COCHLEAR FREQUENCY RESPONSE

Although the hair cells are distributed rather uniformly along the basilar membrane, not all portions of the membrane have an identical frequency

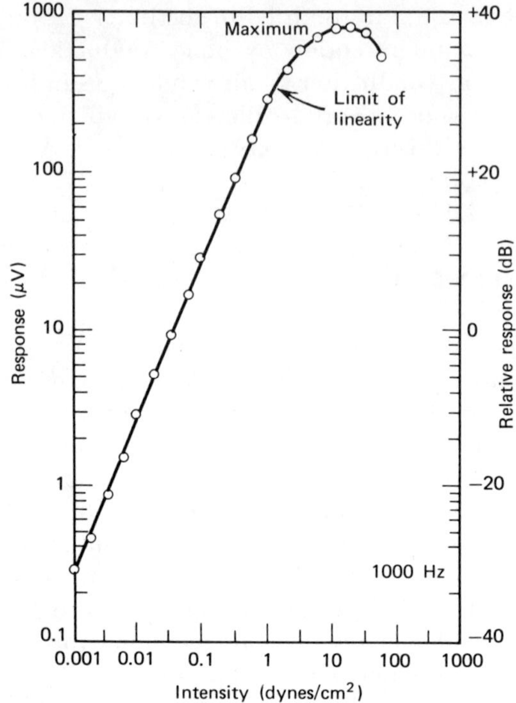

Figure 5   Cohclear potentials observed in a cat as a function of sound intensity at 1000 Hz. At very high intensity the response is nonlinear. (From E. G. Wever, *Theory of Hearing*, Dover Publications, 1970; originally by Wiley 1949.)

response. Figure 6*a* shows the frequency-position dependence of the cochlea for a basilar membrane that has been unrolled.[7] Figure 6*b* shows the approximate shape of the basilar membrane in which the width is narrower near the stapes than it is at the apex; the membrane is only about 0.08 mm wide at the oval window and 0.5 mm wide at the apex. (Note that the basilar membrane width increases as the cochlear cross section decreases).

A qualitative explanation for this frequency response is suggested by Figs. 4*b* and 6. If the stapes moves to the right, the fluid in the cochlea will undergo compression. Because the walls of the cochlea are essentially rigid, and the fluid is only slightly compressible, a vibratory motion of the stapes will yield (a) some deflection of the basilar membrane and (b) an expansion of the round window. If the motion of the stapes is very slow, that is, corresponding to one cycle per second, almost no deflection

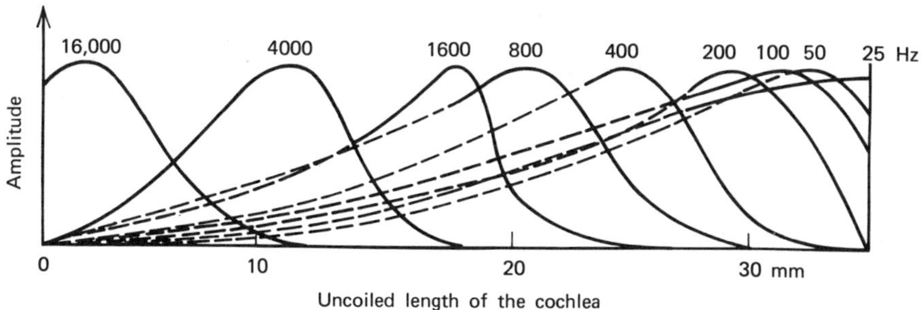

Figure 6*a* Relative displacement of the basilar membrane as a function of frequency. Distances along the cochlea are measured from the oval window. [From L. A. deRosa, *J. Acoust. Soc. Am.*, **19**, 623 (1947).]

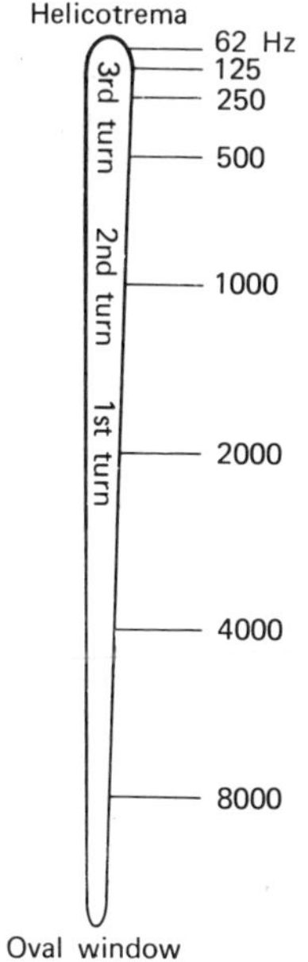

Figure 6*b* Schematic diagram of the basilar membrane. The basilar membrane width increases as the cochlear cross section decreases. High frequency sensitivity is near the oval window. [From L. A. deRosa, *J. Acoust. Soc. Am.*, **19**, 623 (1947).]

will take place in the basilar membrane. Pressure will be relieved by means of the expansion of the round window, as the impedance or constriction to fluid flow at the apex of the cochlea is very small. At somewhat higher frequencies, there will be small deflections of the basilar membrane downward, as the stapes moves inward. As the frequency is increased, the deflection of the membrane will occur nearer to the stapes, and we can consider this deflection as representing a type of hydraulic short circuit. This argument fails to account quantitatively for the specific location of frequency maxima, but it is plausible because the narrower end of the membrane should be most responsive to high frequency stimulation, and the displacement waves would be damped at high frequency near the apex of the cochlea.

THRESHOLD OF HEARING

The threshold of hearing is the minimum sound pressure level that the human ear can detect. Figure 7 is an average threshold curve for young adults with "normal" hearing. The threshold is seen to be markedly dependent upon frequency. The ear requires a minimum signal in the range of 3000 Hz. At lower frequencies the threshold signal for detection must be substantially greater, and at 20 Hz we require a signal whose sound pressure level is 70 dB, with respect to a reference sound pressure level of 2×10^{-5} Pa. At frequencies of 5000, 6000 and 8000 Hz the ear

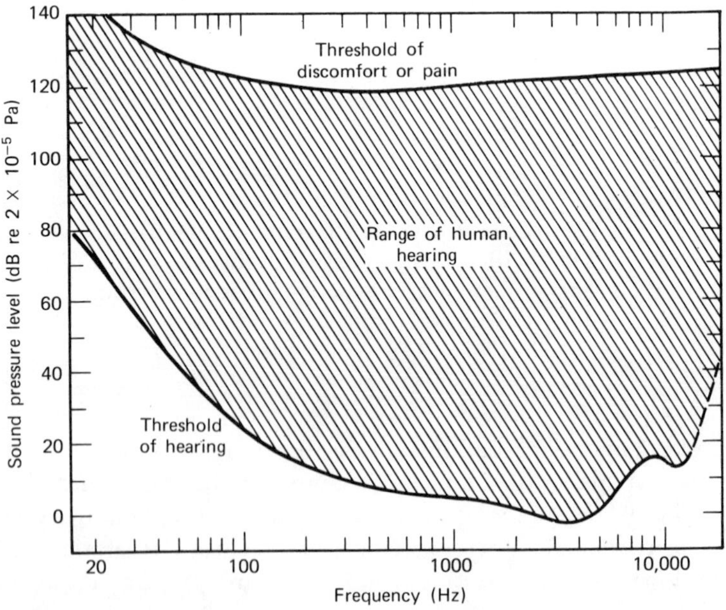

Figure 7   Range of human hearing.

is also less sensitive than at 3000 Hz, although the decrease in sensitivity is not as great. There is some speculation that the ear has adapted itself, in part, to the natural environment. In any event, many sounds in nature have a low energy content in the frequency band where the ear is most sensitive, and many natural sounds are of higher intensity in the frequency bands where the ear's sensitivity is diminished.

The sensitivity of the ear to detect a signal around 3000 Hz is truly amazing. The acoustic intensity is of the order of magnitude of 10^{-16} W/cm². The actual physical displacement of the eardrum at this level has also been found to be somewhat less than 10^{-9} cm (Chapter 2). This displacement distance is much smaller than the "size" of a single molecule, and it is orders of magnitude smaller than the wavelength of visible light. Because of the macroscopic size of the eardrum, it is difficult to envision a displacement of such submicroscopic dimensions. To date, no manufacturer has been able to build a microphone comparable to the human ear.

However, it may be advantageous that the human ear is not any more sensitive than it is, for we would then hear noises at all times. Calculations have been made indicating that we could hear the vast spectrum of noises associated with our own biological system (e.g., blood flow and mechanical noises). At maximum hearing sensitivity we are also close to detecting *Brownian motion,* that is, the statistical fluctuations of random molecular motion in the atmosphere.

Qualifications should be made to the hearing threshold curve of Fig. 7. First, this curve represents hearing for young people having normal hearing sensitivity. A loss in sensitivity always accompanies age. This loss in hearing sensitivity is called *presbycusis.* Audiometric tests show that most persons will have their threshold of hearing shifted over most frequencies. Another type of sensitivity loss is called a *temporary threshold shift* (TTS). When a person is exposed to an intense sound for a few hours, he suffers a temporary loss of hearing sensitivity. After being removed from this noise source (hours or days), he will usually recover normal sensitivity. A *permanent threshold shift* is due to (1) presbycusis, (2) occupational or other exposure to intense sound levels, or (3) physical injury or disease. These effects are discussed further in Chapter 17.

UPPER LIMIT OF HEARING

The upper limit of hearing is somewhat more difficult to determine than the lower threshold. Sometimes this upper limit is called the *terminal threshold.* Some discomfort is apparent at 120 dB, and sensations of tickle and pain are experienced when a 1000 Hz tone is about 140 dB. At this frequency the ear has a dynamic range corresponding to a 10^6 to 10^7 change in pressure or a 10^{12} to 10^{14} in intensity. At low and high frequencies the dynamic range is substantially less. We also note that at both high and low frequencies the sensation of hearing tends to merge into the nonauditory sensation of feeling or pain. Furthermore, this terminal threshold does not vary appreciably with frequency.[8]

It is appropriate also to inquire concerning the upper and lower *frequency* limits of detection. The suggestion has been made that at very high frequencies the ossicular chain cannot respond because of its mass. However, there is probably also a practical limit to the frequencies to which the auditory nervous system can respond. Because of potential danger to human subjects in high frequency and high intensity tests, not much detailed data is available. At very low frequencies ($\sim$15 Hz) sensations of feeling begin to accompany the sensation of sound. At 15 Hz, however, most persons lose all sense of pitch, because the separation in the arrival of pressure waves to the ear is too great, that is, sound loses its tonal quality. This does not mean that frequencies as low as one or two cycles per second cannot be sensed at high sound pressure levels. However, the sensation is usually attributed to tactile sensations or the generation of *aural harmonics* (Chapter 6) arising from an amplitude distortion in the ear at high sound levels.

THE EAR AS A FREQUENCY ANALYZER

The ability of the ear to analyze complex sound into its sinusoidal components was first announced as a formal principle in 1843 by G. S. Ohm, whose name attained prominence through Ohm's law of electricity. Ohm's law of auditory perception postulates that the ear is capable of performing a Fourier analysis of a composite sound comprised of many frequencies and amplitudes. Qualitatively, we know that this is true, for a musician can correctly identify five or more individual notes when a chord is played on a piano. In this sense the ear is quite superior to the eye, as the latter cannot resolve a blend of colors into the separate frequencies of which it is comprised.

We have already noted that when the ear is stimulated by pure tones, the position of maximum vibration of the basilar membrane depends upon the frequency of excitation. However, the frequency resolution of the ear is much better than is suggested by these broad vibrational maxima. For this reason it is necessary to assume a more discrete subdivision of the basilar membrane (and associated hair cells) into regions that are called critical bands.[9]

CRITICAL BANDWIDTH

The concept of a *critical bandwidth* has evolved from a large number of experiments on complex sounds. These experiments have included tests of loudness, masking, threshold sensitivity, musical consonance, and harmonic discrimination. We shall consider a critical band in connection with the sensation of loudness that will be discussed further in the next chapter.

Assume that we are listening to a pure tone of 1000 Hz, which we define as a *center frequency*. We also arrange to have a sound pressure level of 60 dB for this tone. Now let us change this tone so that it includes additional frequencies in the neighborhood of 1000 Hz, but with the constraint that the sound pressure level remains invariant. The tone will, of course, no longer sound like a pure tone, but will the tone give us a louder or softer aural sensation? Figure 8 gives the answer.

For small frequency side bands around 1000 Hz (e.g., 980–1000, 1000–1020) the quality of the tone changes, but we are unable to detect any real change in the "loudness" of the 60 dB signal. When the range of frequencies, Δf, included in the signal gets to a 160 Hz band (i.e., 920–

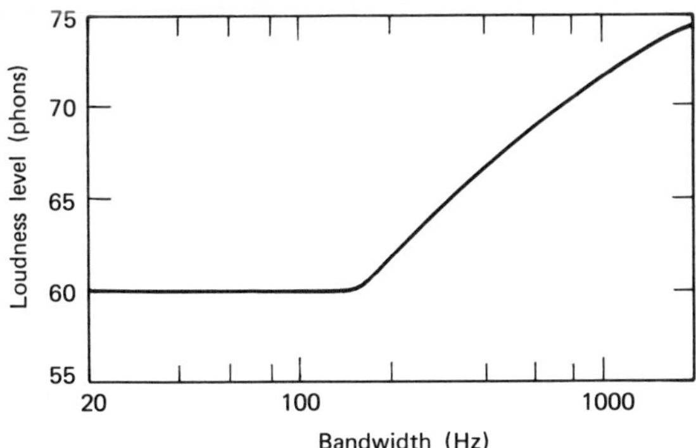

Figure 8   The effect of bandwidth on the loudness of a complex tone of constant sound pressure level (60 dB) with a center frequency of 1000 Hz. (From W. L. Gulick, *Hearing: Physiology and Psychophysics,* Oxford University Press, 1971.)

1080 Hz) a louder sensation becomes evident. As we proceed to a bandwidth, $\Delta f = 1000$, (500–1500 Hz) an even louder sensation is observed. For this particular center frequency of 1000 Hz, we say that there exists a critical bandwidth of 160 Hz.

If we would conduct a similar experiment at a higher center frequency, for example, 5000 Hz, we would observe qualitatively the same phenomenon. There would be a frequency band within which there was no increase in "loudness" and this band would include a greater frequency range (~1000 Hz). Figure 9 shows the increase in critical bandwidth that accompanies an increase in the center frequency.

Critical bands have also been the subject of study in music where it has been found that judgments of pleasantness were fairly constant when two tones are separated by greater than a critical bandwidth.[10] Other studies have also demonstrated that the highest distinguishable harmonic (see Chapter 15) is separated from other harmonics by at least one critical band. These and other tests suggest that the basilar membrane acts like

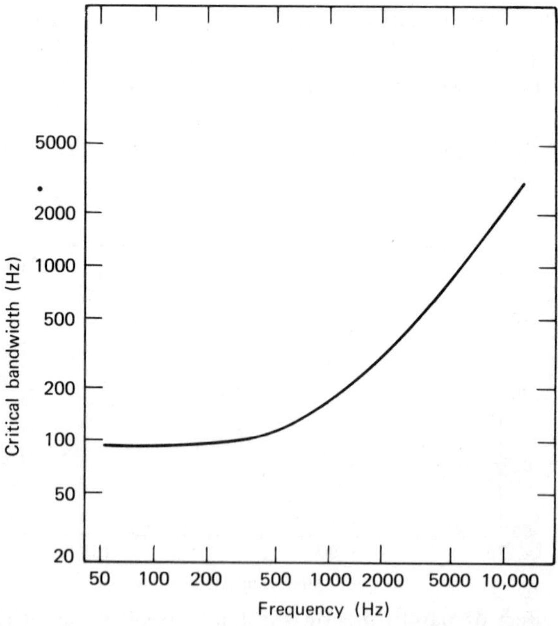

Figure 9   Approximate critical bandwidth as a function of the frequency of the center of the band. [From J. E. Hawkins, Jr., and S. S. Stevens, *J. Acoust. Soc. Am.*, **22**, 6 (1950).]

a distributed filter in which small portions of the membrane respond to the various frequency bands. A critical band corresponds to about a 1.3 mm distance along the basilar membrane, allowing some 24 critical bands (laid end-to-end, up to 16,000 Hz) to exist on a basilar membrane 32 mm long.[11] Unfortunately, little is actually known about the critical band mechanism as a physiological process.

Nevertheless, we know that the ear does function as a constant percentage bandwidth analyzer having a bandwidth of about 1/3 octave (Table 2, Chapter 7). Indeed, it is on the basis of the ear functioning like a 1/3 octave-band analyzer that the analytical instrument described in Chapter 7 was developed.

RESPONSE TO TRANSIENT PHENOMENA

Because hearing involves both mechanical and electrical processes, there will be a minimum duration time if a sound of reasonable intensity is to be detected. At very high intensities, of course, the time required for sensing a single burst of energy approaches zero. As with vision, no stimulus can be made of such a short duration that it will not be perceived if it is sufficiently intense. In practical situations the ear integrates energy according to the relation

$$I \cdot \Delta t = \text{constant} \tag{1}$$

where I is the sound intensity and Δt is the signal duration time. This relationship holds up to about 200 msec; for greater duration times this simple reciprocity relation is not valid.[12]

CHANNEL CAPACITY OF THE EAR

The ear not only has a wide frequency and dynamic range, but possesses other features characteristic of an electrical communication network or a computer. Specifically, it has a "channel capacity" in the sense that it can only accept and process information at a limited rate. This channel capacity is a function of both the sensation level of the sound and the integration time of the ear. Indeed, it has been suggested that the ear has some resemblance to a binary device, if one assumes that neural firings in the inner ear are characteristically bursts of uniform intensity, but whose rate is determined by the parameters of the sound stimulus. Thus, if unit counts are stored in the integrating mechanism of the ear after being triggered by a meaningful signal, then recognition of the signal might depend upon the presence or absence of "one preponderant least count, analogous therefore to 1 or a 0."[13]

Of course, signals such as speech arriving at the ear are not immediately translatable into binary digits. Nevertheless, the very large number of possible speech sounds and the varying sensitivity of the ear for sounds of different frequencies produce statistical effects that allow the problem of speech recognition to be treated as if the ear were functioning as a binary device. This concept has been useful in analyzing speech sounds, and in developing speech-hearing theories that can be compared to experimental speech intelligibility tests.

PERSISTENCE OF HEARING

Up to this point we have tacitly assumed that when an acoustical signal reaching the ear ceases, auditory perception also is terminated. This assumption is not completely valid. There is a finite persistence of an auditory stimulus that is comparable to that of vision. In the optical case visual images persist for about 1/50th of a second. This fact makes it possible for motion pictures and TV presentations to be viewed with an impression of continuous motion, although in reality we are looking at a rapid succession of discrete images. This finite resolving time of the eye is thus advantageous even though it places an upper limit upon high speed reading or the resolution of rapidly moving objects.

In an analogous manner, the ear has a "time lag" that prevents the instantaneous termination of an auditory stimulus. Measurements of this phenomenon for a pure tone have shown that the sound sensation decays to the threshold of hearing in 0.14 sec regardless of the initial intensity.[14] Figure 10 shows the decay curves for a 800 Hz tone where the signal was terminated at 40, 60, 80, and 100 dB above threshold, and the sensation of sound ceased in the 0.1 to 0.15 sec range.

We conclude that our ability to completely resolve sounds is limited by the physiology of the auditory system and that there is an upper limit to sensing the discrete elements of music and speech. We have already noted that a reverberant field will always cause a persistence of sound in any practical enclosure. To this must be added the small but measurable decay time of the hearing process itself. Even under ideal conditions, two tones of equal intensity will not be heard as discrete signals unless they are separated in time by about 50 msec. For shorter times the two signals will merely merge into a single signal whose quality and loudness will depend on their individual signal strengths.

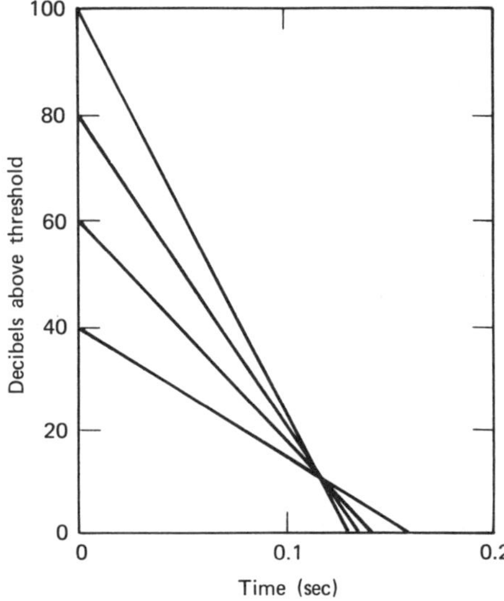

Figure 10   The persistence of hearing phenomenon. Decay· curves are shown for a 800 Hz pure tone. (From S. S. Stevens and H. Davis, *Hearing: Its Psychology and Physiology*, Wiley, 1938.)

AUDITORY FATIGUE

The ear is similar to the eye in that exposure to a prolonged stimulus results in a change in its sensitivity. Auditory adaptation, or auditory fatigue, has been observed to be a function of sound intensity and frequency. At low intensities, that is, 10 to 40 dB, a simple exponential loss of sensitivity occurs over short periods. At high intensities, there appears to be a rather rapid decrease in sensitivity followed by a more gradual fall off in the hearing sensitivity. One method for measuring the degree of "fatigue" has been to expose one ear to a sustained tone for a preset time and intensity, and then test the other unstimulated ear with a comparison tone of the same frequency. Thus, by many successive trials, a match to the loudness decrement of the exposed ear can be estimated.

Figure 11 shows the results of an experiment designed to measure auditory fatigue using this methodology.[15] The stimulus was a 1000 Hz tone of moderate intensity, and the sustained tone was presented to one ear for periods of 2 to 120 sec. Immediately after a given period the other ear of the listener was presented with a tone for matching or balancing. A loss of about 10 dB is noted during the first 40 to 45 sec, with little observable loss during the final minute of exposure. This general type of sensitivity loss has been confirmed by numerous other experiments.

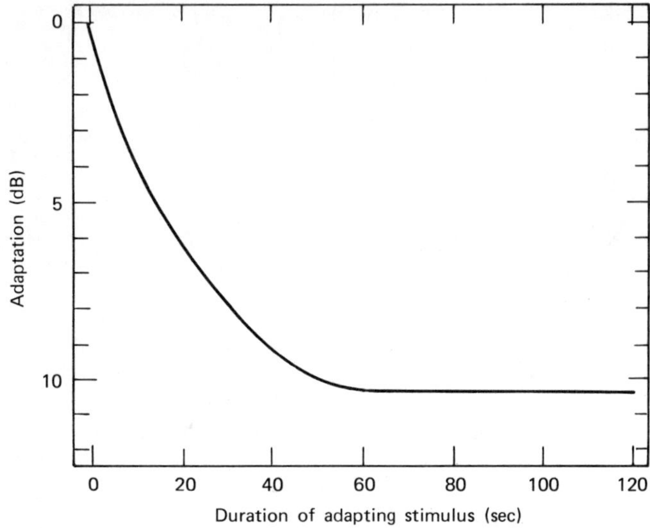

Figure 11   Relative adaptation or fatigue of the ear (in dB) as a function of the adapting stimulus, in seconds, for a tone of constant intensity. (From A. G. Wood, Ref. 15.)

MASKING

Masking is the phenomenon in which one sound is obscured by another.* Intuitively we might expect a signal to be masked by a second signal if (1) the second signal is more intense and (2) both signals have about the same frequency. We recall an analogous situation in radio reception where a strong transmitting station obscures or interferes with the signal of another that has almost the same radio frequency. The phenomenon of masking is common in music. An instrument can easily mask another instrument (e.g., a trumpet can mask a second trumpet playing the same note if the first trumpet is played slightly louder). On the other hand, if the signals are widely different in frequency (e.g., a tuba and flute playing simultaneously in an orchestra) we are able to distinguish both sounds quite easily. Because masking is important for noise, music, and speech intelligibility, this phenomenon has been studied extensively.

We should note clearly the difference between masking and auditory fatigue. In the preceding section we observed that fatigue is a temporary loss in sensitivity to one stimulus *following* exposure to another stimulus,

* The masking of television audio signals by aircraft flyovers is among the most frequent disturbances reported by people living near airports.

whereas masking is a loss in sensitivity to a stimulus *during* exposure to another stimulus, that is, fatigue is a sequential, and masking a concurrent sensitivity loss.[16]

Figure 12 shows the effect of masking resulting from a single pure tone.[17] A masking tone of 415. Hz is presented to a listener at several different intensity levels, at 20 dB, 30 dB, . . 80 dB, above the threshold of hearing (at 415 Hz). The curves of Fig. 12 then show the extent to which other tones must be raised above a listener's normal threshold of hearing in order to be heard in the presence of the masking tone. First, it is clear that the masking effect is greatest near the 415 Hz frequency. Second, the greater the intensity of the masking tone, the broader the band of frequencies for which masking is evident. Third, masking is greater at frequencies above the masking tone. Low frequency tones, in general, are more effective in masking higher frequencies than vice-versa.

Studies have also been conducted on the masking effect of noise.[18] In Fig. 13 the abscissa is the narrow band noise level and the ordinate is the amount of masking, expressed as the magnitude of the signal needed for detection over that required in the absence of any noise. These data

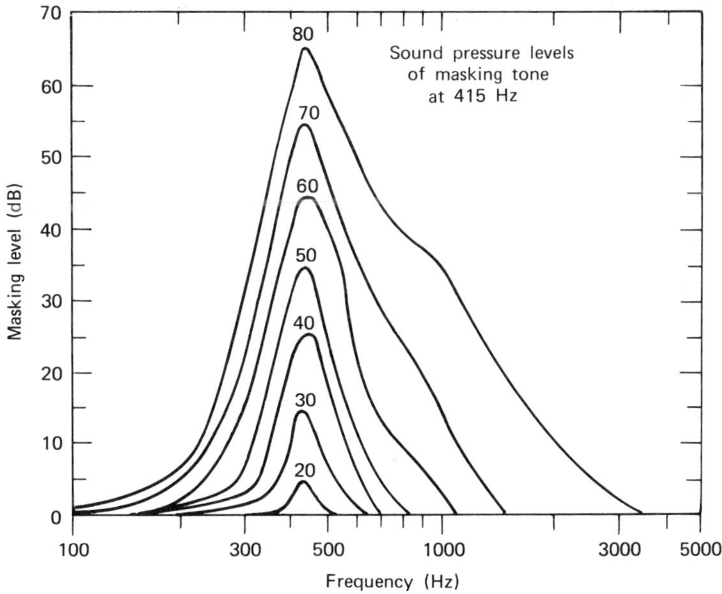

Figure 12   Masking levels corresponding to a pure tone of 415 Hz for various sound pressure levels of the masking tone. [From J. P. Egan and H. W. Hake, *J. Acoust. Soc. Am.*, **22**, 622 (1950.)]

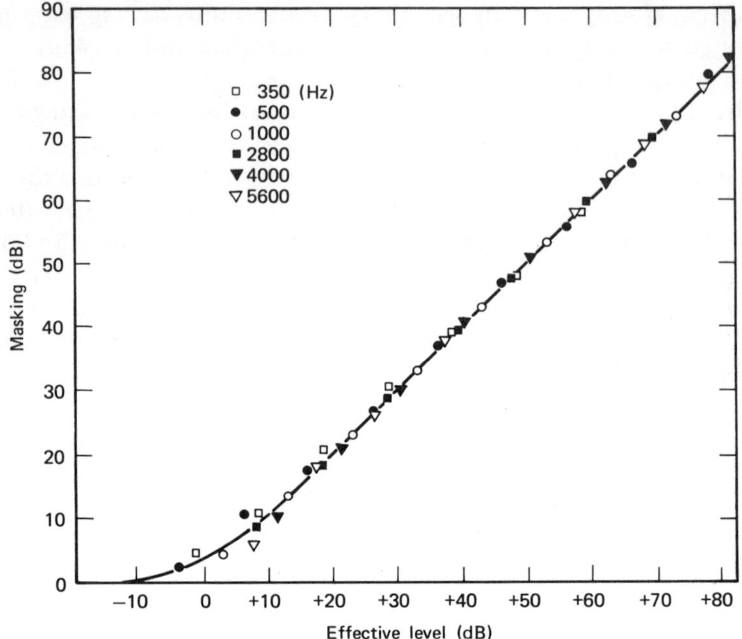

Figure 13   Relation between the effective level of noise and the degree of masking. The effective level is the amount of noise power in a narrow frequency band (the "critical band") centered about the frequency of the masked signal. When the sound intensity is plotted in terms of the "effective level," the graph shows that the amount of masking is independent of frequency. [From J. E. Hawkins, Jr., and S. S. Stevens, *J. Acoust. Soc. Am.*, **22**, 66 (1950).]

show that a linear relationship exists between masking and noise level for narrow (critical) bands whose center frequencies are 350, 500, 1000, 2800, 4000, and 5600 Hz. Qualitatively, masking occurs over the entire range of audible frequencies, and the general level of ambient noise surrounding a listener determines his effective threshold of audibility. As a result, a person with poor hearing may be able to hear speech as well in a noisy environment as a person with normal hearing, because conversationalists will raise their speech power as their own threshold for hearing becomes higher.

Physiologically, masking is believed to be a neurological and not a sensory phenomenon.[19] It is not attributed to the hair cells that generate the cochlear potentials but to the excitation of the cochlear nerve fibers that connect to the brain. Thus, when a group of nerve fibers are vigorously excited by a given frequency band, the fibers cannot convey other impulses to the brain because the "lines" are already loaded.[20]

Furthermore, some evidence exists that masking is not only dependent on intensity and frequency, but is also slightly dependent on time. Masking has been detected at very short time intervals (milliseconds) both prior and subsequent to the arrival of a desired sound signal (termed forward and backward masking, respectively). This extension of the definition of masking is attributed to time-dependent properties of the neural mechanism of hearing.[21]

LOCALIZATION IN SPACE

Because normal hearing is binaural, we are able to locate a sound in space in a manner somewhat analogous to the visual depth perception and angular acuity that we possess with two eyes. Specifically, our two ears permit us to determine the "horizontal" angle between the line joining the two ears and the direction of the incident sound (i.e., in a horizontal plane). Direction perception of the vertical angle is poor.

Direction perception is of considerable importance when a signal is detected in the presence of other interfering sounds. A difference in interaural time, intensity, and phase not only produces a directional sensation, but the intelligibility of the desired sound can be enhanced. Qualitatively, this phenomena can be substantiated by common experience. When speech and an interfering sound are separated widely in direction, the intelligibility of speech is greater than if the speech and noise are received along the same path. Sometimes termed the "cocktail party effect"—when there are many voices being heard at the same time—a person may *focus* his hearing on one particular conversation from a specific direction, and discriminate against the sounds coming from other points. The improvement of hearing ability that can be realized in this manner is sometimes called "masking level difference" (MLD). In one qualitative study,[22] with a 1000 Hz signal, an MLD of 7 dB was noted when the signal was separated by more than 30–40° from the direction of interfering noise.

To gain such an improvement in hearing a signal from a discrete direction when several sources are present, the listener must decide which sound he wants to hear and concentrate upon it. Thus the concept of "attention" has been introduced as a hearing parameter, similar to the notion of "mental set" in psychology.[23] Some unknowns depend on how neural signals from both ears are processed, but it seems clear that hearing ability can be improved when directional information is included in the overall sound spectrum.

THE PRECEDENCE (HAAS) EFFECT

In a previous section we stated that a fusion or blending of two signals into a single sound occurs if these signals reach the ear within a period of 1/20th of a second (50 msec). Additional conditions for detecting only a single sound require that the two signals reach the ear with approximately the same intensity, and that the signals, if complex, have nearly an identical frequency spectrum. This phenomenon of blending or fusion has special significance with respect to the reception of speech or music, and the use of loudspeaker systems.

Consider the listener in Fig. 14 who is positioned at equal distances from loudspeaker A and loud speaker B. If the same signal is fed into both speakers, A and B, the observer will not sense two sound sources, but he will (in an anechoic chamber) hear sound from a virtual source located at a phantom position C. If speaker B is then placed farther away (e.g., to position B'), the apparent source will shift toward speaker A, and no sound will appear to be coming from speaker B. If position B' is

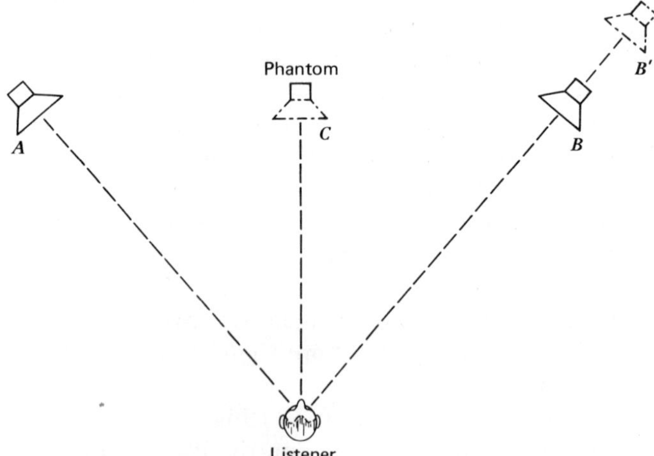

Figure 14   Arrangement of loudspeakers for demonstrating the precedence (Haas) effect. The observer detects sound from the nearer source, from which the sound arrives first. [From M. B. Gardner, *J. Acoust. Soc. Am.*, **43**, 1243 (1968).]

about 3.4 meters more remote from the listener than position B, there will be a delay time of about 10 msec for sound to reach the listener, relative to sound from speaker A.

If the delay time from the more remote source is less than about 50 msec, the nearer source will always be granted "precedence" by the listener.[24] The nearer source, in fact, will be the only one "heard" even though the sound energy from the remote source may be as much as 10 dB greater than from the near source. This *precedence effect* is also called the *Haas effect,* the *"law of the first wavefront, first arrival effect,"* and other names. The physiological explanation is that during the 50 msec after the first arrival of sound the hair cells and nerve centers are preempted and cannot be further stimulated, so a second impulse arriving during this period will not be perceived.[25] A perceptible "echo" will be sensed if the delay time from the second source is greater than 60 to 70 msec.

The importance of this phenomenon in connection with auditorium acoustics is discussed further in Chapter 16.

REFERENCES

1. T. S. Littler, *The Physics of the Ear,* Pergamon Press, London, 1965, p. 22.

2. *Ibid.,* p. 31.

3. F. A. Geldard, *The Human Senses,* Wiley, New York, 1972, p. 174.

4. W. L. Gulick, *Hearing-Physiology and Psychophysics,* Oxford, New York, 1971, p. 55.

5. W. J. V. Osterhout, *Phys. Rev.,* **16**, 216 (1936).

6. W. J. V. Osterhout and S. E. Hill, *J. Gen. Physiol,* **18**, 369 (1935).

7. L. A. deRosa, *J. Acoust. Soc. Am.,* **19**, 623 (1947).

8. Gulick, *op. cit.,* p. 111.

9. R. Plomp, *J. Acoust. Soc. Am.,* **36**, 1628 (1964).

10. R. Plomp and W. J. M. Levelt, *J. Acoust. Soc. Am.,* **38**, 548 (1965).

11. J. V. Tobias, *Foundations of Modern Auditory Theory,* Vol. 1, Academic Press, New York, 1970, p. 188.

12. J. D. Harris, *Some Relations Between Vision and Audition,* Thomas, Springfield, 1950, p. 23.

13. E. L. R. Corliss, *J. Acoust. Soc. Am.,* **50**, 672 (1972).

14. G. V. Bekesy, *Ann. Phys.,* **16**, 844 (1933).

15. A. G. Wood, Master's thesis, University of Virginia, 1930.

16. Gulick, *op. cit.,* p. 167.

17. J. P. Egan and H. W. Hake, *J. Acoust. Soc. Am.,* **22,** 622 (1950).

18. J. E. Hawkins, Jr. and S. S. Stevens, *J. Acoust. Soc. Am.,* **22,** 6 (1950).

19. E. G. Wever, *Theory of Hearing,* Dover, New York, 1970, p. 387.

20. C. A. Culver, *Musical Acoustics,* McGraw-Hill, New York, 1956, p. 70.

21. J. M. Pickett, *J. Acoust. Soc. Am.,* **31,** 1613 (1959).

22. M. Ebata, T. Sone, and T. Minura, *J. Acoust. Soc. Am.,* **43,** 290 (1968).

23. *Ibid.,* p. 297.

24. M. B. Gardner, *J. Acoust. Soc. Am.,* **43,** 1243 (1968).

25. F. Winckel, *Music, Sound and Sensation,* Dover, New York, 1967, p. 63.

6

SUBJECTIVE PARAMETERS OF HEARING

Because the human ear is the final judge of all sounds, we must examine the relationship between physically measurable quantities and subjective sensations. These relations are quite complex. Sensations involve complicated physiological and psychological mechanisms, and much of our knowledge of the hearing process remains empirical. Furthermore, our subjective measurements rely on having a statistically significant number of people compare sound qualities; we then obtain some average of their opinion. In this sense, subjective measurements cannot be precise. Nevertheless, we can approximate such parameters as loudness and pitch, in relation to physical observables, and we can make some qualitative statements about "annoyance," the phenomenon of beats, and the concept of "volume." We will also note that the human ear even "creates" some frequencies that do not exist in an actual sound wave.

LOUDNESS

Loudness can be loosely defined as the magnitude of the sensation experienced by a listener when sound energy impinges on the tympanic membrane. Since the ear is a nonlinear device, we would not expect a simple relationship to exist between sound pressure and the magnitude of a stimulus, especially when we recall that the outer ear, the middle ear, and inner ear have their own individual response characteristics. Thus, in order to provide a practical scale for loudness, the International Standards Organization (ISO) meeting in Paris in 1936 adopted the tone of 1000 Hz as the standard reference tone. Subjective magnitudes of all other tones are then given in terms of the physical magnitude of an "equally loud" tone of 1000 Hz. This unit of equivalent loudness, or loudness level as it was later defined, was termed the *phon,* a nomenclature that had already been in use in Germany to express noise magnitudes. The basis for any loudness scale, however, begins with "loudness contours"[1,2,3] that relate sound pressure level and frequency to the sensation of loudness over the complete audio frequency range.

Loudness Level Contours (Phons)

Figure 1 shows the equal *loudness level contours* that are now accepted as representative of the sensation of loudness vs. frequency for a large group of listeners. These contours show the dramatic difference in hearing sensitivity at the threshold of hearing, and this variable sensitivity remains (to a degree) up to the loudest sounds that the human ear can hear.

In the conditions under which these contours of equality of loudness have been obtained, the listener making a comparative judgment of loudness must listen naturally with both ears to two alternate sounds (one a 1000 Hz tone) while facing the source of sound in a nonreflecting space (i.e., an anechoic chamber). In such an environment we see that a listener would judge a 20 Hz tone at 78 dB to be of the same loudness as a 4000 Hz tone at a 13 dB pressure level. Also, a 125 Hz tone at 90 dB would have the same loudness level as a 4000 Hz tone at 80 dB. These contours are the loudness level contours L_s that we call *phons.* Furthermore, at 1000 Hz every decibel is the equivalent loudness of a *phon* unit.

We restate that these contours have been obtained on the following basis: (1) the standard tone is 1000 Hz or narrow band noise centered at 1000 Hz, (2) the measurement is in a free-field environment, and (3) the primary instrument is the human observer. The reference sound pressure level is also 0.0002 μb (20 μN/m^2 or 20 μPa).

The Sone Scale

The development of the phon scale is fairly straightforward yet it is deficient in the sense that the numbers assigned to this scale do not cor-

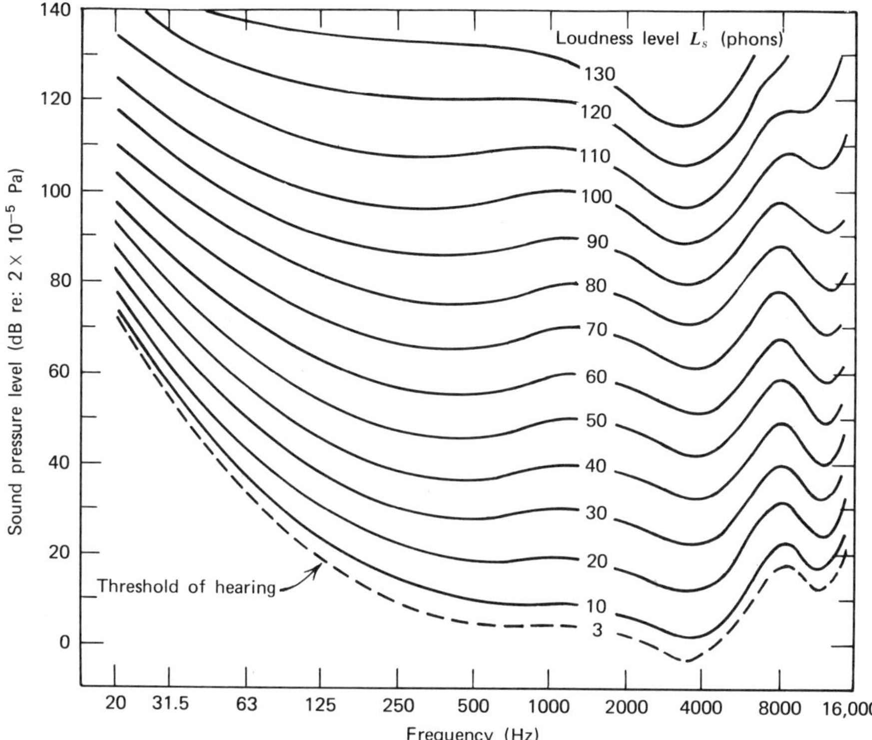

Figure 1   Equal loudness level contours (phons) expressed in decibels relative to 2×10^{-5} Pa, for average listeners for narrow band or pure tones (International Standards Organization).

respond to a listener's intuitive evaluation of relative loudness. Stated another way, apparent loudness bears little relationship to loudness level. For example, a noise of 80 phons is judged by a typical listener to be much louder than twice that of 40 phons. Even a small change in loudness level, from 40 to 50 phons seems to correspond to a doubling of apparent loudness.

To resolve this difficulty the International Standards Organization, in 1947, adopted a new subjective scale called the *sone* scale. The intent of this scale was to establish a more reasonable relationship between loudness level and subjective loudness. The *sone*[4] scale has the following features; first, a sone is defined as the loudness of a 1000 Hz tone of 40 dB intensity level. It is also the loudness of any sound having a loudness level of 40 phons. Second, the sone scale is arranged so that a 10 phon change in the loudness level corresponds to a doubling or halving of loudness.* On the basis of these two prerequisites and for loudness levels of 20 to 120 phons, the relationship between the numerical values of loudness level P (*in phons*) and loudness S (*in sones*) is given by

$$S = 2^{(P-40)/10} \tag{1}$$

or

$$S = \text{antilog} \, [0.03 \, (P - 40)] \text{ sones} \tag{2}$$

Figure 2 is a graphical representation of this relationship.

This sone scale has the distinct advantage of representing loudness in terms of human hearing. Typical values of loudness level and loudness for various sounds are given in Table 1.

Table 1

Loudness Level (phons)		Loudness (sones)
120	Jet aircraft	256
110		128
100	Truck	64
90		32
80	Orator	16
70		8
60	Conversation	4
50		2
40	Quiet room	1
30		0.5
20	Rustling of leaves	0.25
3	Hearing threshold	

LOUDNESS INDEX, ANNOYANCE, AND PERCEIVED NOISE LEVEL

Up to now we have been primarily concerned with establishing a correlation between measurable sound pressure levels (in dB) and the subjective sensation of loudness for pure or narrow band signals. Most real noise sources, however, generate sound over a very wide range of fre-

---

* Except for very low loudness levels. Thus, 40 phons = 1 sone, 50 phons = 2 sones, 60 phons = 4 sones, that is, each time the loudness level is increased 10 dB the loudness in sones is doubled.

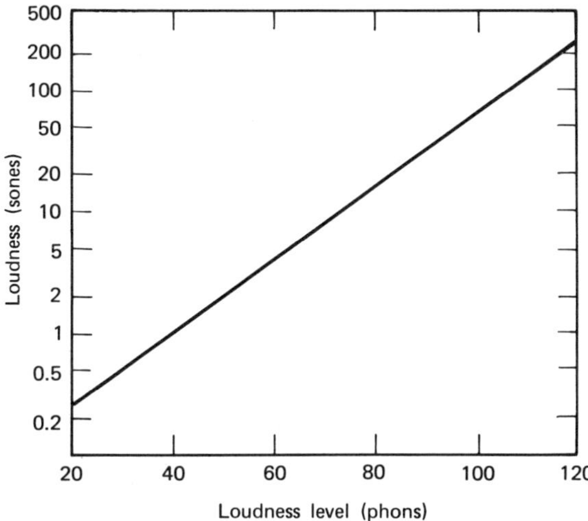

Figure 2   Graph showing the relation between sones and phons.

quencies. Thus, it would be advantageous if we could relate sound intensity and spectral composition (frequency), using *bands* of frequencies, that is, octave, 1/3 octave, and so on. Also several investigators have attempted to determine what make some sounds more "annoying" than other sounds. We shall restrict our discussion to two methods: (1) the loudness index of Stevens[5] based on octave and 1/3 octave-band analysis and (2) the Kryter[6] method for *noisiness* or *annoyance*.

The Stevens method of calculating loudness requires the use of Fig. 3 or tables. Since this loudness index graph is based on loudness judgments of a large number of observers, it is useful in subjective measurements for persons of normal hearing. Here is the procedure for determining loudness.

1. Using an octave-band analyzer* (see Chapter 7), record the sound pressure level (L_p) in each of the eight octave bands having center frequencies of 63, 125, 250, 500, 1000, 2000, 4000, and 8000 Hz.

2. Use the graph to determine the loudness index of the observed noise for each octave band.

3. The *total loudness* of the noise in sones S_t, is given by $S_t = S_m + 0.3 (S - S_m)$, where S_m is the largest of all of the loudness indices and S is the sum of the loudness indices of all of the bands.

* An alternative method is to use 1/3 octave bands, in which case the coefficient of $(S - S_m)$ is 0.15 rather than 0.3.

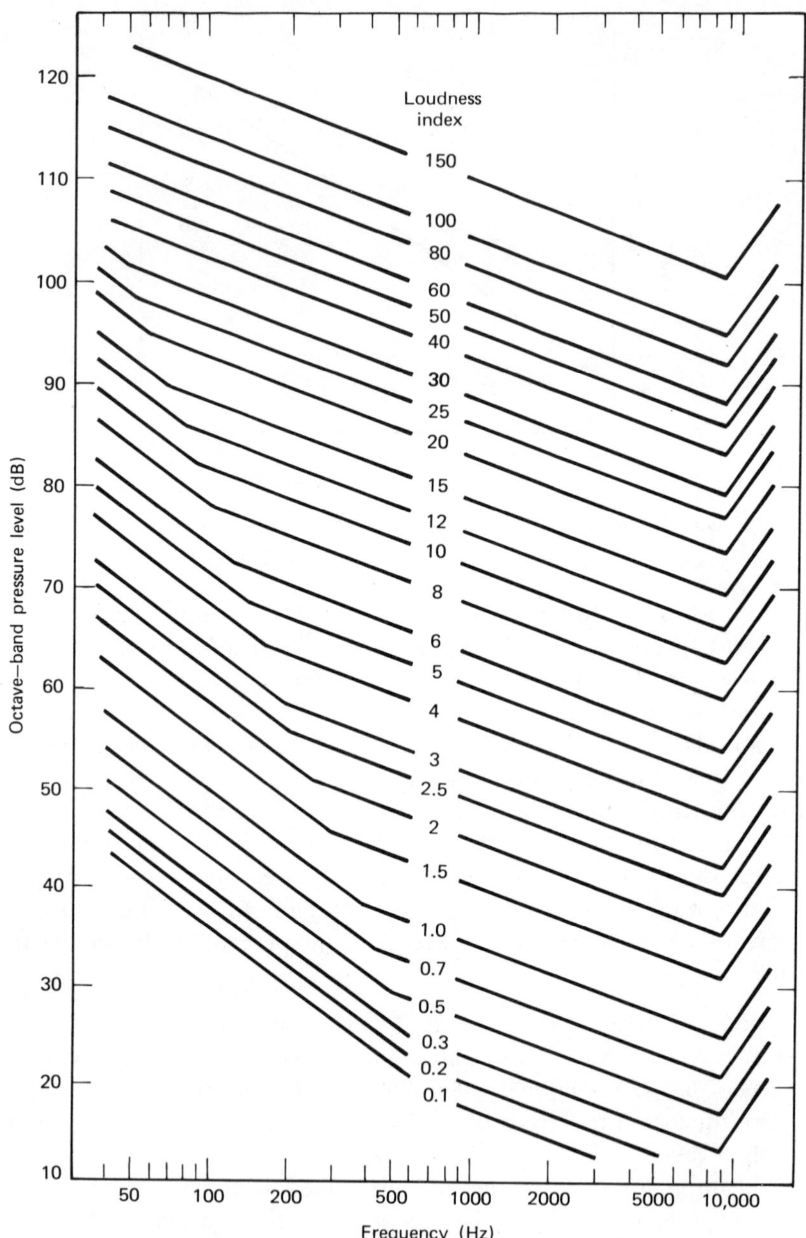

Figure 3   Contours developed by Stevens. A loudness level index in sones can be determined by measuring the sound pressure level of eight octave bands having center frequencies of 63, 125, 250, 500, 1000, 2000, 4000, and 8000 Hz. [From S. S. Stevens, *J. Acoustic Soc. Am.,* **33,** 1577 (1961).]

4. The *total loudness level* is then obtained by converting the total loudness (in sones) to a calculated loudness level, P, (in phons) by means of the equation

$$S_t = 2^{(P-40)/10}$$

or the graph of Fig. 2.

Example

Determine the loudness of a noise whose sound pressure level (L_p), in each octave band has the following values:

Frequency (Hz)	63	125	250	500	1000	2000	4000	8000
L_p (dB)	77	82	78	73	75	76	80	60
Loudness index (sones)	5	10	10	8	12	15	25	8

The loudness index numbers are obtained from Fig. 3.

Total loudness (sones) $= S_t = S_m + 0.3\,(S - S_m)$

$$S_t = 25 + 0.3\,(93 - 25)$$
$$= 25 + 20.4$$
$$= 45.4$$

The calculated loudness level, P (from Fig. 2) is approximately 95 phons.

The degree to which sounds may be really annoying is more complex than determining loudness. Here we are faced with psychological effects. There is general agreement that the following considerations relate to the degree of annoyance.

a. High frequencies are usually more annoying than low frequencies, at comparable sound levels.

b. A continuous noise or one that is repeated periodically with only short intervening times is more annoying than infrequent periods of noise of short duration; for intense sounds there may also be a "startle" effect.

c. People often do not object to noise of moderate intensities unless they can identify the source or at least the direction associated with the sound. (The degree of annoyance will, however, be markedly dependent on the specific situation and the individual.)

d. The annoying character of noise depends on the time of day and the activity of the listener.

e. Noises produced by machines appear to be more annoying than natural sounds of equal loudness (rain, wind, waterfalls, surf).

f. Even loud sounds (e.g., from jet aircraft) are not too annoying provided that the frequency of occurrence is very low (and sleep is not interrupted).

The Kryter method for determining annoyance does not consider all of these factors. It does, however, give a greater weighting to the higher frequencies than does the Stevens criteria. Also, the Kryter scheme was designed to assess noisiness, rather than loudness, particularly with respect to sounds from jet aircraft.

Figure 4 shows Kryter's equal noisiness contours[7] for converting octave or 1/3 octave bands of noise into units that he denotes as *noys*. As in the Stevens' case, these contours were obtained from relative annoyance judgments of narrow bands of noise.

The annoyance N_t of the noise is determined in the same manner as in the Stevens' method, by summing the contributions of the various bands:

$$N_t = N_m + F(\Sigma N - N_m) \tag{3}$$

where N_t is the total annoyance, N_m is the largest contribution in noys of any band, ΣN is the sum of all the contributions, and the function F is given by:

$$F = 0.15 \quad \text{for critical bands}$$
$$F = 0.15 \quad \text{for } \tfrac{1}{3} \text{ octave bands}$$
$$F = 0.2 \quad \text{for } \tfrac{1}{2} \text{ octave bands}$$
$$F = 0.3 \quad \text{for octave bands}$$

The total noisiness N_t (in noys) can also be converted into a rating that is called the *perceived noise level* (PNdB), by using the Stevens' formula for converting sones to phons. That is

$$N_t = 2^{(x-40)/10} \tag{4}$$

hence

$$x = 40 + \frac{100}{3} \log_{10} N_t \tag{5}$$

where x is the perceived noise level in PNdB.

Note the similarity between sones for loudness and noys for noisiness, and between phons for loudness and PNdB for perceived noise level.

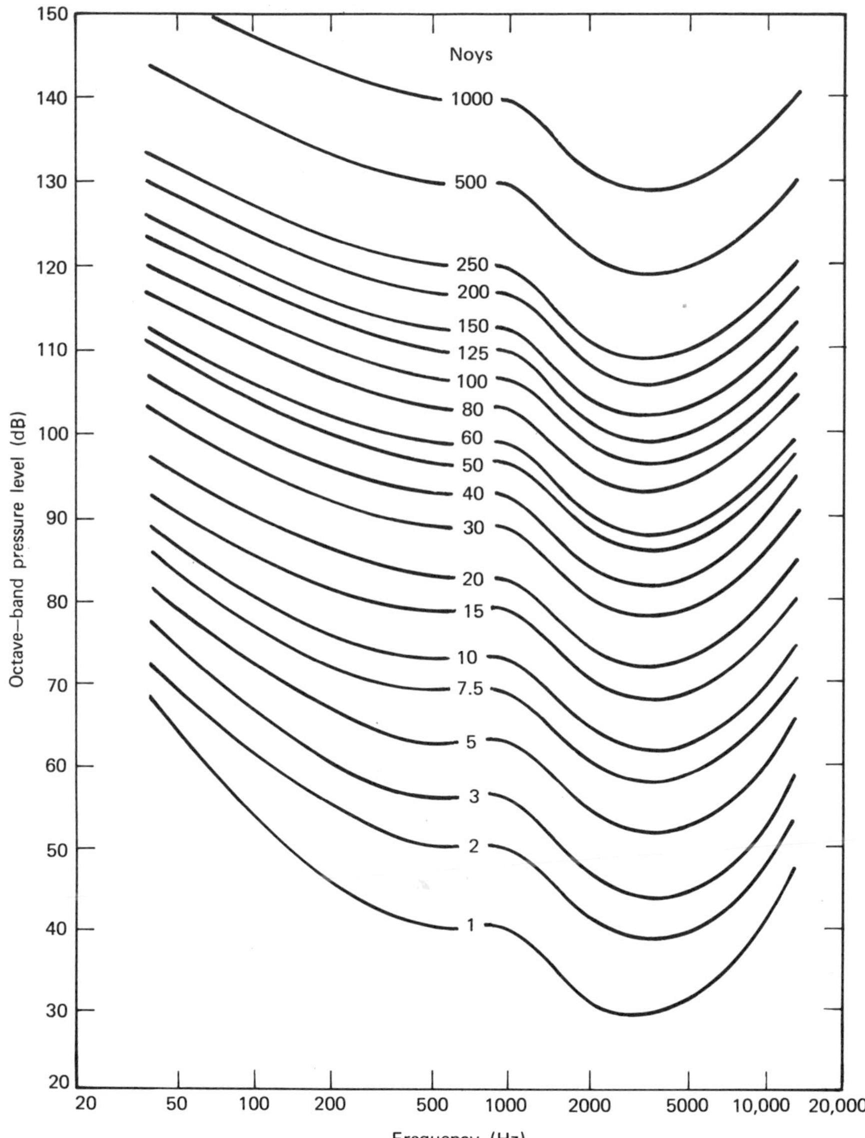

Figure 4   Kryter's contours of perceived noisiness (noys) that are used to determine annoyance. The total annoyance is determined by measuring the sound pressure level in various octave bands. [From K. D. Kryter, *J. Acoust. Soc. Am.*, **31**, 1415 (1959).]

Example

Compute the perceived noise level in PNdB of a noise whose sound pressure level (L_p) in each octave band has the following values:

Frequency (Hz)	63	125	250	500	1000	2000	4000	8000
L_p (dB)	100	95	81	73	73	70	60	75
N (noys)	30	30	15	10	10	15	6	15

Figure 4 is used to obtain noy values corresponding to the L_p for each octave band center frequency.*

The total noisiness N_t is given by

$$N_t = N_{\max} + 0.3\,(\Sigma N - N_{\max})$$

where $N_{\max}$ is the highest noy value and $\Sigma N =$ sum of the noy values in all octave bands. Hence,

$$N_t = 30 + 0.3\,(131 - 30)$$
$$= 30 + 30.3$$
$$= 60.3 \text{ noys}$$

From equation 5 or Figure 2, 60.3 noys corresponds to a perceived noise level of approximately 100 PNdB.

We shall defer until later chapters a discussion of other noise rating schemes such as *traffic noise index* (TNI) and various subjective aircraft noise numbers.

LOUDNESS LEVEL AND DURATION TIME

In previous paragraphs we have discussed steady-state conditions. We shall now inquire into time dependency. Studies by Munson[8] show that the loudness level of a sound does indeed vary for the interval of time that it is detected. The loudness level of a sound does not reach a maximum value at the beginning of a sound signal. Instead, its magnitude is a monotonically increasing function of time, from about 10^{-2} to 1 sec; thereafter, its decreases gradually for periods of 100 sec and greater. Figure 5 shows this relation between loudness level in phons and duration time. (A 1-sec reference tone served as a comparison). Because the mechanical components of the ear are reasonably well damped, the

---

* In the aircraft industry, 1/3 octave bands are employed.

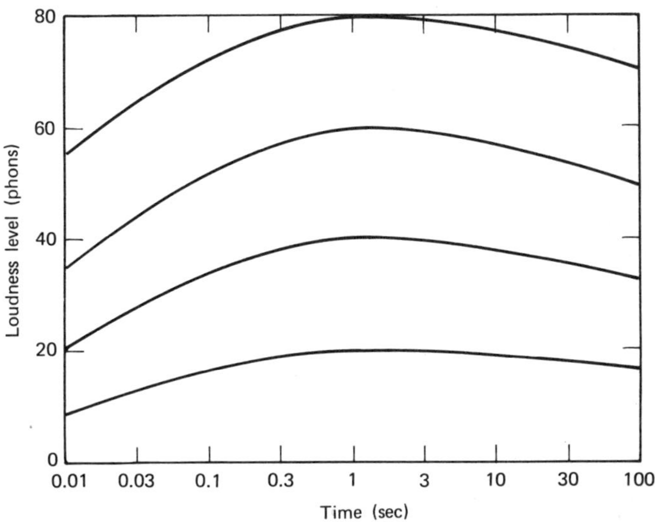

Figure 5   Relation between loudness level and duration time. [From W. A. Munson, *J. Acoust. Soc. Am.*, **19,** 584 (1947).]

growth of loudness level with time cannot be ascribed to their action. Thus it is presumed that this characteristic response reflects the neural action in the auditory system.

Recent research has also indicated that the loudness of a sound of short duration (~1 sec) is somewhat dependent upon the "rise time" of the initial transient.[9] In tests where the sound attained its maximum in the range of 0.03 to 0.5 sec, the signals with a very fast transient (0.03 sec) produced a louder sensation than signals requiring a longer period to attain a given dB level. This observed change in loudness with the *rate* at which the sound level increases has also been related to the complex neural response of the ear.

PITCH

Pitch, like loudness, is a complex characteristic of hearing, and it is dependent upon not one but several physical parameters. The pitch of a musical sound is determined primarily by the frequency of the sound wave, but it is also somewhat dependent upon intensity, the wave form, and the duration of the signal. The subjective sensation of pitch is usually alluded to in terms of "high" and "low" pitches even though an analogous scale is never referred to in optics, that is, the violet-to-red spectrum is never mentioned in terms of higher-to lower colors or hues. Several ideas have been advanced to explain the nomenclature in acoustics.[10] First, investigators report that, psychologically, high frequency tones create an impression of sound localization that is higher in space. Second, the musical scale uses a system for note location that places high frequency sounds at the top of musical clefs. Finally, one could assume that a high pitch refers simply to a high frequency, except that this explanation presumes that listeners are aware of a physical frequency-pitch relationship.

Units of Pitch—Mels

Even though pitch cannot be described with the precision of the physical parameter, frequency, a pitch scale has been developed by sensory physiologists and psychologists.[11] The scale is defined in terms of pure tones.

A tone having a frequency of 1000 Hz is arbitrarily defined as having a pitch of 1000 mels (derived from the word, melody). A pitch with a psychological magnitude that is "twice as high" as the reference pitch is given a value of 2000 mels; a pitch of 500 mels is "half as high" as the 1000 mel pitch. Figure 6 is a graph of the relationship of mels to frequency, based on many experiments with many observers. We note that tones of 500 and 2000 mels have pitches half as high and twice as high as the 1000 reference pitch, but that they correspond to frequencies of 400 and 3000 Hz rather than the 500 and 2000 Hz values.

In tests to obtain this pitch-frequency relationship, it is customary to present alternately to a listener two pure tones of the same loudness level, usually at 40 or 60 phons, and to require that the listener vary the frequency of one of the tones until its pitch seems to be just one half of the other. A repetition of this procedure is followed to cover the entire audible range.

PITCH AND INTENSITY

The pitch of a pure tone is slightly dependent upon intensity. Because of the complexity of the ear this dependency may not come as a surprise, but there is no clear-cut explanation for this phenomenon. Two different hypotheses are advanced by Stevens and Davis.[12] The first

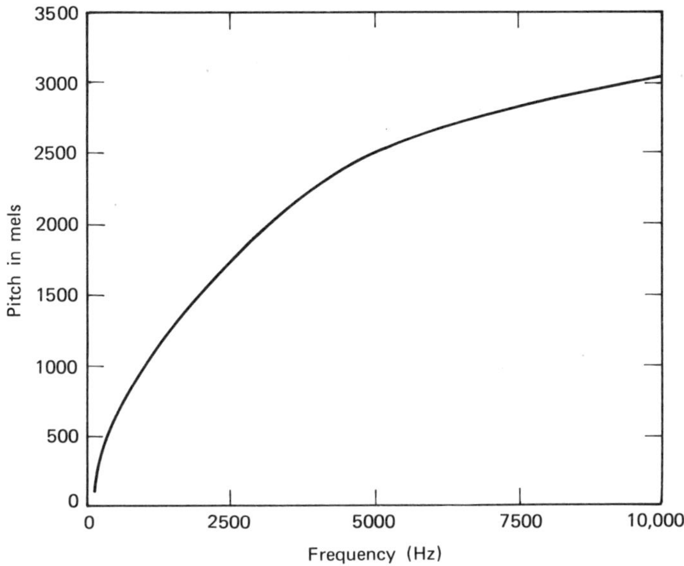

Figure 6   Relationship between pitch in mels and frequency.

hypothesis presumes that the muscles of the middle ear contract at high intensities, resulting in a slight change in the resonance characteristics of the ear as a whole. The second suggests that there may be a shift in the location along the basilar membrane of the peak amplitude of the cochlear potential (CP) because of the nonlinearity of this cochlear potential at high intensities. These hypotheses are currently questioned because the pitch-intensity effect is observable at intensities below those for which muscular contraction is known or the CP is nonlinear. It has been suggested that for low tones, a broader wave pattern exists in the cochlea when either frequency decreases or intensity increases. The net effect could give rise to an illusion of a change in pitch.

In any event, this change in pitch has been documented by Stevens[13] over a frequency range of 150 Hz to 12 kHz. In this test by Stevens, tones of slightly different frequency were presented to a listener who was permitted to adjust the intensity of one tone until it appeared to be of the same pitch as the other, that is, an intensity differential was made to compensate for a difference in pitch. The results are shown in Fig. 7 and are termed equal-pitch contours. It is seen that for tones from 1000 Hz and lower, pitch decreases with increasing intensity, whereas for frequencies of 3 kHz or greater, pitch increases with intensity. Some recent work indicates that the pitch-intensity relationship that is suggested in Fig. 7 may not be representative for the average listener, although, qualitatively, such pitch-intensity relationships have been confirmed by other investigators.

One might suspect that this phenomenon would result in unpleasant sounds in the playing of music. Recall, however, that the data of Fig. 7 are for pure tones, whereas music always involves complex tones that produce pitches of remarkable stability in spite of loudness differences. Some distortion probably occurs, but loud tones (~70 dB) above 3000 Hz are rare; an exception is music that has been amplified electronically.

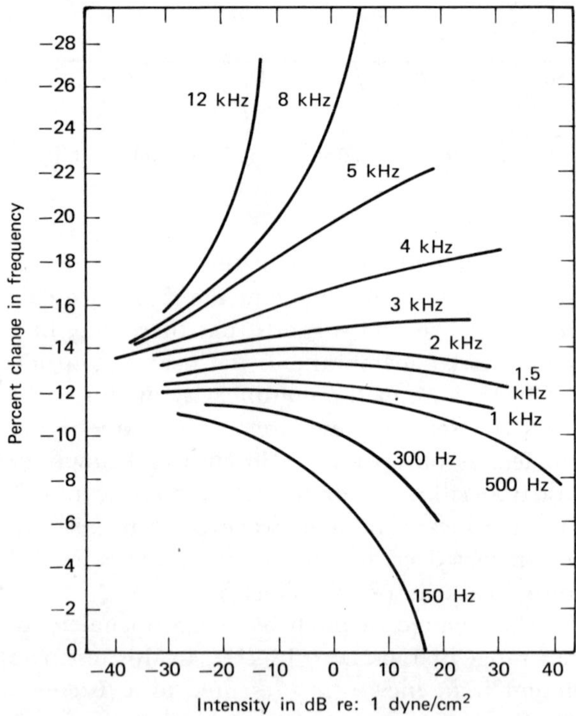

Figure 7   Pitch vs. intensity contours. For tones lower than 2000 Hz it is necessary to raise the frequency of the comparison tone at increasing intensities. For high frequency tones (>3000 Hz), it is necessary to lower the comparison tone frequency, at increasing intensities, to maintain equality of pitch. [S. S. Stevens, *J. Acoust. Soc. Am.*, **6**, 150 (1935).]

MINIMUM PERCEPTIBLE CHANGE IN PITCH

The sensitivity limit for this characteristic of hearing is of considerable importance. It places limits on our ability to estimate velocities by means of the Doppler shift, or to sense the increase or decrease of the rotational speed of motors and other devices that generate frequencies dependent upon their translational or rotational velocities. Skilled mechanics automatically determine the approximate "loading" of motors by detecting small changes in the pitch of rotary saws, gear trains, and lathes. In playing a stereo record even a small change in frequency is objectionable, and the degree that the collective members of an orchestra can tune their instruments depends on their relative acuity of sensing small pitch or frequency increments.

The minimum detectable change in frequency, as a function of frequency, is shown in Fig. 8.[14] Notice that the ear is less sensitive to a percentage change in frequency in the low frequency range, where the relative intensity sensitivity of the ear is less. We also note that at 1000 Hz a percentage change in frequency of about 0.3% can be detected at reasonable intensities. This differential is a factor of about 20 times less

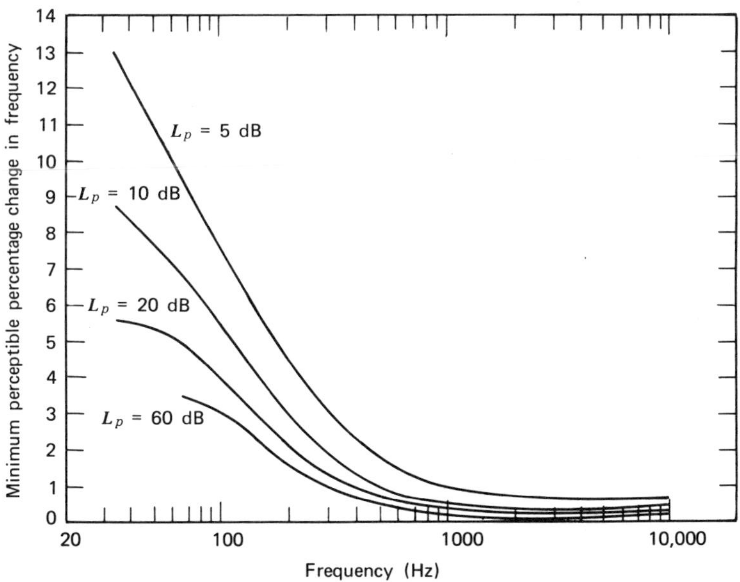

Figure 8   Minimum detectable change in frequency as a function of frequency.

than the interval between two consecutive notes of a piano, where each semitone represents roughly a 6% change in frequency. Although the ability of the ear to sense changes in frequency appears to be quite poor at low frequencies, in practice many low frequency sounds are often rich in higher frequency harmonics. This makes possible a greater frequency discrimination than is evident for the pure tones of Fig. 8.

PITCH AND DURATION TIME

We have seen that time is a parameter affecting the loudness of a sound; it is also important for pitch determination. The time required before the sensation of a definite pitch can be established has been explored by Bürch *et al.*[15] Their results are shown in Fig. 9. There is a substantial variation in the time required for pitch determination from 60 to 10,000 Hz. At 50 Hz, 60 msec are required for a definite pitch determination. At 2000 Hz (close to the maximum hearing sensitivity) only about 12

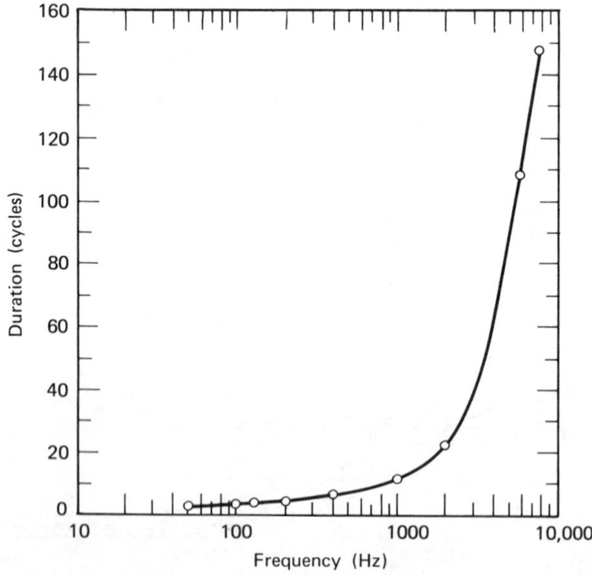

Figure 9   Time required before the sensation of a definite pitch is established. The duration is measured as the number of cycles in the pulse of tone that is just sufficient for pitch determination. (From Bürch et al., Ref. 15.)

msec are needed. At 10,000 Hz, the time for pitch establishment is about 26 msec.

Even though a definite pitch may be sensed in very small intervals of time, this does not imply that pitch is invariant with time. Actually, if one compares the pitch of a short burst of tone with a tone of long duration, it is found that the pitch increases with duration time. For example, with a tone source of 1000 Hz, Ekdahl and Stevens[16] found that it was necessary for the tone to be heard about 40 msec before a steady-state pitch was established (Fig. 10). For very short bursts of tone ($<10^{-3}$ sec), sounds are recognized only as clicks.

There is a general relationship termed the *indeterminacy of tone perception*. This states that the exact pitch of a sinusoidal vibration is limited by[17]

$$\Delta t \cdot \Delta f = 1 \tag{6}$$

where Δt is the time span for which a frequency band Δf is detected. The more precisely we fix one magnitude, the more inexact is the determination of the other. Clearly, if Δt is very short the uncertainty in pitch or frequency will be very large. Conversely, in principle it would require a period of infinite duration in order that a sound have a discrete vibration of frequency f (i.e., $\Delta f = 0$). Nature seems to have constructed the ear as a compromise between perfect sensing of time and frequency, for the optimum hearing of speech and music. One need only play a tape-recording of speech or music at twice its recorded speed to show the qualitative trade-off of duration time vs. pitch. With a fast playback (Δt decreased) the uncertainty in pitch Δf always increases.

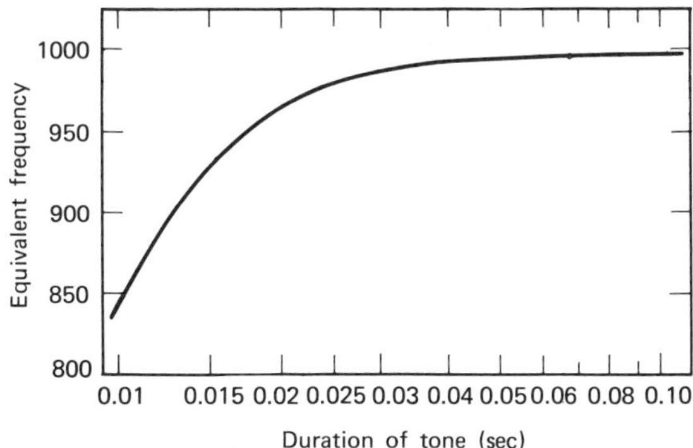

Figure 10   The variation of the pitch of a tone as a function of its duration. [From A. G. Ekdahl and S. S. Stevens, *J. Acoust. Soc. Am.*, **10**, 261 (1939).]

PERCEIVED PITCH OF A MUSICAL SOUND

Musical sounds are rich in *harmonics* or integral multiples of a fundamental frequency (Chapter 15). Indeed there is no voice or orchestral instrument that is capable of emitting a sound with no harmonic content whatsoever. In music there will also be transients, but most musical sounds will be sustained for a large number of cycles, that is, 200 msec or greater. In this case, what will be the perceived pitch of the sound? If the fundamental (first harmonic) and higher harmonics are present, the perceived pitch will be that of the fundamental frequency. Suppose, however, we generate a musical note having a fundamental frequency of 100 Hz and harmonics of 200, 300, 400, and 500 Hz, but we listen to this note through a filter that completely suppresses the 100 Hz frequency (Fig. 11). What will be the perceived pitch?

The answer is that we will still identify the note as being 100 Hz.[18] Essentially the ear senses the pitch of a complex wave on the basis of its overall harmonic structure (i.e., its periodicity) so that even if the first harmonic is very weak or totally absent the ear will subjectively supply it. Of course the tonal quality will be different if one or more of the lower harmonics are absent, but the sense of pitch will be the fundamental that is associated with the harmonic series. Because of this phenomenon of hearing, it is possible for radio manufacturers to produce reasonably acceptable portable radios with small output speakers. These speakers cannot reproduce the lowest bass notes, but the ear tends to "fill in" the low frequency sound on the basis of the harmonics that are actually present. Above about 1400 Hz, however, pitch matches of complex tones are more likely to be determined by the lowest note actually present in the stimulus.[19]

AURAL HARMONICS AND COMBINATION TONES

When an intense pure tone impinges upon the ear, it can generate new tones or frequencies within the ear itself. These supplemental tones are termed *aural harmonics*. Also, when two pure tones are incident upon the hearing mechanism, it is sometimes possible for these two tones to generate additional tones that are called *combination tones*. One combination tone is called a *difference tone* because it consists of a frequency that is equal to the frequency difference of the two components. The other combination tone is called the *summation tone* because it corresponds to a frequency that is the sum of the two component frequencies. Aural harmonics and combination tones are distinctly different phenomena. The feature common to each, however, is that new tones are

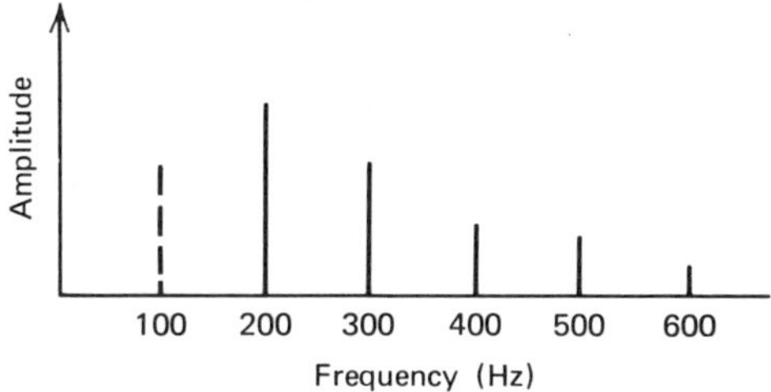

Figure 11   Harmonic frequencies of a tone having a 100 Hz fundamental. The pitch will be perceived as 100 Hz even when the fundamental is removed.

created or perceived by the ear that are not actually present in the original sound source.

The fact that we should be able to hear frequencies that do not exist in the original sound wave should not come as a complete surprise when we realize that the ear is a complex transducer. If it is overdriven by an intense sound, we might expect that it would respond in a nonlinear fashion. We know from experience that harmonics (multiples of the fundamental frequency) are generated when a bow is drawn over a violin string and that many modes of vibration are excited when a membrane, such as a drum, is hit with substantial force. In the case of the ear, new cochlear potentials are actually generated, and these potentials can be detected. Wever and Lawrence,[20] in a classic experiment, stimulated a cat's ear with a 1000 Hz pure tone of high intensity and then probed the cat's cochlea. Using a wave analyzer, they discovered not only the fundamental signal (1000 Hz) but multiples of this frequency. Figure 12 shows their results in which cochlear potentials corresponding to harmonics or overtones were distinctly present. These aural harmonics are also known to occur in the human ear, and they account for some of the real stimuli produced that are nonexistent in an actual sound source.

Combination tones were first noted circa 1745 by the German organist, W. A. Sorge, who reported hearing a third "deeper tone" when playing two notes simultaneously. In 1754 an eminent Italian violinist independently discovered this phenomenon; he called such tones "terzi suoni" or "third sounds" and they are sometimes now referred to as *Tartini's tones*. For example, consider two tones being produced (at moderate intensities) by oscillators at $f_1 = 350$ Hz and $f_2 = 500$ Hz. Experiments show that a trained listener will hear, in addition to these primary tones,

$$2f_1 = 700 \text{ Hz}$$
$$2f_2 = 1000 \text{ Hz}$$
$$f_2 - f_1 = 150 \text{ Hz}$$
$$f_1 + f_2 = 850 \text{ Hz}$$

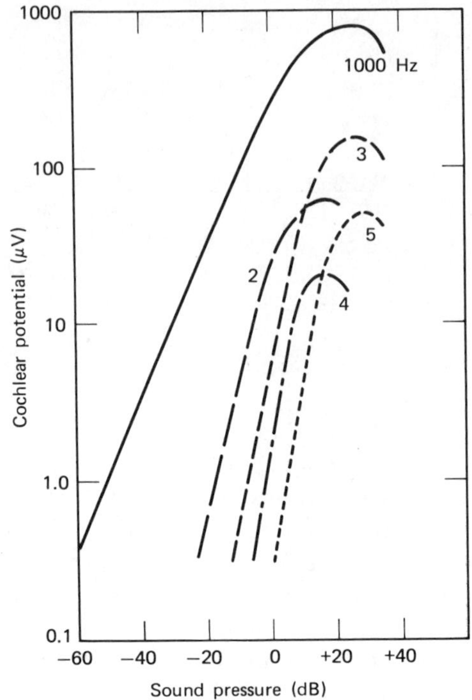

Figure 12   Cochlear potentials generated by a 1000 Hz pure tone. Harmonics are generated that do not exist in the sound source. (From E. G. Wever and M. Lawrence, *Physiological Acoustics*, Princeton University Press, 1954.)

The frequencies of 700 and 1000 Hz will be recognized as harmonics of the generating tones and would be expected on the basis of aural harmonics. But the *difference* and *summation* tones, 150 and 850 Hz, respectively, bear no simple relationship to the generating frequencies.

Theoretically, there should also be other summation and difference tones. A general expression for the formulation of the combination tones f_c for two primary tones with frequencies f_1 and f_2 is:

$$f_c = m \cdot f_1 \pm n \cdot f_2 \tag{7}$$

in which m and n represent the series of whole numbers.[21] Thus, if $f_2 > f_1$, combination tones will be $(f_1 + f_2)$, $(f_2 - f_1)$, $(2f_1 + f_2)$, $(2f_1 - f_2)$, $(2f_2 - f_1)$, $(2f_2 - 2f_1)$, and so on. These tones can also be explained on the basis of the nonlinear response of the ear, or on the basis of *beats* at very low frequencies.[22,23,24] For example, at frequencies of 350 and 500 cycles/sec the *beat frequency* would be 150 cycles/sec (see the next section). This frequency is too fast to usually be identified as a beat, but it is the modulation frequency of the wave pattern, and the ear hears this new frequency as a difference tone.

In practice, difference tones are more easily discerned than summation tones, as these latter tones are often masked by higher primary frequencies or their harmonics. The most prominent difference tone $(f_2 - f_1)$ is actually of value in the construction of pipe organs. A *resultant bass* is designed into the pedal division of some organs, with selected low-pitched pipes connected in pairs according to specific frequency differences. This scheme permits using smaller (and less costly) pipes to be used than would be required if the lowest tones were generated as fundamental frequencies of very large pipes.

THE PHENOMENON OF BEATS

The phenomenon of beats has been heard by almost everyone at some time or another. A piano that is out of tune will produce beats, and two sound sources that have frequencies close together may yield a noticeable fluctuation in sound intensity. Beats are usually evident as a periodic wavering or change in sound intensity. The frequency of such a modulation may range from less than one cycle per second to many cycles per second.

For simplicity, let us consider two pure sinusoidal tones, of frequencies f_1 and f_2 and amplitudes A_1 and A_2.[25] Let the phase difference between vibrations be denoted by ϵ. The superposition of these two tones can be represented by the vector diagram of Fig. 13. The resultant vector can be represented by the equation

$$x = A_1 \cos 2\pi f_1 t + A_2 \cos (2\pi f_2 t - \epsilon) \tag{8}$$

We can consider the two variations as projections of the rotating vectors OP_1 and OP_2, of amplitudes A_1 and A_2. Then the resultant will be the projection of the vector OP in Fig. 13. This vector OP will rotate at an angular velocity that fluctuates between $2\pi f_1$ and $2f\pi_2$; its amplitude will fluctuate also and have limits of $(A_1 + A_2)$ and $(A_1 - A_2)$. Now if the angular frequencies of both vectors are quite close together, the angle P_1OP_2 will be nearly a constant during a single rotation, and we can consider the resultant vector as describing a simple harmonic motion, with an amplitude varying between the limits $(A_1 + A_2)$ and $(A_1 - A_2)$.

The period of this *amplitude fluctuation* will be the period during which the vector OP_2 gains 360° on vector OP_1 that is $2\pi/2\pi(f_1 - f_2)$.

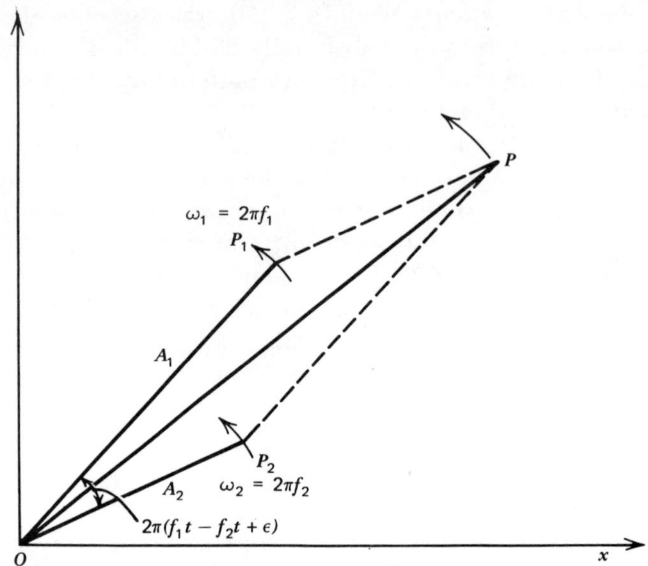

Figure 13   Vector diagram to illustrate the production of beats. The resultant vector of A_1 and A_2. OP will have an angular velocity that fluctuates between ω_1 and ω_2.

This is the same as saying that *the frequency of the amplitude fluctuation is the difference of frequencies of the two tones or vibrations.* This amplitude fluctuation is what the ear hears as a *beat.* Beats are most likely to be sensed when the two individual sources (Fig. 14) generate sound waves that are equal in amplitude, for then the contrast between maximum and minimum will be the greatest. For this case where $A_1 = A_2 = A$, and with the further simplification that $\epsilon = 0$, we obtain

$$x = A \cos 2\pi f_1 t + A \cos 2\pi f_2 t$$

$$= 2A \cos 2\pi \tfrac{1}{2}(f_1 - f_2)\, t \cdot \cos 2\pi \tfrac{1}{2}(f_1 + f_2)\, t \tag{9}$$

This equation represents a simple harmonic vibration whose amplitude varies between (0 and $\pm 2A$ and whose frequency is $\tfrac{1}{2}(f_1 + f_2)$. The amplitudes of $+ 2A$ and $- 2A$ would represent the same intensity; hence, the interval between successive maxima or minima is $1/(f_1 - f_2) = 2\pi/\Delta\omega$ as shown in Fig. 15. Thus the *beat frequency is equal to the difference in frequencies of the two vibrations.*

We should now recognize that this beat is somewhat fictitious: if we employed a frequency analyzer to detect this new frequency, we would

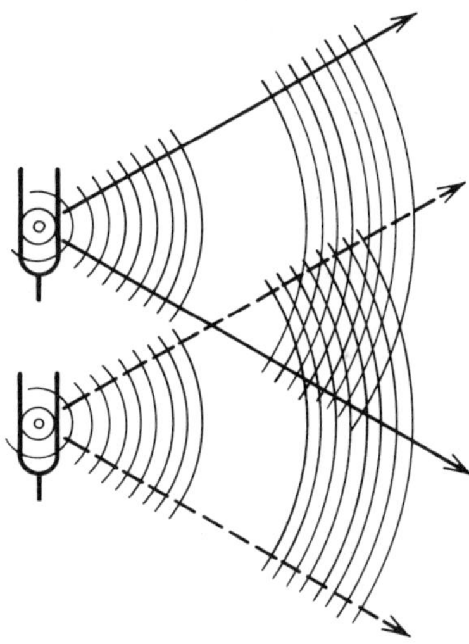

Figure 14   Beats are generated by the superposition of sound waves from two tuning forks that have nearly the same frequency.

not detect it, although a meter having a band pass from f_1 to f_2 would respond (at suitable intensities) by fluctuating at $(f_1 - f_2)$ times a second. We hear beats because the ear is able to sense fluctuations in the intensity of a signal, provided these fluctuations are within the time limits that the ear can sense. When beats occur about several times a second, we sense them as an intensity fluctuation or a *vibrato* of a fixed pitch that corresponds to an average frequency of the two con-

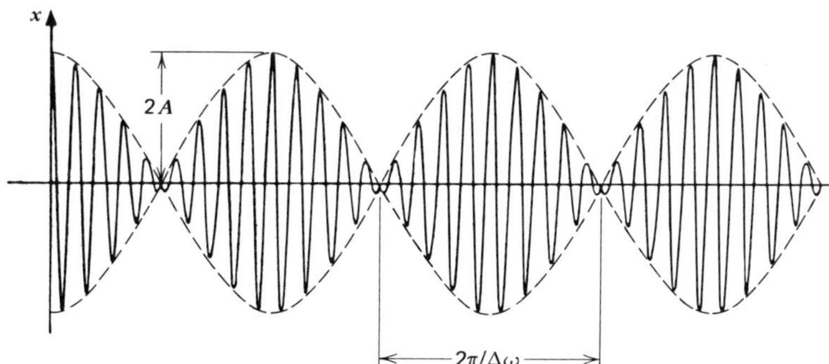

Figure 15   Amplitude vs. time graph showing beats resulting from two tones of slightly different frequency. The ear detects a beat as a periodic wavering of tone (pressure modulation) with a frequency of $(f_1 - f_2)$ times a second.

tributing tones. However, if two sound sources are emitting 10,000 Hz and 10,100 Hz so as to produce 100 beats per second, the ear will be unable to distinguish such rapid alterations in intensity and no beats will be heard. Also it is impossible to detect beats for two signals outside the audio range. For example, two ultrasonic signals of 35,000 and 35,005 Hz would produce 5 beats per second. However, because the primary frequencies cannot be heard, variations in their intensity cannot be sensed and the ear will hear no beat.

Applications have been made of beats in both music and industry. Piano tuners are able to bring various strings into tune by eliminating beats. In mining, two identical pipes are blown simultaneously, one with pure air from a container, the other with the ambient "mine air." If the mine air is pure, the two pipes will emit the same frequency and no beats will be heard. If methane is present, however, the velocity of sound will be changed slightly in one pipe (the velocity of sound in methane is about 30% greater than for air) and beats can be detected. In fact, it is possible to measure quantitatively, the amount of methane in mine air on the basis of beats.[26]

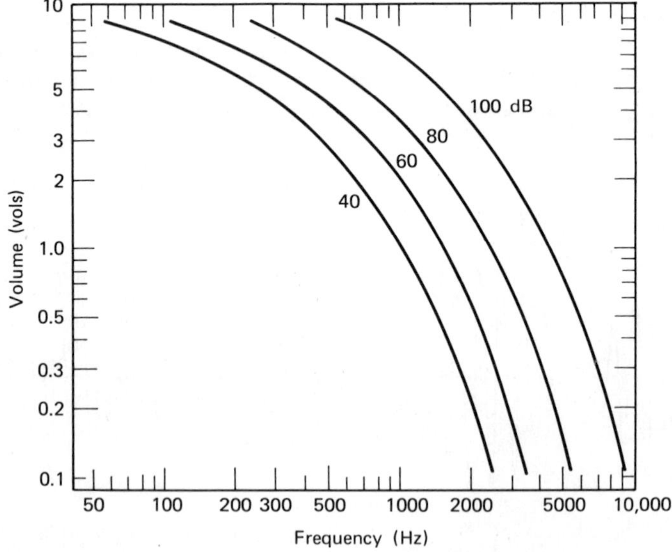

Figure 16   The subjective parameter of "volume" as a function of frequency. [From H. S. Terrace and S. S. Stevens, *Am. J. Psychol.*, **75**, 596 (1962).]

THE CONCEPT OF VOLUME

There are several other subjective aspects of auditory sensation, one of which is termed *volume*. The nomenclature is a psychoacoustic one that attempts to quantify whether tones appear to be "big or little," "massive or thin." High-pitched tones seems localized; they do not seem to encompass a large volume. The high-pitched sounds from a piccolo belong to this category. Conversely, the bass notes of a tuba or even low-pitched noises generate an auditory sensation of depth, expanse, that is, *volume*. There is not complete agreement concerning this auditory sensation among investigators, and studies are still being conducted to correlate this subjective sensation with respect to frequency and sound pressure level.

Figure 16 shows results of a study by Terrace and Stevens[27] that plots tonal volume as a function of frequency at 40, 60, 80, and 100 dB. The proposed volumic unit, the *vol*, is defined as the apparent volume, extensity, or spatiality of a 1000 Hz tone at 40 dB. The volume of a tone is clearly shown to decrease rapidly from low to high frequencies at all sound pressure levels.

REFERENCES

1. H. Fletcher and W. A. Munson, *J. Acoust. Soc. Am.*, 5, 82 (1933).

2. B. G. Churcher and A. J. King, *J.I.E.E.*, 81, No. 487, 57 (1937).

3. D. W. Robinson and R. S. Dodson, *Brit. J. Appl. Phys.*, 7, 166 (1956).

4. S. S. Stevens, *Psychol. Rev. No. 5*, 43 (1936).

5. S. S. Stevens, *J. Acoust. Soc. Am.*, 33, 1577 (1961).

6. K. D. Kryter, *J. Acoust. Soc. Am.*, 31, 1415 (1959).

7. K. D. Kryter and K. S. Pearson, *J. Acoust. Soc. Am.*, 35, 866 (1963).

8. W. A. Munson, *J. Acoust. Soc. Am.*, 19, 584 (1947).

9. K. Gjaevenes and E. R. Rimstead, *J. Acoust. Soc. Am.*, 51, 1234 (1972).

10. W. L. Gulick, *Hearing*, Oxford, New York, 1971, p. 135.

11. S. S. Stevens and J. Volkman, *Am. J. Psychol.*, 53, 329 (1940).

12. S. S. Stevens and H. Davis, *Hearing*, Wiley, New York, 1938.

13. S. S. Stevens, *J. Acoust. Soc. Am.* 6, 150 (1935).

14. E. G. Shower and R. Biddulph. *J. Acoust. Soc. Am.*, 3, 275 (1931).

15. W. Bürch, P. Katowski, and H. Lichte, *Elektrotech. Nachrichten-tech*, 12, 355 (1935).

16. A. G. Ekdahl and S. S. Stevens, *J. Acoust. Soc. Am.,* **10,** 261 (1939).
17. F. Winckel, *Music, Sound and Sensation,* Dover, New York, 1967, p. 49.
18. H. Fletcher, *Phys. Rev.,* **23,** 427 (1951).
19. R. Plomp, *J. Acoust. Soc. Am.,* **41,** 1526 (1967).
20. E. G. Wever and M. Lawrence, *Physiological Acoustics,* Princeton University Press, Princeton, 1954.
21. F. Winckel, *op. cit.,* p. 136.
22. C. A. Taylor, *The Physics of Musical Sounds,* Elsevier, New York, 1965, p. 117.
23. C. H. Wenner, *J. Acoust. Soc. Am.,* **43,** 77 (1968).
24. R. Plomp, *J. Acoust. Soc. Am.,* **37,** 1110 (1965).
25. T. S. Littler, *The Physics of the Ear,* Pergamon Press, London, 1965, p. 129.
26. M. Y. Colby, *Sound Waves and Acoustics,* Henry Holt, New York, 1938, p. 172.
27. H. S. Terrace and S. S. Stevens, *Am. J. Psychol.,* **75,** 596 (1962).

7

INSTRUMENTATION FOR SOUND ANALYSIS

MICROPHONES

All acoustic measurements begin with the transducer that converts sound
pressure variations into electrical signals. Many types of microphones
have been described in the literature.[1] The condenser microphone, however, comes closest to meeting the specifications of the "ideal" microphone for use in environmental applications* because it has these
characteristics:

1. Flat frequency response over the audio range.
2. Low self-noise.
3. Stability with respect to time, temperature, and humidity.
4. High sensitivity.
5. Small dimensions so that diffraction of the sound field is negligible.

The condenser microphone is so named because it makes use of a condenser (capacitor) principle to convert acoustic signals to an output voltage. Figure 1 shows the general features of this device. A thin metallic
membrane or diaphragm is one element of a parallel plate condenser.
When sound is incident upon this membrane, the alternating pressure

* The piezoelectric microphone is also widely used, but its high frequency response is not comparable to the condenser type.

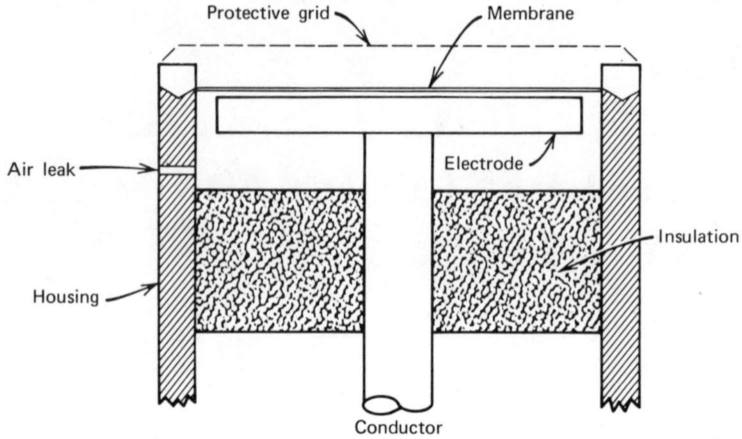

Figure 1   Schematic diagram of a condenser microphone.

of the sound wave causes small displacements of this electrode, resulting in a small change in condenser capacitance. The other parallel plate is called the polarization electrode. This electrode remains fixed, and it is maintained at a potential of about 200 V. A suitable insulator supports this polarization electrode and insulates it from the housing at ground potential. A grid protects the delicate membrane, and a small capillary hole in the housing of the microphone allows an air leak to the interior, so that the static pressure inside the microphone assembly is always equal to the ambient atmospheric pressure.

The equivalent circuit for the condenser microphone is shown in Fig. 2.

Here

$$C = \text{microphone capacitance}$$
$$C_s = \text{stray capacitance}$$
$$E_p = \text{polarization voltage}$$
$$R_p = \text{polarization resistor}$$
$$R = \text{preamplifier input resistance}$$

A constant charge is maintained by the polarization voltage to the condenser through a very high resistance, R_p. Thus, the change in capacitance of the condenser will be due to the minute deflection of the diaphragm only. Let Δx be the displacement of the diaphragm from its equilibrium position x_0, Q_0 the initial charge supplied to the condenser

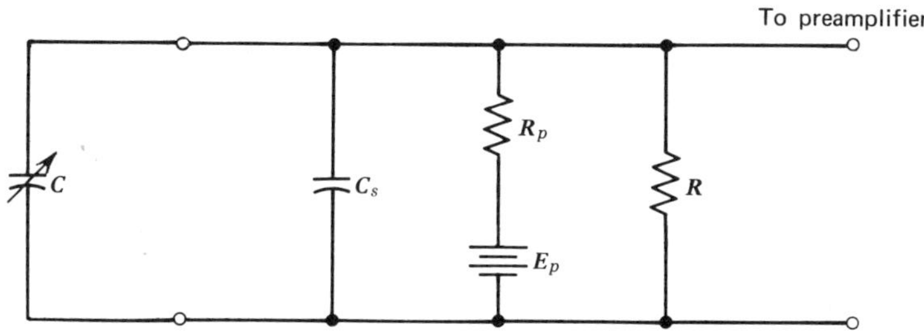

Figure 2   Equivalent circuit of a condenser microphone.

in coulombs, C the microphone capacitance in farads, $C_s = $ stray capacitance in farads, and k a constant of calibration for the device. Then, the change in voltage across the output terminals in Fig. 2 due to a deflection of the diaphragm will be

$$\Delta E = -kQ_0\,\Delta x\,\frac{C}{C + C_s\,[1 + (\Delta x/x_0)]} \tag{1}$$

It is customary to rate the sensitivity of a microphone on the basis of the magnitude of the voltage generated, ΔE, for a given sound pressure level, L_p. The sensitivities are often expressed as dBV/μb, where dBV is the output voltage level ΔE, in decibels *re* 1 V [i.e., $20 \log_{10} (\Delta E$ V/1 V)] for an incident sound pressure at the microphone of 1.0 μb. A 1 μb sound pressure is equal to a sound pressure level of 74 dB *re* 2×10^{-4} μb, which is close to the sound pressure level of street noise. The range of typical microphone sensitivities is between -50 dBV/μb and -125 dBV/μb.[2] These values correspond to output voltages of about 3 mV and 0.5 μV, respectively.

In order to achieve high sensitivity and linearity of the output voltage, it is essential that C_s in the equivalent circuit be kept as small as possible. This means that interconnecting cables to output circuitry should be very short, or that a preamplifier should be connected directly to the microphone housing. This latter option is chosen for all microphone systems where high sensitivity is required.

A recent development in condenser microphone design has been that of the "electret." This type uses a thin plastic sheet as a diaphragm that moves with respect to a backplate, but the plastic (suitably treated) maintains its own polarization voltage.[3] As with the usual condenser microphone it has a linear response over most of the audio frequency range.

SOUND LEVEL METER

The sound level meter is the basic instrument for the measurement of environmental noise. Its essential components are indicated in the block diagram of Fig. 3. In some instances the microphone will be attached directly to the instrument. In others, the microphone and its preamplifier will constitute a separate unit with a shielded cable connecting it to the instrument proper, which contains (1) a "weighting" network, (2) an output to external filters, (3) a main amplifier, (4) a detector circuit, and (5) an indicating meter having a "fast" and "slow" response. Some instruments may have a preamplifier (or input amplifier) equipped with an attenuator control that permits an accurate attenuation of the input signal in steps of 10 dB, and peak and impulse response circuitry. Most sound level meters also contain batteries for remote operation and a means for calibration.

The frequency response of the *A, B,* and *C* weighting networks is shown in Fig. 4. These weighting networks conform now to international standards. Originally they were designed to approximate the loudness level sensitivity of the human ear for pure tones, (*A* at 40 phons, *B* at 70 phons, and *C* at 100 phons). The *A scale* is reasonably matched to the human hearing sensitivity for a wide range of sounds, and this weighting is stipulated for virtually all governmental and industrial regulations. The *B scale* is used only occasionally. The *C scale* is fairly "flat", with only a small attenuation at both low and high frequencies. Comparative readings of *A* and *C* scales are helpful in making a qualitative assessment of the low frequency contribution to noise. The *linear scale* is used when the instrument is connected to a tape recorder output, or when attached to an external set of filters.

The *peak* and *impulse* features of an instrument are important. A *peak* or "hold" reading shows the maximum sound pressure level sensed by the microphone during a monitoring period; the meter will hold until reset. In a sense, it is analogous to a ballistic galvanometer having a very

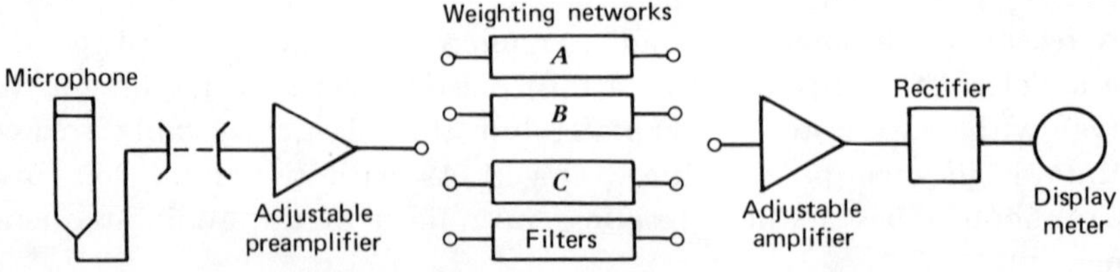

Figure 3   Block diagram of a sound level meter.

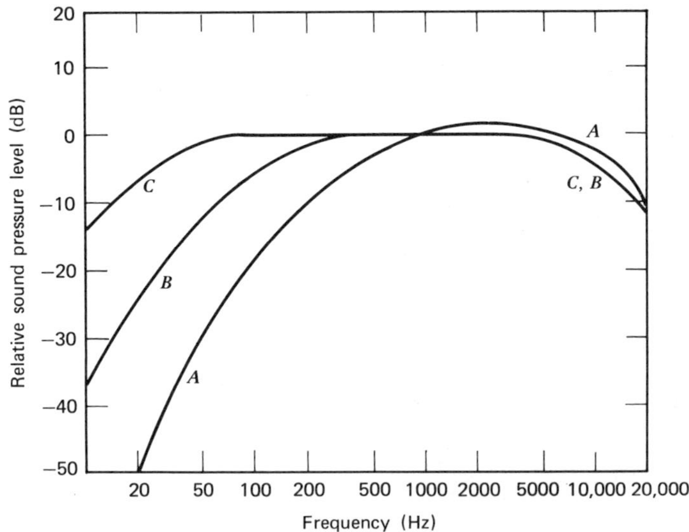

Figure 4   International Standard *A, B,* and *C* weighting curves for sound level meters.

long time constant. The *impulse* response performs an integration for a very short period (50 μsec). "Fast" or "slow" responses of the meter are available for convenience. A "fast" response will indicate a true sound pressure level within 250 msecs; a slow response provides a time constant of about 0.5 sec.

Figure 5 is a photograph of a sound level meter with most of the features described above.

NOISE EXPOSURE MONITOR

While a sound level meter is an ideal instrument for survey work, a device for continually monitoring an industrial area is often desirable. A device that permits detection of "on-again, off-again" noises that are typical of many industrial work areas is shown in Fig. 6.

This instrument enables safety engineers and industrial hygienists to record intermittent noise and thus to verify compliance with federal or state regulations, or with corporate hearing-conservation programs. In the model shown, noise is sampled about twice a second. Input signals are then categorized and weighted according to sound level, and a panel display indicates percent of noise exposure or percent of test time as

Figure 5   A sound level meter. (Courtesy of General Radio Co.)

selected by panel pushbuttons. Noise exposure data are then registered through recorders or printouts so that the instrument can be used directly to determine compliance with the Walsh-Healey Act or OSHA standards (see Chapters 11 and 12).

PERSONNEL NOISE MONITOR

Although survey type instrumentation or a noise exposure monitor is often adequate to monitor industrial noise levels, a variation in an actual worker's exposure may be so great as to require an auxiliary measure-

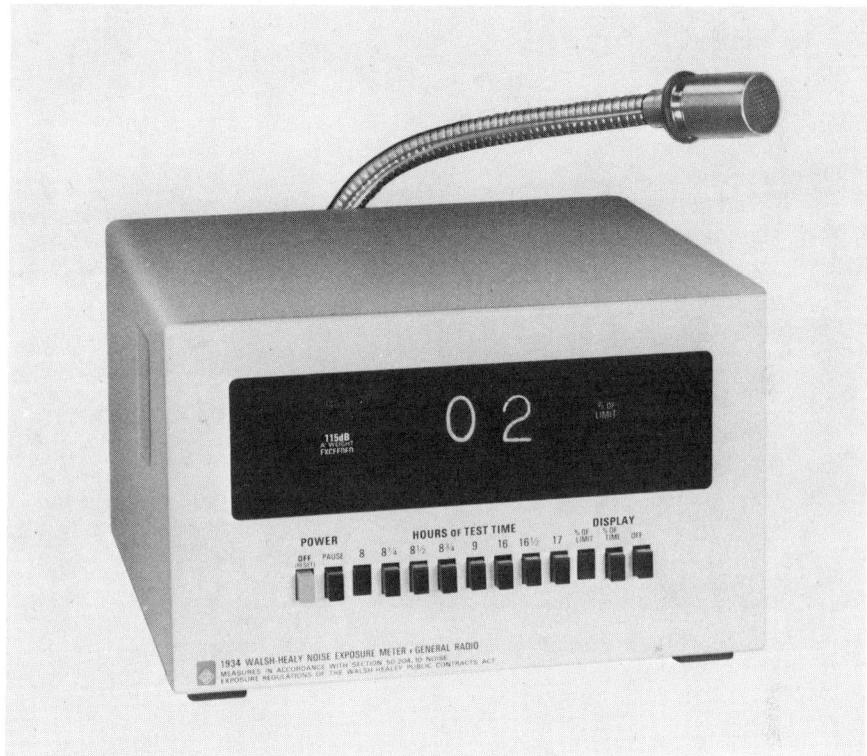

Figure 6   A noise exposure monitor. (Courtesy of General Radio Co.)

ment. Recently, small portable noise exposure monitors have been developed that can be worn by individual workers throughout a working day. These units are pocket-sized, weigh less than 8 oz, include a small microphone and circuitry to indicate an integrated exposure, based on governmental health and safety regulations. In some models, a digital display provides a direct reading, in percent, of the allowed exposure, and a high level detector (with a flashing lamp) indicates when a person has been subjected to any noise level greater than 115 dB A. Figure 7 shows a pocket-size noise exposure monitor being worn by a machinist.

OCTAVE-BAND ANALYZERS

In the field of noise measurements, as well as in measurements relating to speech and music, it is often essential to ascertain the frequencies and

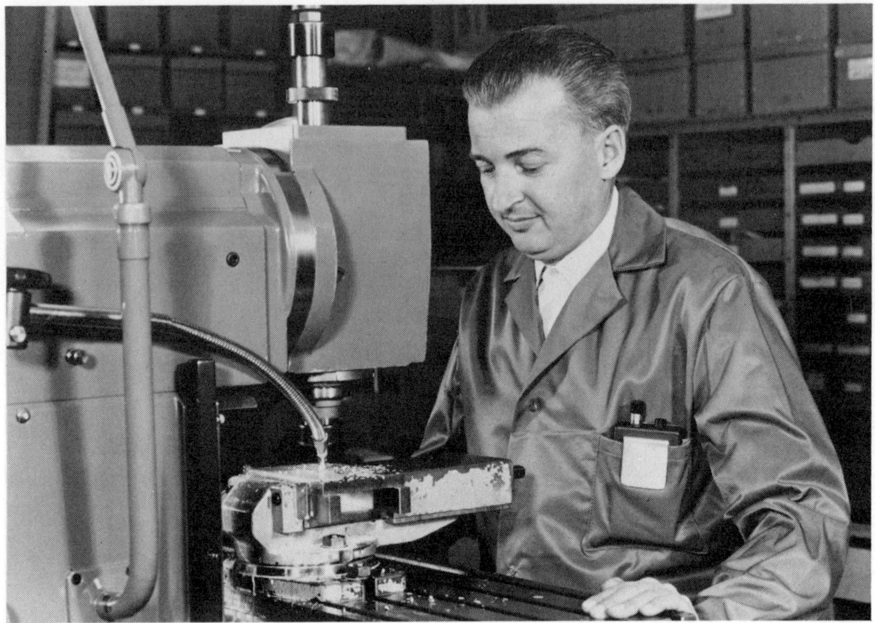

Figure 7   A personnel noise monitor. (Courtesy of B&K Instruments, Inc.)

intensities of sound with greater resolution than is available by the "weighting" scales of a sound level meter. Accordingly, instrumentation has been designed to divide the audio spectrum into 10 bands. By definition, an octave band includes a range of frequencies in which the highest frequency, f_2, of the band is twice the lowest frequency, f_1. Also, the center frequency, f_c, is defined as

$$f_c = \sqrt{f_1 \times f_2} = \sqrt{f_1 \times 2f_1}$$
$$= f_1\sqrt{2} \tag{2}$$

If the center frequency is given, the upper and lower frequency limits are 0.707 f_c and 1.414 f_c, respectively. Because sounds with frequencies greater than 8000 Hz are not commonly encountered at high intensities, commercial instrumentation includes only nine octave bands. The center frequencies[4] and range of these bands are listed in Table 1.

These bands are used as filters in connection with sound level meters or other instrumentation. The ideal filter would have a sharp "cut-off," so as to pass all frequencies within the octave band without attenuation, but such a filter is impractical to design. A standard set of octave-band

Table 1   Internationally Preferred Octave Bands

Center Frequency (Hz)	Effective Band (Hz)
31.5	22–44
63	44–88
125	88–177
250	177–354
500	354–707
1000	707–1414
2000	1414–2828
4000	2828–5657
8000	5657–11314

filters will have zero attenuation at the center frequencies, but with a transmission characterized by Fig. 8.

1/3 OCTAVE-BAND ANALYZERS

When a more detailed analysis of a sound spectrum is required, a 1/3 octave-band analyzer may be used. Center frequencies of thirty 1/3 octave-band filters are given in Table 2. For the third octave band whose center frequency is f_c, the lower frequency limit is $(1/2)^{1/6}$ or 0.891 f_c, and the upper frequency limit is $(2)^{1/6}$ or 1.1225 f_c. Thus, the extent of the 1/3 octave-band with a center frequency of 500 Hz will be from 445 to 561 Hz.

NARROW BAND ANALYZERS

The octave and 1/3 octave-band analyzers are ideal for many types of measurements, but their limited frequency resolution (octave, $\Delta f \approx 70\%$; 1/3 octave $\Delta f \approx 23\%$) restrict their desirability in certain situations. Hence, several narrow band analyzers have been designed having the following features:

Very narrow band analyzer

This type of analyzer usually has a constant bandwidth of about 3 Hz. The instrumentation is basically a heterodyne type analyzer that "scans"

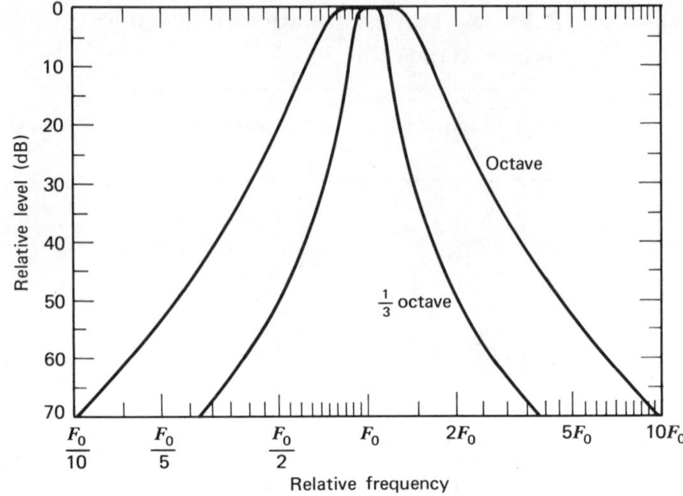

Figure 8   Characteristic response of 1/3 octave and octave-band filters. (Courtesy of General Radio Co.)

a large frequency range by means of a mechanical or electronic sweep control.

Constant bandwidth analyzer

The constant bandwidth analyzer has a constant frequency band, Δf, of 10 Hz, 50 Hz, or some other selected frequency range. In some instances the sweep rate is optional, that is, a frequency scan may be linear, loga-

Table 2   Center Frequencies of Internationally Preferred 1/3 Octave Bands

Center Frequency (Hz)	Center Frequency (Hz)	Center Frequency (Hz)
10	100	1000
12.5	125	1250
16	160	1600
20	200	2000
25	250	2500
31.5	315	3150
40	400	4000
50	500	5000
63	630	6300
80	800	8000

rithmic, or specially programmed so as to spend the most time in a freqency range where the maximum amount of data is desired.

Constant percentage analyzer

This analyzer is designed so that Δf is always a *constant percentage* of f_c. In other words, $\Delta f / f_c =$ constant, and the absolute bandwidth, Δf, (in Hz) thus depends upon the frequency f_c, to which the analyzer is tuned. Octave and 1/3 octave analyzers are in this general class and constant percentage analyzers are built as narrow as 1%.

Figure 9 illustrates the distinction between constant bandwidth and constant percentage bandwidth analysis. If the measurement problem consists in determining a large number of stable, harmonic frequency components, a constant bandwidth analysis may be preferred. For example, if many high harmonics of the same fundamental frequency are to be identified, a constant percentage analyzer might fail to resolve them. For a 6% constant percentage instrument, however, no practical "overlap" would be encountered until one approached the 15th harmonic.

REAL TIME ANALYZERS

In octave or 1/3 octave-band analyzers, sounds can be analyzed according to frequency, but the analysis for each band must be done sequentially.

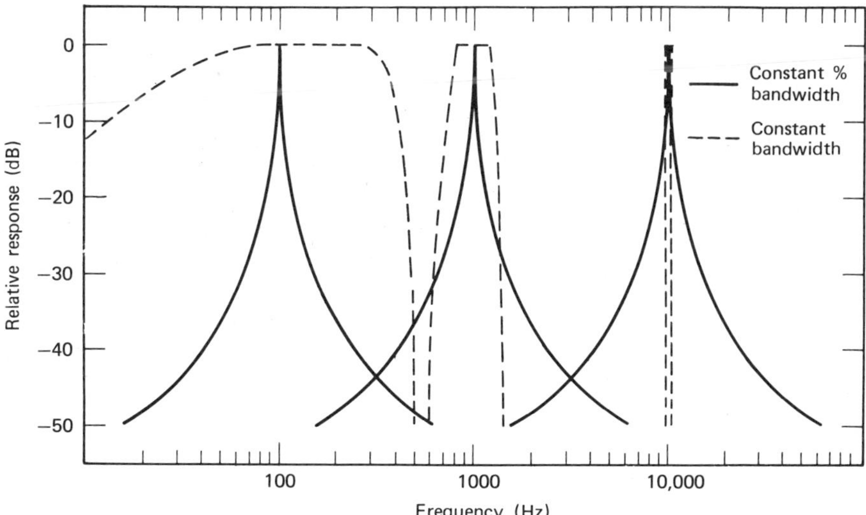

Figure 9   Distinction between constant bandwidth and constant percentage bandwidth analysis.

A *real time analyzer* provides a readout of many channels simultaneously, thus increasing markedly the information that can be acquired from non-steady-state noise sources. Figure 10 shows a modern real time analyzer consisting of 30 1/3 octave-band filters, operating in parallel and covering a 70 dB dynamic range. It permits the frequency spectrum to be determined instantly and displayed on high intensity neon-readout tubes that show the levels of the various bands of the frequency spectrum. The

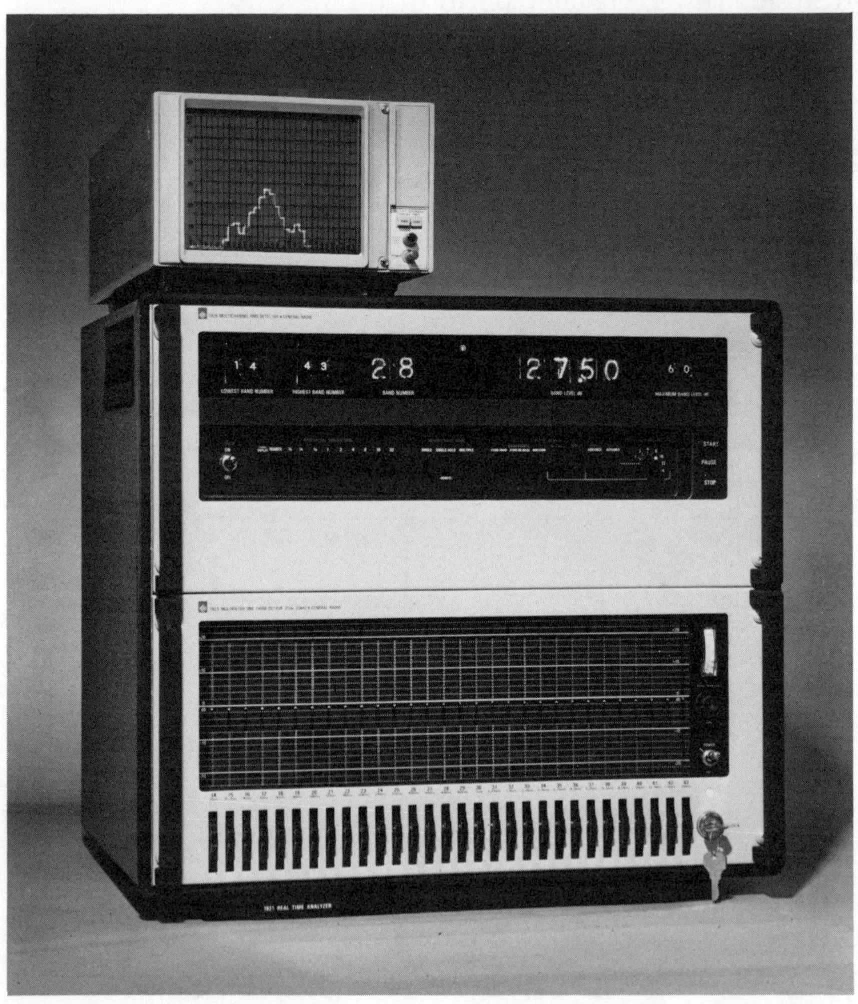

Figure 10   A real time analyzer. (Courtesy of General Radio Co.)

analyzer also provides both digital and analog outputs, and various models cover a frequency range from subsonic to 80 kHz.

In situations involving on-line production control, machine-noise, or any vibration where a readout as a function of frequency and time is desired, the real time analyzer is indispensable. Practical environmental situations for which this instrumentation is useful include the analysis of factory noise, traffic, airport operations, and building acoustics.

For the so-called "signature" analysis of gears and bearings, the study of short transients (involving wide frequency bands) and the detailed analysis of signals containing discrete frequency components that vary in a significant manner with time (e.g., speech and jet noise), even greater resolution is needed than can be provided by 1/3 octave-band instrumentation. Hence, real time analyzers have been developed with as many as 500 equal percentage bandwidth elements. The 500 "lines" can be applied over a frequency range of 10 Hz up to 100 kHz so that very high resolution is achieved. Such instrumentation also includes circuitry for "data averaging" whereby a discrete frequency that is usually buried in noise can be enhanced and identified. Memory circuits also permit a continuous comparison to be made (on a dual display oscilloscope) of previous data with new instantaneous spectra. The real time analyzer shown in Fig. 11 is also equipped with a vertical line cursor that can be superposed on any "peak" of interest, and the absolute frequency of such a signal is automatically presented on the digital display.

TAPE RECORDERS

Tape recorders are essential for survey and field type measurements as it is not practical to bring extensive analytical equipment to many environmental locations. Levels of traffic, airport, and community noise most generally are taped, unless only spot checks are to be made with a sound level meter. Studies of machine noise, transformers, vibration problems, time-dependent and intermittent sounds also require recording in order that a subsequent detailed analysis can be made in the laboratory.

When measurements must be made with great precision, it is necessary to utilize rather costly multichannel recording equipment that has a high tape speed, a large dynamic range, and low distortion. For many environmental investigations, however, two-track magnetic tape recorders* of the

* Some small recorders designed for entertainment use are not satisfactory, however, as they have an automatic gain control that makes it impossible to calibrate the tape.

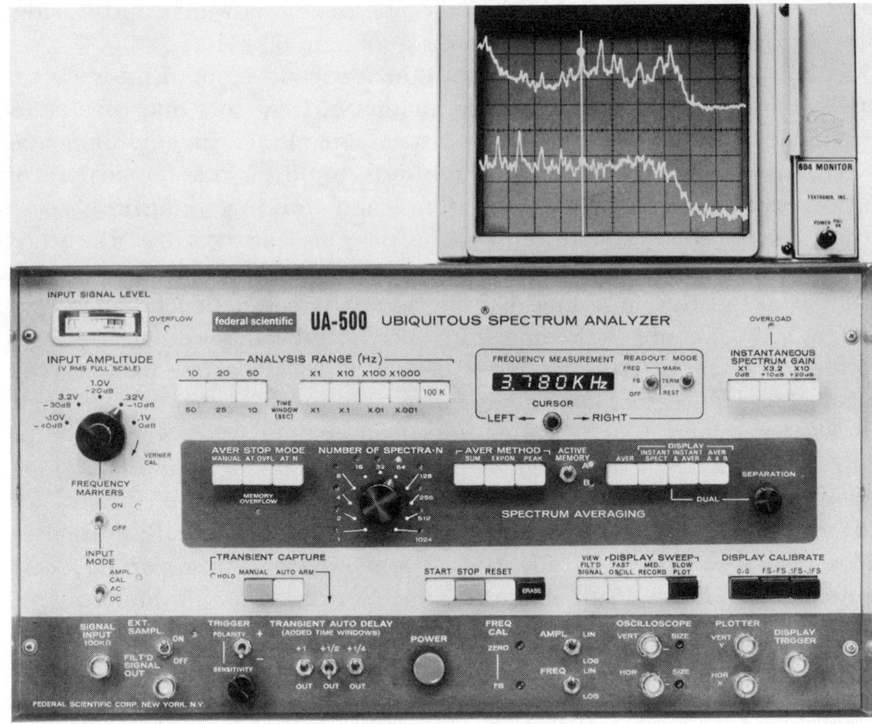

Figure 11   A real time analyzer of 500 channels. A digital display provides a direct frequency reading of any peak that is located by a vertical-line cursor. (Courtesy of Nicolet Scientific Corp.)

cassette type are adequate. Such recorders are now available that have been specifically designed for use with sound level meters. Commercial models typically have a frequency response of ± 2 dB from 50 Hz to 12 kHz, and a signal-to-noise ratio of 50 dB. One recording track registers the desired sound signal; the second track normally records the setting of the sound level meter range control. Voice notes can also be recorded in this second channel, so as to include essential descriptive information concerning the recorded event.

An important ancillary device to both the recorder and the sound level meter is a *sound level calibrator* or *acoustic calibrator*. Such calibrators generate a known sound pressure level at a given frequency. These acoustic signals are fed directly to the microphone that is used with the recorder or sound level meter. In this manner a known reference signal can be registered on magnetic tape permitting other recorded signals to be subsequently normalized.

DIGITAL DATA SYSTEMS

Because the objectives in making noise measurements are so diverse, it is desirable to have a means for not only recording data over extended periods, but for processing the data in convenient form. For example, output data should be available that is most meaningful for evaluating community noise, highway noise, baseline noise at proposed construction sites, plant noise, and so on.

A recently developed system (B&K Instruments, Model SP-321) is composed of two major parts. The first is a field unit comprising a microphone and a digital data recorder. With this recorder, short sampling intervals of 1, 3, and 10 sec can be selected so that the cassette tape data capacity can be extended up to 10 days. The second part is an office system consisting of (a) a digital data translator and (b) a programmable calculator. An accelerated playback of the complete cassette tape can be made in one hour, digital and analog outputs are available from the translator, and several printouts can be programmed from the calculator. These latter include L_N (noise level exceeded $N\%$ of the time); L_{dn} (day-night average sound level—see Chapter 12); the standard deviation; and a pseudohistogram of the statistical data.

OSCILLOSCOPES

A cathode-ray oscilloscope is the most universally used laboratory instrument for observing the waveform of sound. It presents the waveform instantaneously and continuously on a screen, and estimates can be made quickly with respect to peak amplitudes and rates of decay. With experience an observer can also determine something about the frequency components of sound by adjusting the sweep frequency. *Dual-channel* oscilloscopes are also available that permit a sound or vibratory signal to be compared directly with a reference signal (e.g., a sine-wave generator) on the same screen. Two channels also permit an interesting visual comparison to be made of sound signals that have been fed through various filter networks. A *storage* oscilloscope represents another type of oscilloscope that is important in observing transient and nonperiodic phenomena. Such a scope has the capability of retaining virtually any waveform (e.g., the waveform from impact noise, the overall decay of sound in a room after cessation of a sound source, and the starting transient of a musical instrument) on the display screen as long as desired. With suitable output circuitry such a signal can also be fed into a computer for analysis.

The field or mini-oscilloscope is a recent development. Operated with batteries and a direct microphone or vibration sensor input, it provides information that heretofore could only be obtained in the laboratory. Some models have dual channel inputs, making them even more versatile. Such instruments are especially useful in connection with field studies of musical and architectural acoustics, the detection of standing waves, and vibrations of machines and structures. Figure 12 is a photo of a dual channel mini-oscilloscope.

REVERBERATION TIME APPARATUS

The development of portable instrumentation capable of measuring reverberation time is of increasing interest. The industrial safety engineer may require such apparatus in his survey program in order to discern the relationship between reverberation and excessive noise levels in a fabrication plant. The architect may also employ such instrumentation to investigate the properties of auditoria and concert halls. Reverberation times can, of course, be measured by observing the decay of sound via an oscilloscope. The enclosure must first be excited by a sound source. And if frequency vs. reverberation data is required, either the sound source or the microphone-amplifier circuitry must have appropriate filters.

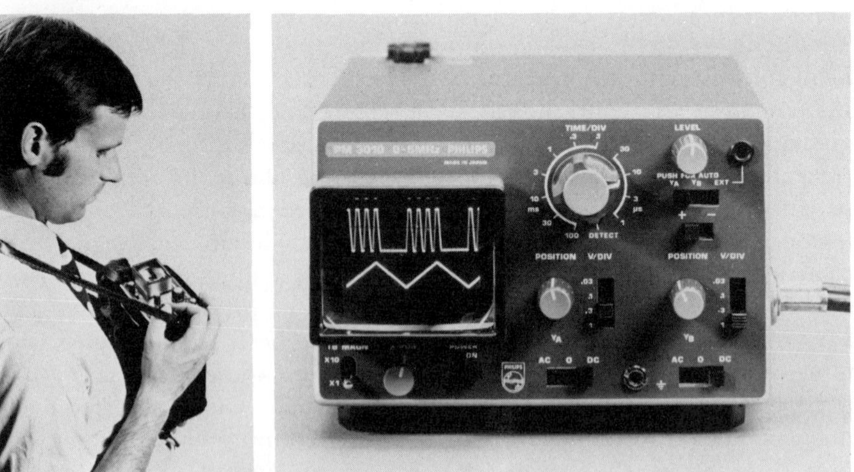

Figure 12   A field type or mini-oscilloscope. (Courtesy of North American Philips Co.)

Commercial instrumentation, however, has been made available that facilitates the direct reading of reverberation times. Such a "Reverberation Processor" (B&K Type 4422) is shown in Figure 13, although even with this device ancillary equipment is needed. Essentially, the Reverberation Processor generates pulses through a filter and power amplifier to a loudspeaker. The decay of sound in the enclosure is then detected by a microphone that feeds the reverberant signal to a microphone, through another filter and back to the Reverberation Processor. Decay rates can then be monitored on the built-in meter or with a digital voltmeter.

FIELD-TYPE ACOUSTIC WATTMETER

Some specialized types of instrumentation have been recently engineered that are useful for making detailed measurements relating to the sound output of loudspeakers, the noise output of machines, the sound absorption of walls, and the acoustic performance of sound-reproducing systems in rooms. Such measurements often require the determination of the intensity of a sound field or sound power density vector, in which both the magnitude and direction of the transmitted sound are quantified. Olson[5] has described a "field type acoustic wattmeter" consisting of a pressure microphone responding to the sound pressure in a sound wave, and a velocity microphone responding to the particle velocity in the wave. The voltage output of the pressure microphone is amplified and fed to one transformer, and the voltage output of the velocity microphone is fed to a second transformer. Then the transformer outputs are connected through band pass filters to power converters that yield dc output voltages that are proportional to the square of their ac input signals. The outputs of these power converters are then fed to a suitable logarithmic meter for the direct reading of sound power density in decibels, referenced to 0 dB at 10^{-12} W/m^2.

This type of instrumentation has applicability in locating standing waves and measuring the ratio of direct to generally reflected sound in a room, or in the measurement of sound power densities in outdoor environments.

VIBRATION TRANSDUCERS

In some instances, it is more convenient to measure the displacement, velocity, or the acceleration of a sound source, than to monitor airborne

Figure 13   A reverberation time processor. (Courtesy of B&K Instruments, Inc.)

sound radiating from it. Vibrating structures and machinery components are especially amenable to monitoring by commercially available vibration sensors. A *displacement transducer* generates an output voltage proportional to the relative displacement of the two elements of the device. A *velocity transducer,* develops a voltage proportional to the relative velocity of the two elements of the sensor, that is, the output is proportional to both the displacement and frequency of the vibrating surface to which it is fixed. *Accelerometers* produce a voltage output proportional to the acceleration of the vibrating element; hence the output signal varies directly with the displacement and with the square of the frequency.

The usual reference levels for velocity sensors and accelerometers are 10^{-6} cm/sec and 10^{-3} cm/sec² , respectively. Most structure-borne vibration specifications are given in terms of the latter unit and called *acceleration decibels,* or c dB.[6]

STROBOSCOPES

Stroboscopes have been used for studying rotating or reciprocating objects for many years. If a rotating vane is revolving at 2000 rpm and it is viewed with a light that flashes at a precise 2000 Hz frequency, the vane will appear to "stand still." At 1999 flashes per minute the vane will seem to be rotating at 1 rpm, and at 2001 flashes per minute it will appear to rotate in the opposite direction at 1 rpm. Because the eye has a "time constant" of about 1/50th of a second, no flicker will be evident except at slow speeds.

This principle is utilized in modern electronic stroboscopes.[7] The light source is mounted in a parabolic reflector and the flash frequency is determined by an electronic pulse generator. The pulse generator can be varied from 110 to 25,000/min in one model, and up to 150,000/min (2500 Hz) in another. Figure 14 is a photo of such a stroboscope, equipped with a photoelectric pickup that serves to synchronize the stroboscope with repetitive mechanical motion.

Such instrumentation, while essentially designed for the visual observation of rotation or vibration phenomena is also valuable for locating sources of noise. An analysis can be made of the deformation of helicopter blades, the vibrating modes of turbine blades, or the detailed behavior of other objects that give rise to noise. Furthermore, this analysis can be done without affecting the vibration. Because the reduction of noise at the source is of paramount importance, the stroboscope is an important analytical instrument for noise detection.

AUDIOMETERS

An audiometer is an apparatus for determining a person's hearing sensitivity to pure tones as a function of frequency. It consists of an audio oscillator, an attenuator, an amplifier, and an earphone. The oscillator generates an alternating current of small magnitude that is amplified and fed into the earphone. The amplitude of the signal to the earphone is controlled by the attenuator that is calibrated directly in decibels. A frequency control indicator provides for standard frequencies of 500, 1000, 2000, 3000, 4000, 6000, and 8000 Hz. The basic function of the instrument is to indicate the difference (in decibels) of a person's threshold of hearing to the "normal" threshold sensitivity curve (see Figs. 3 to 8, Chapter 17).

If the auditory response of the person being tested is below "normal" at any frequency, the attenuator will have to be adjusted so that a more intense signal is supplied to the earphone. When the signal is just audible to the subject, the attenuator dial reading will indicate the decibel increase required above "normal."

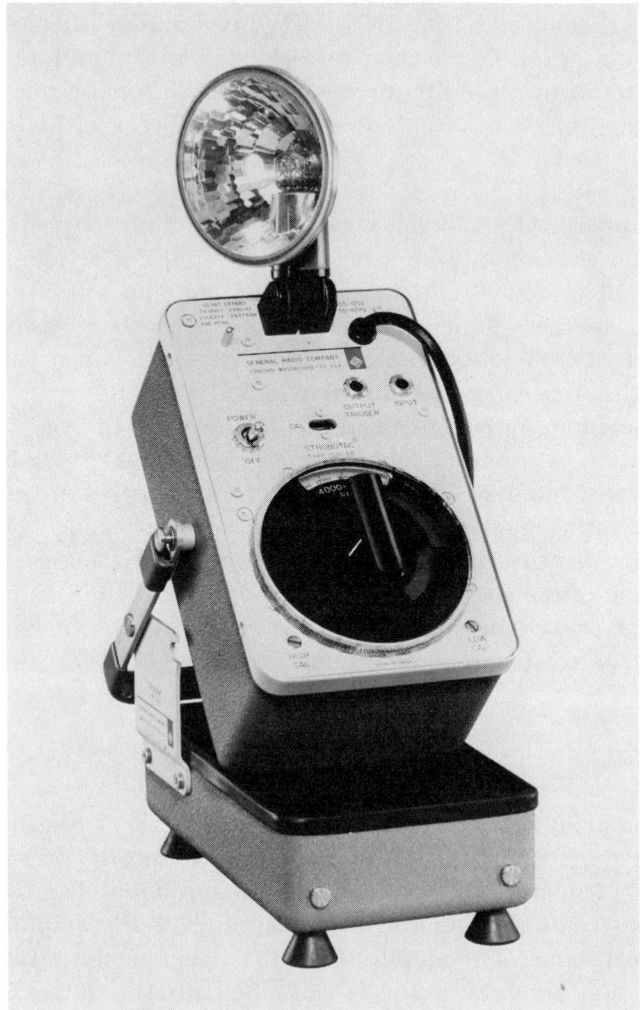

Figure 14   An electronic stroboscope. (Courtesy of General Radio Co.)

After readings at the standard frequencies have been obtained, a graph can be constructed showing any deviation from normal hearing. Such a graph is represented in Fig. 15. Today such data are routinely obtained (1) during the medical testing of prospective employees, (2) to determine occupational hearing loss, (3) for general audiometric diagnosis, and (4) to test a person's auditory response to musical or other sounds.

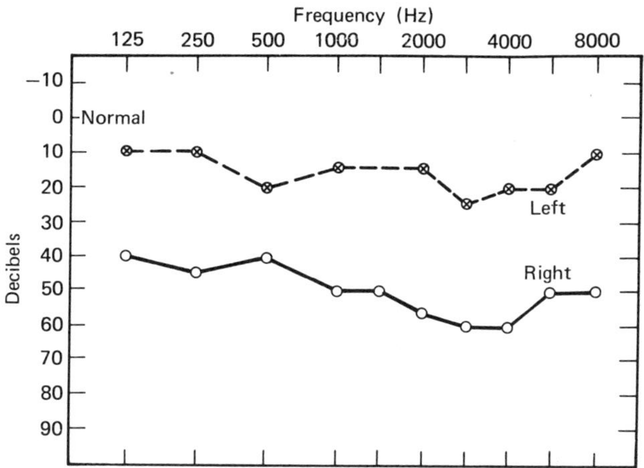

Figure 15   Hearing response curves determined with a standard audiometer.

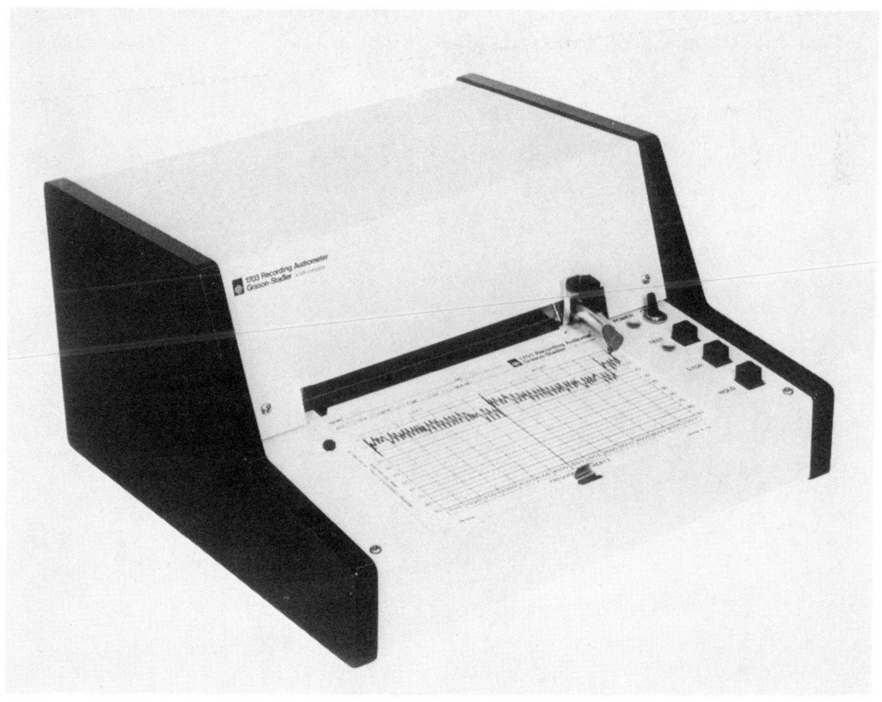

Figure 16   A modern recording audiometer. (Courtesy of Grason-Stadler Co.)

Figure 16 is a photo of a modern audiometer that is self-recording, and that is especially useful when a large number of persons are to be tested. The unit presents seven standard audiometic frequencies sequentially for 30 seconds each, varies the intensity of each frequency until the hearing threshold is reached, and records the results of the entire test on a chart.

REFERENCES

1. L. L. Beranek, *Noise and Vibration Control*, McGraw-Hill, New York, 1971, p. 46-58.
2. *Ibid.*, p. 46.
3. S. V. Djuric, *J. Acoust. Soc. Am.*, **51**, 129 (1972).
4. B. F. Day, R. D. Ford, and P. Lord, *Building Acoustics*, Elsevier, London, 1969, p. 5.
5. H. F. Olson, *J. Audio, Eng. Soc.*, **22**, 321 (1974).
6. G. M. Diehl, *Machinery Acoustics*, Wiley, 1973, p. 49.
7. A. P. G. Peterson and E. E. Gross, Jr., *Handbook of Noise Measurement*, General Radio Corp., Concord, Mass., 1972, p. 137.

II

NOISE CONTROL

8

NOISE AS A SOCIETAL PROBLEM

Noise as an environmental pollutant has recently become an object of serious social concern. However, it cannot be isolated from other social concerns, and is inextricably linked with the many facets of contemporary civilization—accompanying man's essential, as well as nonessential, activities and his way of life. Noise differs significantly from air pollution and also from thermal pollution, in the sense that it decays almost instantaneously and leaves no residue. Yet noise is a persistent pollutant in the many practical situations where noise sources cannot be suppressed, and it degrades the quality of life.

Continued exposure to high levels of noise can, of course, result in permanent hearing loss, and there is increasing evidence that noise may have other physiological and psychological effects (Chapter 17). Even low levels of noise can create an environment that is unsatisfactory for human communication, work, sleep, and recreation. Furthermore, if one accepts the definition of health adopted by the World Health Organization—"a state of physical, mental, and social well-being, and not merely the absence of disease"—then freedom from the annoyance of noise is, indeed, a desirable amenity.[1] Perhaps the most immediate improvements in noise control will be experienced in occupational environments where federal legislation is specific and regulations can be strictly enforced. In addition, one can expect some industrial managements to seek occupational noise levels substantially below federal standards in order to achieve increased employee satisfaction and productivity. But a few of the constraints that will delay the early advent of a "quiet society" include:

1. The impact of urbanization
2. The growth of noise sources
3. Conflict in societal goals
4. Legal considerations

THE IMPACT OF URBANIZATION

Noise levels increase with increasing urbanization, just as other forms of pollution are aggravated by increases in population density. Primary factors that have caused people to abandon spacious rural environments in the United States have included (1) industrial growth, which provided jobs in the city, (2) recreation, cultural, and sport facilities in the city, and (3) an agricultural revolution in which the farmer who in 1850 could feed four is now able to feed 30 people. In 1850 the United States population was approximately 23 million; in 1970 the population was approximately 204 million. More important, of course, has been the dramatic increase in the percentage of the population living in the nation's urban areas,[2] as shown in Fig. 1.

GROWTH OF NOISE SOURCES

Population growth, coupled with the increasing demand for transportation, and a wide variety of powered products, is reflected in Table 1.

Noise sources have increased at a much greater rate than the population. The total power consumption of the United States has also followed a pattern of approximately doubling every 15 years.

The most important environmental noise increment, of course, has accompanied the advent of a very large turbofan jet fleet and increased truck traffic. It is estimated that 9,000,000 people living in homes covering an area of 2000 square miles are currently exposed to aircraft and highway noise levels that are deemed to be incompatible with residential living.[4] In the Kennedy Airport environs alone, some 700,000 persons, and 220 schools attended by 280,000 pupils are included in such an area.[5] The population affected by aircraft noise in the Boston area has also been reported, and is indicated in Table 2.

The overall impact of noise on society cannot be assessed in simple terms. It is estimated that in the United States, 80 million people, or

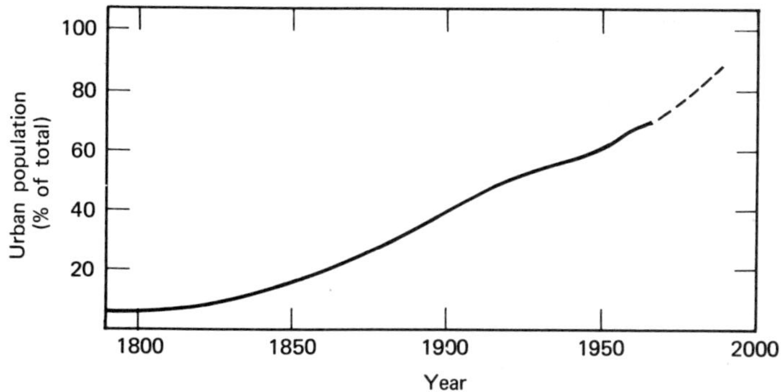

Figure 1   Percentage increase of the United States population living in urban areas since 1790.

Table 1   Growth in Noise Sources (M = Million, TH = Thousand)[3]

	Year: 1950	1960	1970
Population (M):	151	181	204
Transportation Vehicles			
Cars, buses, trucks (M)	49.2	73.9	106.3
Motorcycles (M)	0.45	0.51	3.0
Powered boats (M)	2.6	4.7	5.8
Snowmobiles (TH)	0	2	1600
Commercial aircraft (turbofan)	0	202	1989
Private aircraft (TH)	45	76.2	136
Outdoor appliances (approximate)			
Lawn mowers (M)		10	17
Chain saws (M)		0.5	1.2
	1953	1960	1970
Home Appliances			
Dishwashers (M)	1.3	3.2	14.9
Clothes washers (M)	32.2	42.0	57.6
Clothes dryers (M)	1.5	9.0	25.3
Air conditioners (M)	0.6	6.5	23.0
Food mixers (M)	12.6	27.0	51.2
Food waste disposers (M)	1.4	4.8	14.4

Table 2   Estimated Impact of Noise at Logan Airport (Boston)[6]

	1967	1975
Area "not compatible with residential living" (mi²)	25	80
Population in area	177,000	556,000
Schools	93	272
Hospital beds	1,391	3,158

roughly 40% of the population, are subject to noise levels that are annoying and that frequently interfere with communication, relaxation, sleep, or some other social activity. A much smaller number are subject to a potential health hazard, in terms of hearing loss, attributed solely to high level noise exposure. Solutions to the noise problem will be concerned with (1) the noise source (2) the path of sound propagation, and (3) the receiver. Figure 2 suggests that any major noise prevention or abatement program must address the interrelationship of these three elements, and that both technical and legal measures are involved.

CONFLICT IN SOCIETAL GOALS

Despite the increase in annoyance from noise sources, it is questionable if significant noise reductions will be achieved in many fields in the imme-

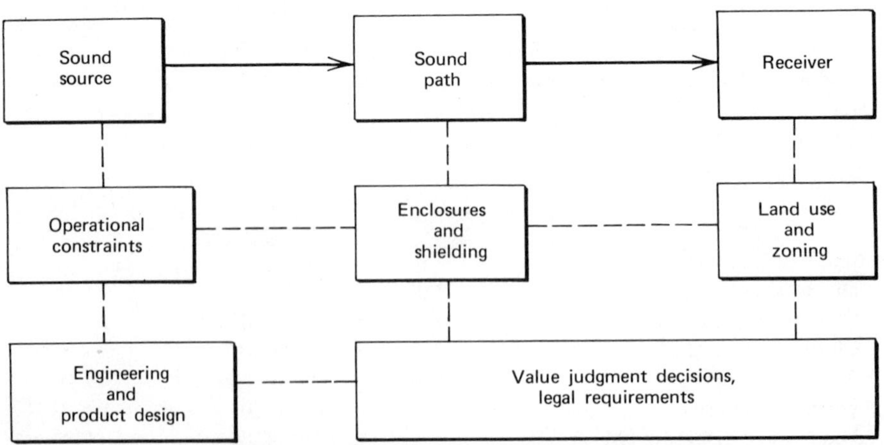

Figure 2   Basic elements of a noise abatement program.

diate future. Sound reduction in residential dwellings cannot be realized by the lightweight construction techniques employed currently by builders who are trying to reduce costs, although research is underway to devise a new generation of building materials.[7] Indeed, economic considerations (as well as building codes) will play an important role in all future plans to attenuate aircraft or traffic noise. The argument has also been advanced that it might be unfair to insist that household and recreational products be "quiet" if in doing so a large number of persons are priced out of the market by a more expensive but quieter product. This argument has limited validity because often it is possible to design products that are not only quieter but that have improved efficiencies, greater reliability, and a longer life. Thus, a slightly greater initial cost may be more than compensated by an extended operating period. Nevertheless, a higher quality product will usually carry a higher initial price tag, a factor that is not trivial in the real marketplace. Furthermore, the demands for federal support in other areas of social concern—energy, education and welfare, medicine, housing and transportation, plus other environmental fields—suggest that noise control will not have an unchallenged priority.

The conflict in societal goals over transportation is especially hard to resolve. In the nineteenth-century a railroad was a welcome addition to virtually any area, despite the noise. It opened up new markets to farmers and cattlemen, and a railroad was the vital link in industry and commerce. To some degree, air and surface transport represent an analogous situation today. Air freight has opened up vast new markets for perishable produce, and a modern city cannot exist without an air terminal. Admittedly, airport planning has not always been optimum, and convenience of the air traveler has often been given a priority over environmental concerns. The miles of expressways and the ever-increasing delivery of products (including mail) by trucks obviously contribute to the noise problem, but they provide us with the necessities and comforts we are loath to do without in today's society. So for a city, an enhanced "quality of life" is not an acceptable substitute for the closing of an airport or the exclusion of surface transport. An individual may move and leave the noise behind, but an entire community cannot do so.

With respect to aircraft noise, it is especially important to interpret fairly the meaning of *health and welfare* that underlies all environmental goals and relevant legislative measures. Noise abatement goals must indeed be pursued vigorously, but they will have to be constrained by realistic considerations of safety and technology. In addition, aviation is an important national asset and a vital link in national and international commerce and trade. Regulations that would severely limit the viability of aviation could conceivably be in opposition to the "public health and welfare" in its broadest sense. Thus, noise control options must be exercised within the overall social context.

Any national noise abatement effort must also reckon with the wide latitude in the social preference of individuals. The choice of a reasonably large percentage of the population is not for quiet, at least during leisure hours and vacation periods. Many remote areas are now attracting tourists who genuinely prefer power boats, snowmobiles, and motorcycles in contrast to less mechanized and quieter forms of recreation. Indeed, there are relatively few vacation areas that remain characterized by the sounds of nature, and noise levels at small inland lakes often approach and sometimes exceed the noise levels of the city. In other recreational spots the loud broadcast of recorded music is intentionally used to create an exciting carnival-type atmosphere; many people "expect this noise" and enjoy it. And for a large percentage of the United States population, television, the radio, and the record player have replaced books as the dominant force in American culture. The result is that among millions of people living in close proximity, what is entertainment to some is intrusive noise to others.

Because we live in a democratic society, the type of environment and the quality of life that can be realized will ultimately depend on voter consensus. Engineers can locate the sources of noise, make quantitative physical measurements of many of them, suggest acceptable standards, and estimate the cost of attaining these standards. In some instances information relative to the physiological and psychological benefits that accompany lower noise levels can be made available to the public. But communities must be aware that noise measurements and the adoption of standards alone will not solve the noise problems and that any meaningful solution will have an economic impact. For some communities the price tag for a dramatic and immediate environmental improvement will be too high, when balanced against other concerns. Legal considerations must also be factored into any overall noise abatement goals.

LEGAL CONSIDERATIONS

Throughout the history of the common law, courts have attempted to reconcile the conflicts of interest of adjacent property owners. For example, one person insists that because he owns property, he is entitled to use it as he wishes; another believes that his ownership entitles him to use his property without annoyance from his neighbor. There is also the larger question of an individual's right vs. the common good. These two conflicts have now been extended to the question of noise, and cases associated with noise can be expected to increase in the courts. People are refusing to accept noise that they consider unnecessary, and courts are currently willing to hear arguments that would not have been considered in the past. Citizen's and community groups usually first request voluntary compliance with neighborhood standards, but they are playing a more active role in initiating complaints against noise sources, in cases of noncompliance. Even individuals are filing class-action suits in regard to noise. Municipalities and administrative agencies have few precedents, so more legal action is inevitable. For instance, the question of "air rights" has not been resolved.

Only a few examples are needed to point up the dilemma of reconciling the freedom of some to undertake noise-producing activities, with the rights of others to be free from annoyance.

Case I

A suit was initiated to bring a total halt to jet traffic at the Washington National Airport on the basis of constitutional rights to health and privacy.[8] Defendants in the case were the Federal Aviation Administration and nine private airlines. In ruling against the plaintiffs, the U.S. Circuit Court judge agreed that the noise "is extremely annoying to and genuinely interferes with the comfortable enjoyment of the property of that segment of the population represented by some of the plaintiffs." But he concluded that he could not order the cessation of jet operations under common law. "Below the level of constitutional right—the interest of the plaintiffs must be balanced against other interests; convenience to the traveling public, economic interest of the community and airlines, and the role of the airport and aviation in national transportation" . . . "Burdensome as it may be—it is no more than felt by that segment of the public that lives adjacent to other methods of transportation, such as the heavily-traveled interstate highway or frequently-used railroad tracks."

Case II

In the case of a private nuisance action against a ready-mix concrete plant,[9] the plaintiff was successful. The operation of the plant was accompanied, for example, by early morning and late evening loud noise of trucks, the banging of tailgates, and there were no other noise-producing plants in the plaintiff's area.

The Court determined that "noises may be of such character and intensity as to unreasonably interfere with the comfort and enjoyment of private property as to constitute a nuisance and in such cases injury to the health of the complaining party need not be shown."

"A fair test of whether the operation of a lawful trade or industry constitutes a nuisance has been said to be the reasonableness of conducting it in the manner, at the place and under the circumstances in question. . . . Thus the question whether a nuisance has been created and maintained is ordinarily one of fact, and not of law depending on all the attending or surrounding circumstances. Each case of this nature must depend on its own facts."

Case III

A homeowner built an expensive home directly across a small river from an existing commercial plant where cranes, barges, and other heavy equipment were routinely used. The equipment was not obsolete, poorly maintained, or excessively noisy compared to comparable equipment used elsewhere. The Court upheld the plant against the action of the homeowner:[10]

"The locality, the occupation of the inhabitants of the neighborhood, the environment, is what determines whether an establishment which uses necessarily noisy machinery, is a nuisance. A manufacturing enterprise that would be a nuisance in one locality might not be so in another. The inhabitants of large cities that are sustained by manufacturing and commercial enterprises must bear the unavoidable discomforts and annoyances thereof. Noises and vibrations from the operation of machinery cannot be a nuisance, subject to injunction, unless they are excessive and unreasonable, depending upon the location of the establishment, its relation to other property, and particularly to other sources of noise or vibration."

Case IV

A group of defendants was charged with disturbing the peace by creating loud and unnecessary noises with motorcycles in violation of a city ordinance.[11] The defendants were found guilty by the

Court: "The protection of the well being and tranquility of a community by the reasonable prevention of disturbing noises is within the city's power."

Case V

A state appeals court upheld the city of Trenton, N.J. in its noise control ordinance against sport parachutists and their noisy aircraft.[12] When the sporting group involved raised legal arguments that only the federal government could regulate air traffic the court disagreed, ruling that a municipality may regulate noise from aircraft "hovering overhead."

Case VI

An instance in which local authority was overruled arose out of the adoption, in 1952, by Cedarhurst, N. Y. of an antiflyover ordinance prohibiting overflights that were less than 1000 ft above the ground. In *Allegheny Airlines* vs. *The Village of Cedarhurst,* the village claimed that the ordinance was necessary because Cedarhurst was within some 4000 ft of the eastern end of the J. F. Kennedy International Airport. Cedarhurst was then sued to prevent enforcement of the altitude ordinance by both the air carriers using the JFK airport, and the Port of New York Authority. In sustaining the air carriers and the Port Authority, the Court of Appeals noted "that the predecessor to the FAA had been directed by the existing Federal law to prescribe air traffic rules regulating safe altitudes of flight and that in complying with these rules aircraft landing or taking off at JFK were required to fly as low as 450 ft over Cedarhurst under certain adverse weather conditions." As a result, the Court found it was not possible for an aircraft at once to comply with both the federal rule and the Cedarhurst ordinance. Given the existence of such a direct conflict, the Court sustained the federal air regulation under the supremacy clause of the Constitution. In the *Cedarhurst* opinion the Court also went on to rule that, without regard to the existence of a conflict, the federal air regulations had completely preempted the field of air traffic regulations and had left no room for any other kind of regulation.[13]

Case VII

For years, residents of Canton, Ill. (population of 14,000) had been hearing seven loud blasts at 6 a.m. each day from a major industrial plant.[14] The sound level at the plant measured 89 dB(A), so that much of the community was exposed to a noise level exceeding that of a pending regulation. Upon receipt of a letter from the Illinois Environmental Protection Agency, suggesting that the whistle might be incompatible with the new law, the plant manager took action by shutting off this early morning work signal. However, within 48 hours after the shutdown, 7000 persons had signed petitions protesting its silencing. Men were late for work, school children were tardy, and housewives were irritated that their dependable alarm clock was no longer available. Accordingly the whistle was turned back on in response to popular demand. The situation was succinctly described by one government official: "There is obviously not much of a nuisance if the majority of the people like the whistle."

Thus, it appears that a large number of cases will have to be resolved on their individual merits. Prohibition of noise, per se, is too vague to be sustained in court, but the Supreme Court has upheld many specific statutes and ordinances limiting noise (Chapter 12). Certain zoning laws, however, based on objective sound measurements, have not been sufficiently tested in the courts to assess their ultimate impact. Also, in some instances where an "essential public facility" has clearly deprived private parties of "private rights," the courts have awarded financial compensation. In conflicts that do not involve a public facility, judicial consideration is still given to the "social value" of the noise or "the impracticality of preventing the annoyance." Overall, the legal resolution of the noise problem—and other environmental concerns—will require (1) a reappraisal of past political and economic axioms and (2) a compromise between the seemingly incompatible ideals of distributive justice and individual freedom.[15]

REFERENCES

1. Proceedings of the International Congress on Noise as a Public Health Problem, Dubrovnik, Yugoslavia (May 1973).

2. F. R. Pitts and R. L. Metcalf, *Advances in Environmental Sciences,* Wiley, New York, 1969, p. 5.

3. U. S. Environmental Protection Agency, *The Social Impact of Noise,* Washington, D. C., 1971, p. 5.

4. *Ibid.,* p. 3.

5. *Ibid.,* p. 124.

6. *Ibid.,* p. 24.

7. Sponsored, in part, by the U. S. Department of Housing and Urban Development.

8. *Noise Control Report,* Business Publications, Silver Spring, Md., May 29, 1972, p. 25.

9. G. M. Betz, *Pollution Engineering: Noise and Its Control,* Technical Publishing, Greenwich, Conn., 1972, p. 15.

10. *Ibid.*

11. *Ibid.*

12. *Ibid.*

13. U. S. Environmental Protection Agency, "Legal and Institutional Analysis of Aircraft and Airport Noise and Apportionment of Authority between Federal, State, and Local Governments," Section 2, p. 47, July 1973, Washington, D. C.

14. *Time,* **103,** No. 3, January 21, 1974, p. 10.

15. J. L. Hildebrand, *Columbia Law Review,* **70,** 652 (1970).

9

TRAFFIC AND VEHICULAR NOISE

TRANSPORTATION NOISE

Highway vehicles that contribute a substantial portion of environmental noise include automobiles, trucks, utility vehicles, buses, and motorcycles. Their use in urban areas account for about 52 percent of the total vehicle miles[1] so that noise patterns throughout the country follow closely the population density. The general growth in the number of highway vehicles, exclusive of motorcycles, is shown in Fig. 1a.[2] As of 1970, there were approximately 106.5 million vehicles operating in the United States, or roughly one vehicle per two persons. Estimated per annum passenger car travel now exceeds 1000 billion miles and truck mileage is over .200 billion miles. Highway and city buses account for less than 30 billion passenger-miles. Most of the total truck operating hours are expended in population centers; about 86% of the time is required for pickup and delivery service, with the remainder being used for long haul service.[3]

The general noise levels attributed to such vehicles is also shown in Fig. 1b. The shaded areas in the bar graphs indicate the substantial range in noise levels observed at a distance of 50 ft from the vehicle at highway speeds. Sources contributing to the overall noise levels include (1) the engine, (2) cooling fan, (3) the drive train and gearing, (4) tires, (5) aerodynamic turbulence, (6) body vibrations, and (7) intake and exhaust systems. Collectively these noise sources combine to produce broadband noise that includes frequencies from the infrasonic to about 10,000 Hz.

190

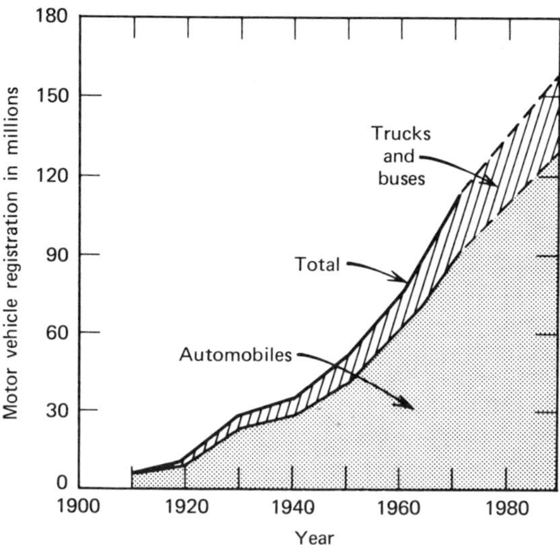

Figure 1*a* Growth in the number of automobiles, trucks, and buses.

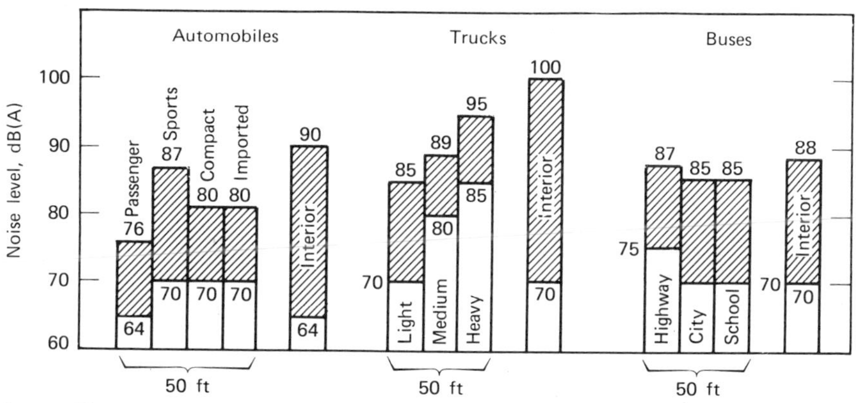

Figure 1*b* Noise levels measured at a distance of 50 feet for various highway vehicles. Shaded areas indicate the range of dB(A) measurements.

Measurements of the attenuation of this broad band noise with distance in typical environments are presented in Fig. 2. The graph shows the relative noise levels generated by the passing of a single passenger car, truck, motorcycle, and a diesel train. It substantiates common experience that in a typical outdoor environment, ~50 dB(A), an automobile can be detected at a distance of about a half a city block, a truck

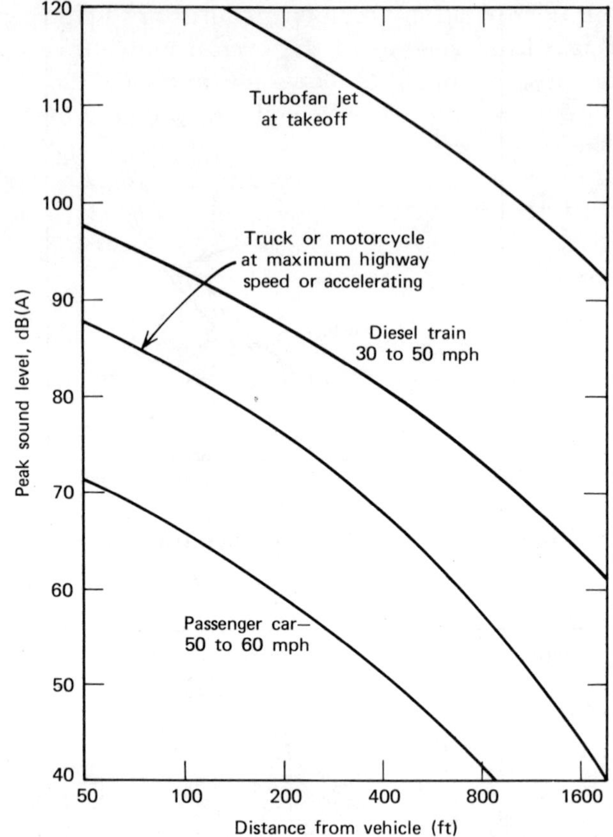

Figure 2   Typical noise levels produced by surface vehicles and aircraft as a function of distance.

much farther distant, and a train is often heard at a half a mile. We shall now consider briefly the characteristic noise spectra of automobiles, trucks, and other vehicles and indicate a few of the problems associated with their noise reduction.

AUTOMOBILES

The major noise sources of the conventional automobile are exhaust, intake, engine and fan, and tires at high speeds. Actually, the modern automobile is a relatively quiet vehicle when operated at low speeds, but the noise output from all car components increases with speed. Figure 3 shows an octave-band analysis of the overall noise from a single automobile from a distance of 50 ft.[4] The solid lines show the sound pressure levels at cruising speeds of 30 to 39 mph and 60 to 69 mph. The dotted lines show the noise spectra when the car is undergoing maximum acceleration (full throttle), in these two speed ranges. At cruising speed there is almost a 10 dB difference at 200 Hz and 2000 Hz between 30 and 60 mph.

At low speeds (~30 mph) the exhaust system constitutes the primary noise source. Engine noise is low at these speeds; the engine is essentially idling, and the massive engine block damps out most of the noise generated from the explosions within the individual cylinders. Mechanical engine and carburetor intake noise is also low and, in addition, the hood and fenders prevent some noise from radiating directly to an observer. Fan noise can approach that of the exhaust system. Characteristic fan and airflow noise depends on the blade configuration and spacing, rotational velocity, the proximity of other accessories, and the radiator-fan distance.

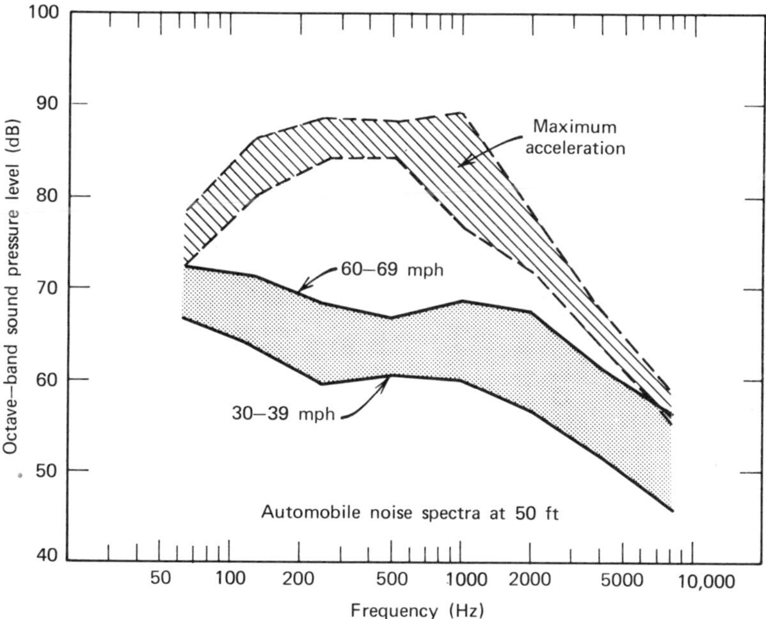

Figure 3   Octave-band spectrum of passenger car noise at a distance of 50 ft. (Courtesy of the U.S. Environmental Protection Agency, Ref. 4.)

The wide range of intake and exhaust noise levels as a function of vehicle speed is explained rather simply.[5] The intake noise results from the air and gasoline mixture that is drawn through the carburetor with the engine acting as an air pump. At low speeds the air may be drawn in at 30 ft^3/min; at high speeds this intake rate may be 300 ft^3/min, and the air velocity may exceed 10,000 ft/min. Eddy currents are formed, and a loud hissing noise results. The exhaust noise results from gas impulses that are fed into the exhaust system under considerable pressure. A four-cycle, eight-cylinder engine will exhaust 240 times/sec at 3600 rpm, but only 48 times/sec at 720 rpm (engine idle). Gas that is forced from the exhaust valves generates broad band noise, and a pulsating gas flow results at the end of the exhaust pipe. The harmonic response of the exhaust system may also contribute to additional noise and vibration.

Figure 4 shows the relative contribution of engine and fan, intake, and exhaust noise for two typical cars undergoing full throttle acceleration

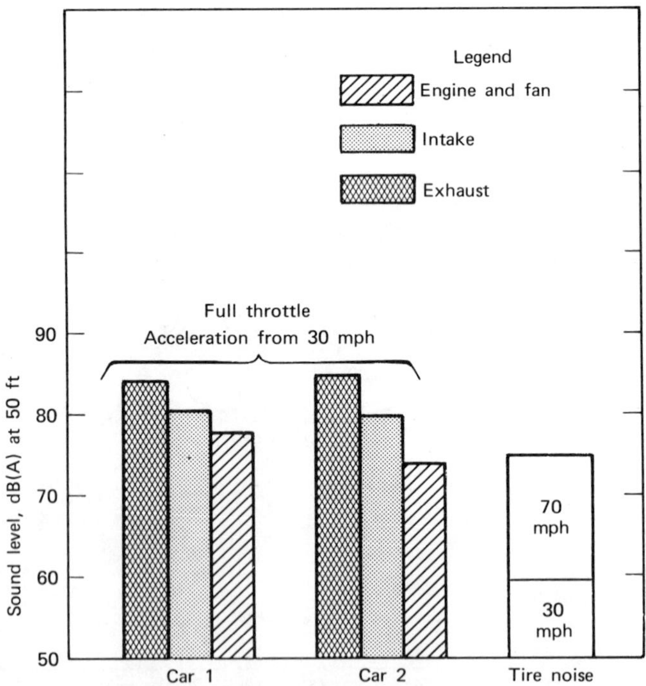

Figure 4   Typical noise levels generated by automobile components, under full throttle acceleration at a distance of 50 ft. (Courtesy of the U.S. Environmental Protection Agency.)

from 30 mph. Under maximum acceleration, exhaust noise is predominant for both vehicles.

Tire noise from passenger cars is important at high speeds. Figure 5a shows the variation in sound levels at 25 ft for two representative tires— "continuous rib" and "discrete block." Figure 5b is a 1/3 octave-band spectra of a "rib" type tire as a function of speed on an asphalt surface. In practice the noise level will also depend on tire pressure and the vehicle loading. At cruising speeds (65 mph) the standard tire may produce from 65 to 75 dB(A). Snow tires have a tread design comparable to truck tires and noise levels at high speeds may reach 85 dB(A).

Considerable engineering effort has been devoted to isolating noise from the passenger compartment.* Sound-absorbent materials are used extensively, and the effective sealing of all windows is important. There is no strict correlation between interior noise level and weight or size of the vehicle, although the heavier and more costly vehicles are generally provided with a greater degree of sound-proofing than lighter and cheaper models. Luxury passengers cars are also equipped with large displacement engines that operate at a lower rpm than sports car engines for the same developed horsepower. Sound measurements within a number of passenger cars has been reported in road tests. Table 1 lists values obtained near the driver's left ear, with windows up, and at 60 mph on smooth roads.[6] For sports cars under full throttle acceleration, interior noise levels have been measured between 85 and 90 dB(A).

Figure 6 shows a 1/3 octave-band spectrum of a domestic sedan [70 mph and 72 dB(A)] and a popular economy import [70 mph and 82 dB(A)]; it also shows the range of noise spectra for a total of 15 cars.

However, neither the dB(A) levels of Table 1 nor the spectral measurements of Fig. 6 reveal the magnitude of sound power that is present at very low frequencies. A recent study[7] has shown that a substantial amount of acoustic power exists below the threshold of hearing (2 to 16 Hz) and extending to about 100 Hz (Fig. 7). These data were obtained on highways at operating speeds between 40 to 70 mph with all windows closed. When tests were made on vehicles with one window open (about 6 in.), levels as high as 110 to 120 dB were observed in the 2 to 32 Hz octave bands. Investigators suggest these high levels are due to air turbulence of a random nature. The slight peak in Fig. 7 at 64 to 125 Hz

---

* Studies at the General Motors Proving Ground Noise and Vibration Laboratory suggest that at moderate operating speeds, passenger compartment (in-car) noise is predominantly tire noise, and that above 400 Hz, in-car tire noise is transmitted principally by airborne paths.

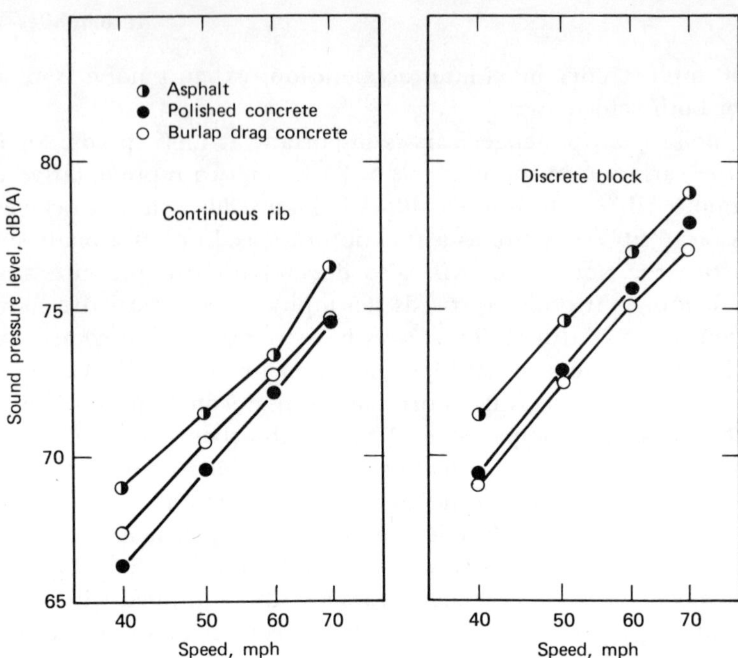

Figure 5*a*   Variation in sound levels at 25 ft as a function of speed, road surface texture, and tread pattern for two representative tires. (Courtesy of General Motors Corp.)

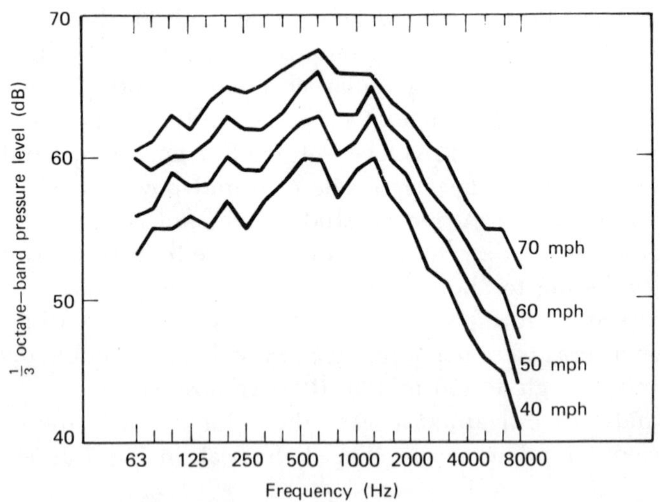

Figure 5*b*   ⅓-octave band spectra for a representative continuous-rib tire as a function of speed on an asphalt surface. (Courtesy of General Motors Corp.)

196

Table 1   Automotive Interior Noise Levels (60 mph)[a]

Model	dB(A)
Ford Thunderbird	67
Chevrolet (Monte Carlo)	68
Mercury (Colony Park)	70
Ford (Torino)	72
Mercury (Cougar)	75
Plymouth (Satellite)	74
Chrysler (Town and Country)	75
Dodge (Monaco)	75
Ford (Mustang)	77
Renault 16	77
Toyota Corolla	77
Ford (Pinto)	78
Dodge (Challenger)	80
Chevrolet (Vega)	83

[a] Data originally reported in *Popular Science* between February 1970 and January 1971.

is possibly due to an engine-exhaust noise component related to the firing rate of the engine.

Low frequency sound at high sound pressure levels is responsible for substantial passenger discomfort and fatigue. It has also been suggested that intense low frequency sound may have some effect on the sense of balance—and thus upon driver safety.

TRUCK NOISE

Noise from trucks originates from the same vehicular components as automobiles. However, truck engines are used in the wide-open throttle mode a greater portion of the time, and in large trucks, the engines are more powerful. A greater sound intensity results. Figure 8 shows the principal contributing components of truck noise (engine, fan, and exhaust) at speeds of about 35 mph for 12 heavy trucks[8] (nine diesels and three gas models).* There is considerable variation in the intensity of the

* In a standard test a 100-ft course is laid out on a lane which is 50 ft from the microphone. The truck accelerates through this 100-ft course under wide-open throttle.

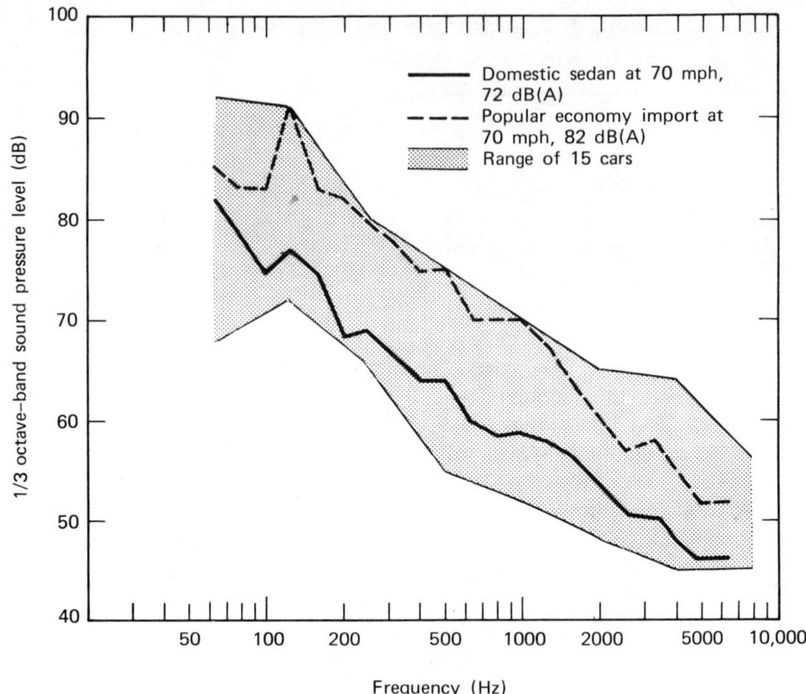

Figure 6   Interior sound pressure levels of a standard passenger car (solid line) and a popular economy import (dotted line) at 70 mph. The shaded band indicates the range of interior noise levels for 15 different cars. (Courtesy of the U.S. Environmental Protection Agency.)

various noise sources for these 12 distinct models. At speeds of about 40 mph, a substantial amount of noise is also generated from tires. There is a significant difference in the noise spectrum generated by gasoline-powered and diesel engines of equal horsepower. Diesel trucks, which now comprise about 3% of all trucks in operation (and a sizable percentage of heavy vehicles) are generally 8 to 10 dB(A) noisier than gasoline-propelled trucks and some 12 to 18 dB(A) noisier than passenger cars. The higher noise levels from diesel engines is a result of the ignition. In gasoline engines combustion is triggered in the cylinder by means of a spark. The spark then generates a flame that spreads rather gradually throughout the cylinder until the fuel-air mixture is completely ignited. This combustion process thus occurs over a reasonable fraction of the power stroke. In a diesel engine, however, the compression ratio is very high, and the fuel-air mixture is spontaneously ignited

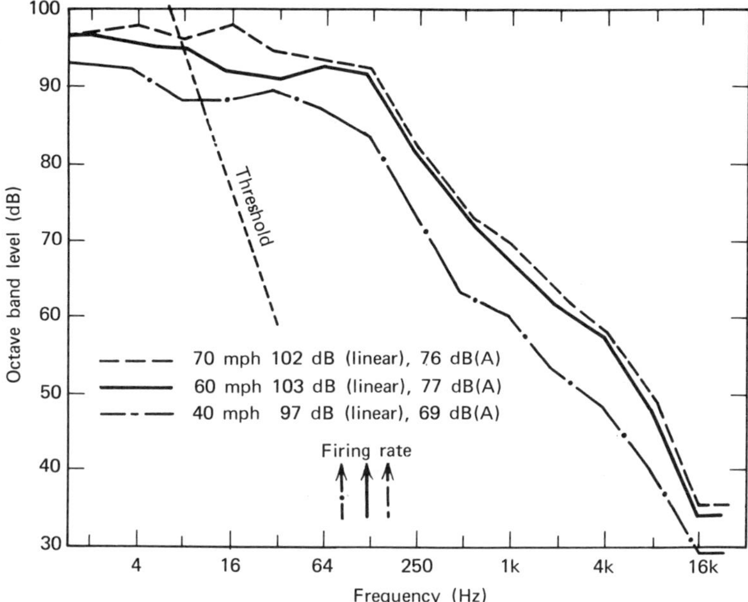

Figure 7   Octave-band spectra of interior noise levels in an automobile travel-ing at 40, 60, and 70 mph. A substantial amount of acoustic energy is below the 15 Hz hearing threshold. [From W. Tempest and M. E. Bryan, *Appl. Acoust.*, **5**, 133 (1972).]

by the compression stroke of the piston. The net effect is a sharper pres-sure rise within the cylinder, a more intense explosion during combus-tion, and more noise radiated from the engine block. Some differentia-tion, of course, must be made between older and newer trucks; for the same gross weight vehicles produced after 1974, there is a smaller differ-ence in the overall noise levels of diesel vs. gasoline-powered vehicles.

More powerful fans are required on trucks than on passenger cars to provide adequate cooling. Present engineering efforts to reduce fan noise include (1) a variable-speed fan clutch, (2) fan shrouds, and (3) slower-turning larger fans. The variable-speed fan clutch provides reduced fan speed when less cooling is required; such fans may also contribute to improved fuel economy. Close-fitting fan shrouds can be designed to direct the air flow for more efficient cooling, and such shrouds permit the use of lower rotational velocity fans, and lower noise.

Exhaust noise in trucks is made up of outlet noise and mechanical "ringing" of the exhaust pipes and muffler body. Current engineering

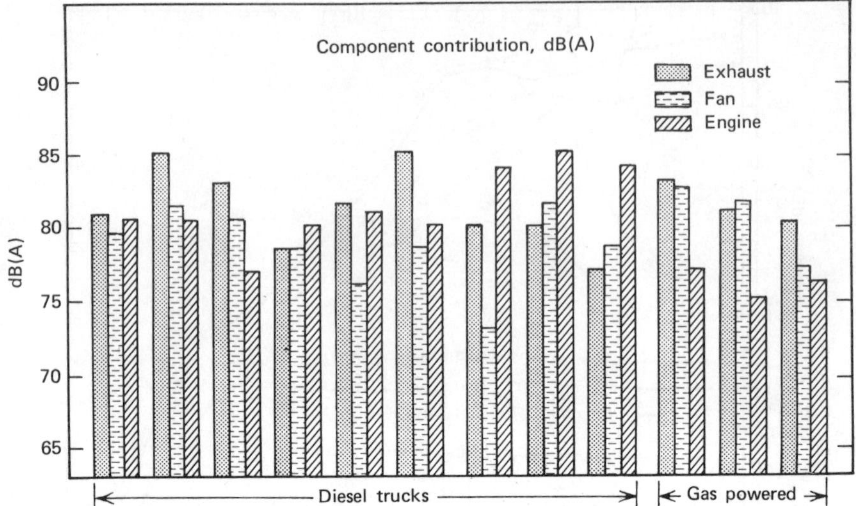

Figure 8   Relative contribution of exhaust, fan, and engine noise, from nine diesel and three gasoline-powered trucks. (Courtesy of General Motors Corp.)

studies to reduce exhaust noise include (1) the use of "premufflers" that provide a reduction in sound energy before the gas flow covers a significant length of the exhaust system, (2) double-walled pipe construction, (3) special pipe and muffler wrappings, and (4) new muffler and pipe designs.*

Figure 9 shows the octave-band spectrum of overall diesel truck noise as determined using procedures outlined by the Society of Automotive Engineers (SAE).

Figure 9 also shows an octave-band comparison of the diesel truck with a passenger car. Both truck and car are being operated at full throttle acceleration at 35 mph, with the microphone located at a distance of 50 ft. The noise output of the truck is actually less than the car at 1000 Hz, but at the very low and high frequencies the diesel is a substantially noisier vehicle. The low frequency noise is very difficult to attenuate, so that diesel trucks can be heard inside buildings that are otherwise quite adequate in providing shielding from the noise of automotive traffic.

* *Note.* Mufflers are usually classified as (1) reactive, or (2) nonreactive, absorbent, or dissipative. A reactive muffler is comprised of one or more expansion chambers and pipes that tend to reflect sound waves and cause some backpressure at the engine exhaust port. Nonreactive mufflers are designed to convert sound energy into heat through the use of sound-absorbent materials.

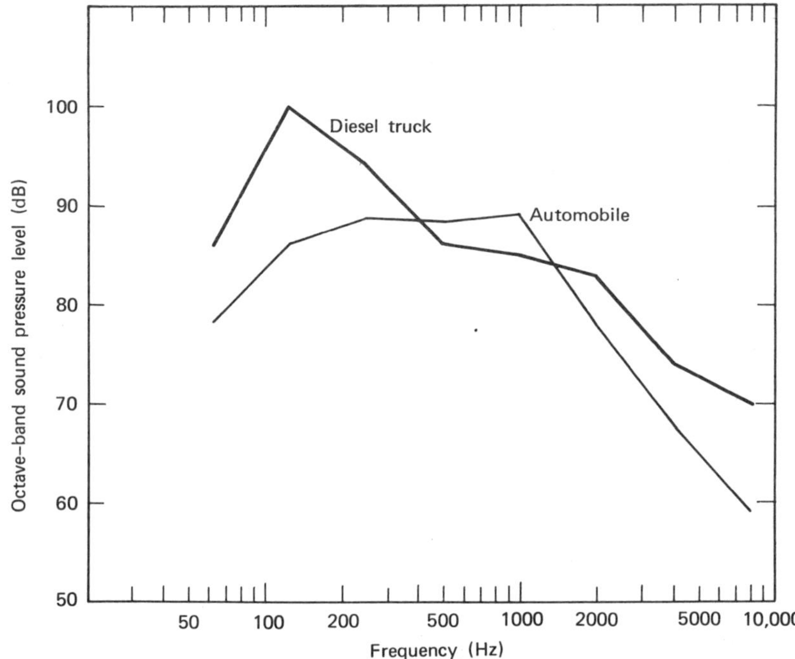

Figure 9   Typical octave-band spectra for a diesel truck and automobile under full throttle acceleration at 35 mph, at 50 ft. (Courtesy of the U.S. Environmental Protection Agency.)

It should be recognized, of course, that there is a marked difference in the acoustic output from any diesel or gasoline-powered truck, depending upon its general condition, horsepower rating, operating load, and tire tread design, as well as upon its speed and acceleration. The general noise output from trucks increases somewhat with age, and at least part of this increase will occur if the components that are replaced are noisier than the original equipment, for example, mufflers and recapped tires. Gasoline engines show a significant reduction in noise with decreasing load (up to 10 dB), but for diesel engines the noise differential will be rarely more than 3 dB for full load to no load.

A number of highway studies have been conducted to determine the distribution in noise levels generated by a large number of trucks.[9] Figure 10 contains plots of the percentage of truck vehicles exceeding a given sound pressure level vs. the peak dB(A) sound level observed at a distance of 50 ft from the vehicle centerline, for truck speeds of 35 mph or lower. Curve A is data for a general mix of 652 vehicles in California;

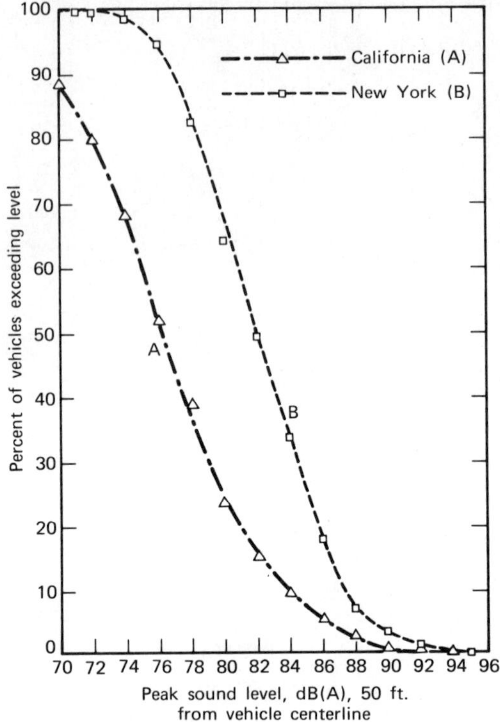

Figure 10   Truck population noise levels at low speeds ($v \leq 35$ mph). [From W. H. Close and T. Atkinson *Sound Vib.*, October 1973, p. 28.]

curve *B* is a mix of 425 trucks in New York. It has been suggested that the lower noise profile of curve *A* is probably due to California's stricter enforcement program.

At high speeds, tire noise becomes a significant factor so that truck noise is highly dependent on tire type, vehicle weight, number of tires, and road surface. One explanation of tire noise suggests that the tire-road contact results from compressing small pockets of air, and that the subsequent tire-road separation is analogous to the breaking away of many suction cups. This phenomenon is called *air pumping,* and it explains why tire noise increases on wet pavements. With modern tires this effect is of lesser importance, however, and the noise is attributed to a complex tread vibration—that is, a repetitive, oscillatory deformation of the tread. Figure 11 indicates the large range of sound levels generated by different tire types and conditions on a single chassis vehicle operating on a concrete road, at a distance of 50 ft. The "crossbar" design is gen-

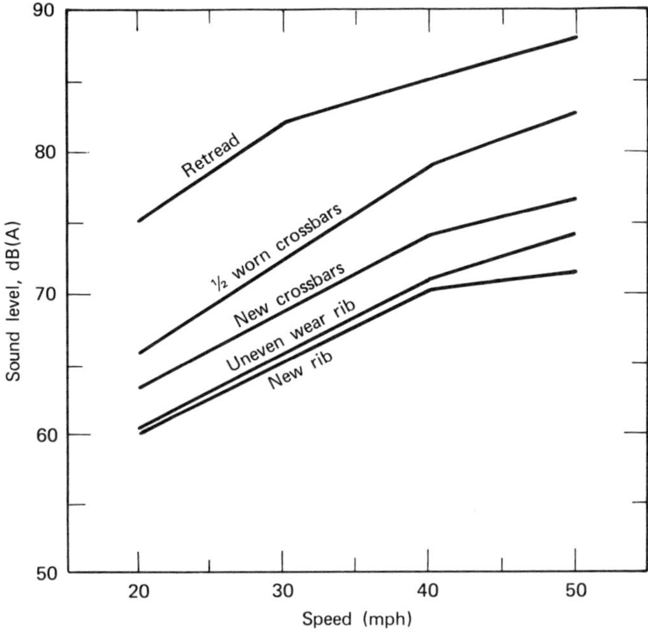

Figure 11   Sound pressure levels, in dB(A) of various tires mounted on the drive axle of a loaded single chassis vehicle, operating on a concrete road, at 50 ft. (Courtesy of General Motors Corp.)

erally used in the drive wheels of trucks; this design exhibits superior traction on both wet and dry pavements, and the tire mileage is greater than the "rib" type configuration used for passenger car use. Retreaded truck tires, using old molds that took little account of noise output, are much noisier than the new treads, which are attempting to provide both traction and low noise. Increased noise from worn tires (not recaps) is ascribed to a change in tread curvature, as the tread thickness decreases. Axle loading, as well as truck speed, is also a factor in tire noise; an unloaded tire does not have as much tread contact with the pavement.

VEHICLE SPEED AND WEIGHT

Both vehicle speed and vehicle weight are important parameters affecting noise levels. The recent work of Olson[10] suggests that if vehicles are divided into appropriate general classifications, the average noise levels from these vehicles can be predicted on the basis of vehicle speed. The study included (1) passenger vehicles, (2) dump trucks, (3) intercity buses, and (4) tractor trailers. Octave-band spectra in the range from 30 to 8000 Hz are shown in Figure 12.

The spectra were taken with the microphone located 50 ft from the center of the traffic lane and 4 ft above street level. The five speed ranges monitored were 20 to 29 mph, 30 to 39 mph, 40 to 49 mph, 50 to 59 mph, and 60 to 69 mph. All spectra indicate higher noise levels at higher speeds. An interpretation of these spectra also confirms that engine noise is the major direct source of noise for heavy vehicles and that dominant frequencies are determined largely by the engine firing rate. Thus, combustion noise appears as direct radiation from the engine block and as exhaust noise. Engine noise from buses is somewhat attenuated, because of a more complete enclosure and muffling system than in trucks. Tire-road interaction adds to noise levels in the 250 to 500 Hz range, and

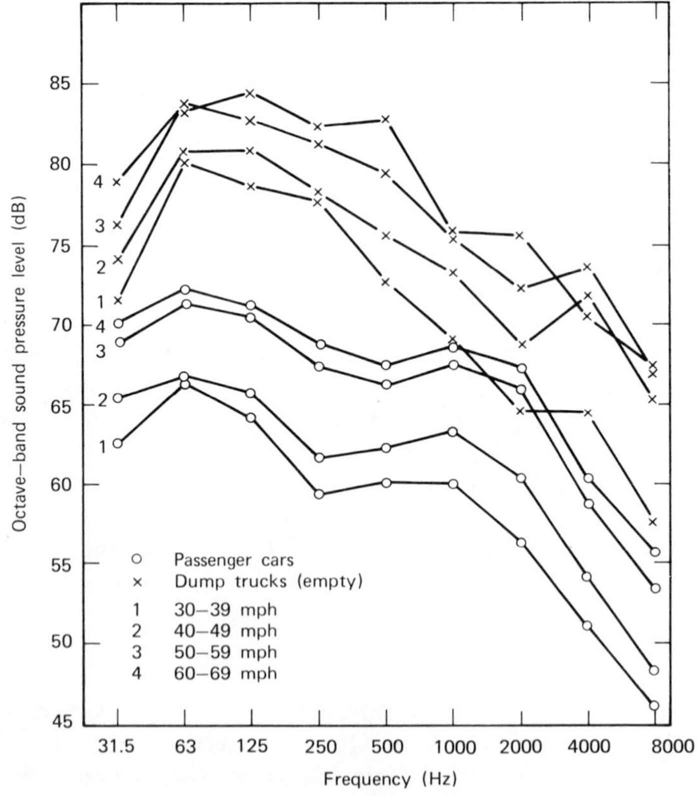

Figure 12   Noise levels generated by vehicles as a function of speed. [From N. Olson, *J. Acoust. Soc. Am.*, **52**, 1291 (1972).]

broad band noise from the carburetor intake is superposed upon discrete frequencies for all vehicles.

A further generalization has been made by Olson[11] on the basis of vehicle gross weight. On the basis of sound level measurements of some 215 passenger cars, 181 light trucks, 125 medium trucks, 102 empty dump trucks, 58 loaded dump trucks, and 32 tractor trailers, he has obtained a noise level vs. weight category shown in Fig. 13. Light trucks are in the 6000 to 15,000 lb gross vehicle weight category; tractor trailers have gross vehicle weights exceeding 40,000 lb. The slope of the curve corresponds to a 3.5 dB(A) increase in noise level for each doubling of gross vehicle (loaded) rating, at a 30 to 39 mph average speed.

EXPRESSWAY NOISE PROFILES

While a study of the noise generated from individual vehicles is essential, it is the total noise from a highway that is of concern to people living nearby. In a recent study made for the State of New York and the

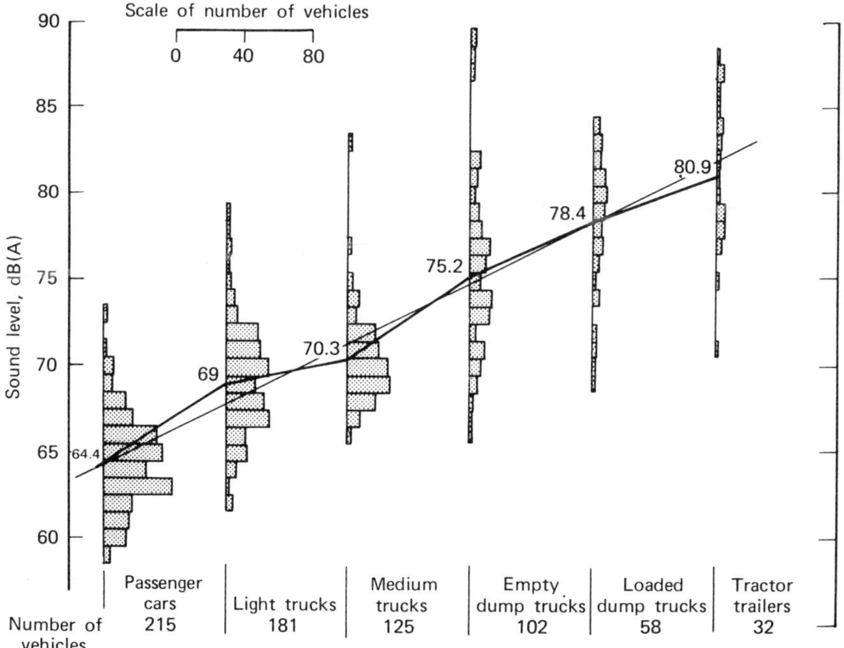

Figure 13   Noise levels generated by vehicles as a function of vehicle weight. [From N. Olson, *J. Acoust. Soc. Am.*, **52**, 1291 (1972).]

Federal Highway Administration,[12] varying noise levels were determined in the proximity of high traffic density expressways. Here are the salient results of this study.

1. A logarithmic relationship exists between traffic volume and noise generated *on* the highway, for a constant mixture of truck and automobile traffic.

2. Variations in the traffic mix will substantially alter the "on highway" noise for a given traffic volume.

3. Site configuration is important; embankments and depth of the roadbed, for example, will affect the noise spectrum for a given traffic density and mix.

4. The "off highway" noise, in proximity to the expressway, will be attenuated differently, depending on expressway design and site (i.e., ravine-type depression, "open" roadway, or elevated span).

5. When the expressway site is such that noise is reflected back to the roadway, the "on highway" noise is increased, the noise attenuation between the highway and the urban area is increased, and the resultant noise level imposed upon the urban area is lowered. For open sites, the noise level both of "off highway" and "on highway" will vary inversely as a function of the distance between the expressway and the nearest reflective surfaces.

6. While site configuration will influence the noise level imposed on the inhabitants of an urban area, the level is predominantly controlled by the traffic characteristics, and increases with hourly traffic volume and the percentage of trucks traveling on the expressway.

These conclusions are supported by data taken for five expressways and a "typical" busy street in the New York City area. Figure 14 shows the weekday diurnal noise level (measured with a B-weighted scale),* at the middle of the roadway for each site.

The road "configuration," daily traffic volume and average speed, and percentage of trucks are listed in Table 2. The Cross Bronx Expressway site is 57 ft below grade, with sloping concrete walls to the six-lane roadbed. The Grand Central Parkway site is a six-lane cantilever-covered highway about 30 ft below ground level. These two configurations yield very high "on highway" noise levels, but provide substantial shielding for the adjacent areas. Figures 15*a* and 15*b* suggest their general contour.

The substantial difference in noise level of an urban street during weekdays and weekends is portrayed in Fig. 16.

* A sound level meter would register slightly lower with an A-weighted network.

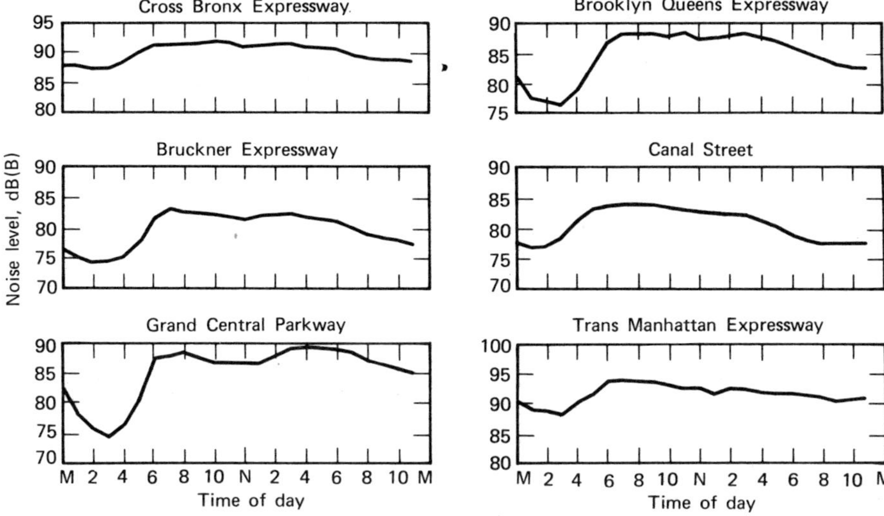

Figure 14   Noise levels measured during a 24-hr period for various expressways and a street in New York City. Measurements were taken at the middle of the roadway with a B-weighting network. (Ref. 12.)

Table 2   Road Configurations and Traffic Volume in the New York City Area[12]

Roadway	Configuration	Daily Traffic	Speed	Cars/Trucks
Cross Bronx Expressway	Deep cut	116,000	48	Mixed
Bruckner Expressway	Grade road	55,000	48	10% trucks
Grand Central Parkway	Cantilever cover	110,000	45	Mostly cars
Brooklyn Queens Expressway	Elevated span	93,000	38	15% trucks
Canal Street	Midcity street	21,000	—	25% trucks
Trans Manhattan Expressway	Sectional "tunnel"	167,000	—	10% trucks

ROADS AS LINE SOURCES

Surveys such as those cited above have permitted analytical models to be developed for freely flowing traffic and large numbers of vehicles.[13] At sufficiently low traffic densities, of course, the noise of an individual vehicle is observed, and the sound pressure level decreases about 6 dB per doubling of the vehicle-observer distance. For high traffic densities

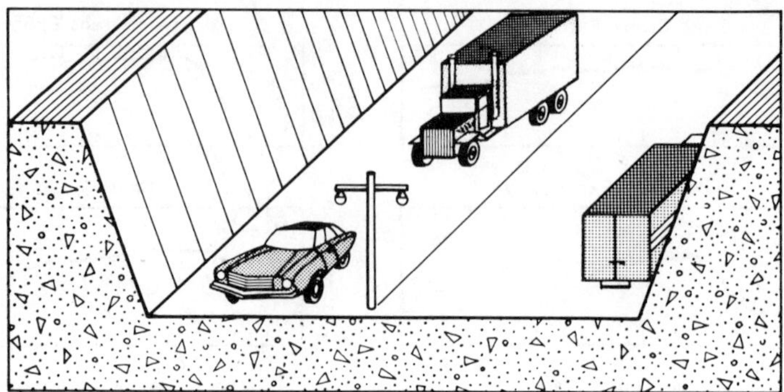

Figure 15*a* Cross Bronx Expressway "deep-cut" roadway with sloping walls. This configuration provides noise shielding for adjacent areas, but "on-highway" noise levels are increased because of reflections from walls.

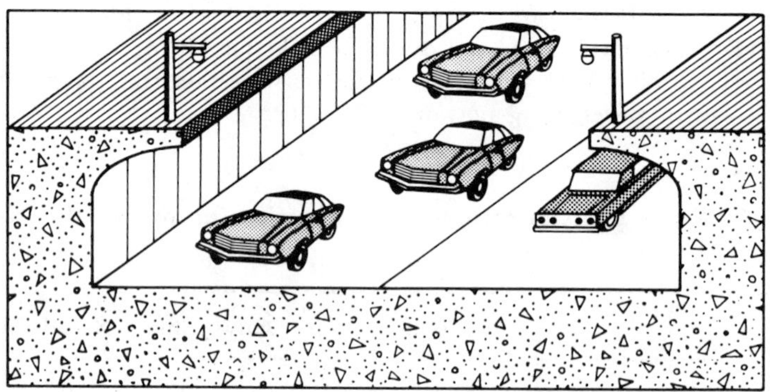

Figure 15*b* Sketch of a portion of the Grand Central Parkway with a "cantilever-covered" six-lane highway some 30 ft below grade level.

and at points beyond the immediate vicinity of the roadway, the noise of individual vehicles tends to blend into a line source for which sound levels decrease at 3 dB per doubling of the road-observer distance.

The simplest model for predicting noise from freely flowing traffic is suggested in Fig. 17. It is assumed that vehicles are spaced sufficiently close together so that the acoustic power is distributed uniformly along the center line of the highway. The source strength can then be described as a constant power per unit length, Π_l, and an expression can be obtained for the mean square acoustic pressure $\langle p^2 \rangle$ at some observation point, 0. Let D be the perpendicular observer—highway distance, $\rho_0 c$

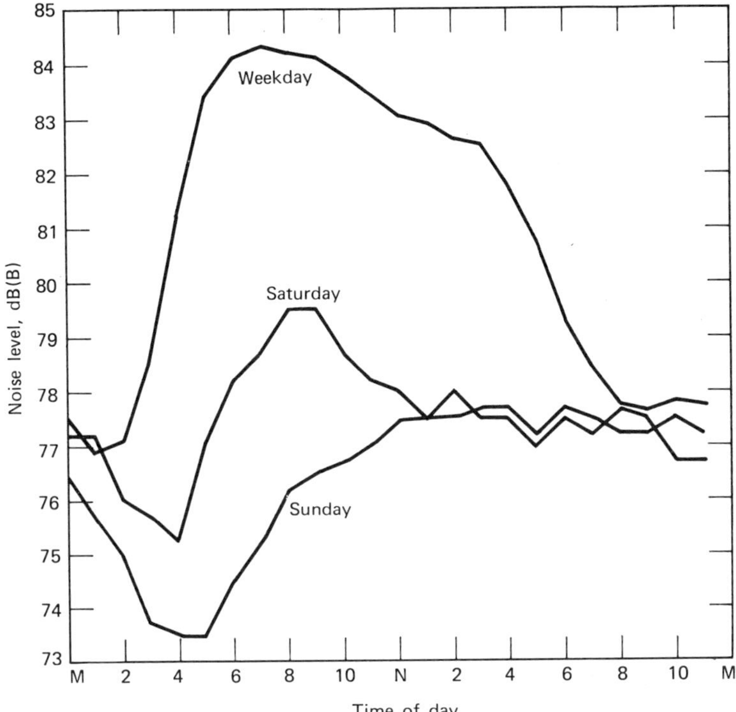

Figure 16   Diurnal noise levels in the middle of Canal Street, New York City.
(Ref. 12.)

the characteristic impedance of air, $d\varphi$ the angle subtending the highway
element, dx. From Fig. 17,

$$x = D \tan \varphi$$

$$dx = D \sec^2\varphi \, d\varphi$$

$$r = D \sec \varphi$$

Then the acoustic power contributed by the element dx is

$$dW = \Pi_l \, dx = \Pi_l \, D \sec^2 \varphi \, d\varphi \qquad (1)$$

and the acoustic intensity from this element is

$$dI = \frac{dW}{2\pi r^2} = \frac{\Pi_l \, D \sec^2 \varphi \, d\varphi}{2\pi \, D^2 \sec^2 \varphi} = \frac{\Pi_l \, d\varphi}{2\pi \, D} \qquad (2)$$

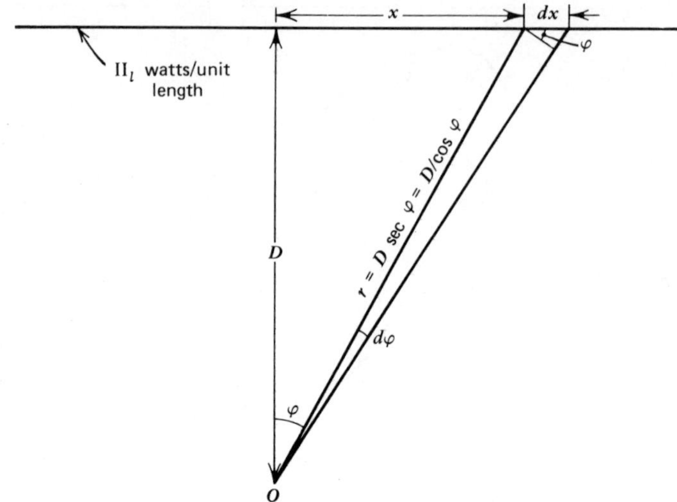

Figure 17   Schematic diagram for computing the acoustic pressure generated from freely flowing traffic. (Ref. 14.)

if hemispherical spreading of acoustic energy from the element dx is assumed. The differential mean square acoustic pressure is then

$$d(\langle p^2 \rangle) = \left(\frac{\rho_0 c \Pi_l}{2\pi D}\right) d\varphi \tag{3}$$

Integrating equation 3 from limits of $-\pi/2$ to $+\pi/2$, corresponding to a line of infinite length, gives the total acoustic pressure:

$$\langle p^2 \rangle = \frac{\rho_0 c \Pi_l}{2D} \tag{4}$$

With the further assumption that all vehicles are automobiles, and using an average *acoustic* power for U.S. automobiles, equation 4 can be expressed in terms of L_{50}, the sound level in dB(A) that is exceeded 50% of the time; the vehicle volume, V, in vehicles per hour; the distance D, measured in feet between observer and roadway centerline, and S, the average speed of vehicular flow in miles per hour:[14]

$$L_{50} = 20 + 10 \log V - 10 \log D + 20 \log S \tag{5}$$

This equation is useful for estimating traffic noise from a level roadway when the vehicle (automobile) flow rate is 1000 vehicles per hour or more.

NOISE vs. TRAFFIC DENSITY

Other attempts have been made to correlate noise levels vs. traffic density with simulated computer models. These models are somewhat more realistic than pure analytical models for several reasons, First, traffic flow is not characterized by a uniform spacing of vehicles. Second, account is taken of multiple lane highways. Third, at large distances a correction can be made for air and ground absorption. Other parameters, which can be adjusted in a computer model, include the mix of vehicles, corrections for low density traffic flow, road surface condition, and highway topography.

A simulated model, which provides good agreement with measured data, is represented by the expression:[15]

$$L_{50} = 10 \log V - 15 \log D + 20 \log S$$
$$+ 10 \log \left[\tanh \left(1.19 \times 10^{-3}\, VD/S\right)\right]$$
$$+ 29 \text{ dB(A)} \tag{6}$$

This equation, for automobiles only, is plotted in Fig. 18 for a fixed distance of 100 ft from the traffic lane and for various automobile speeds, S_A.

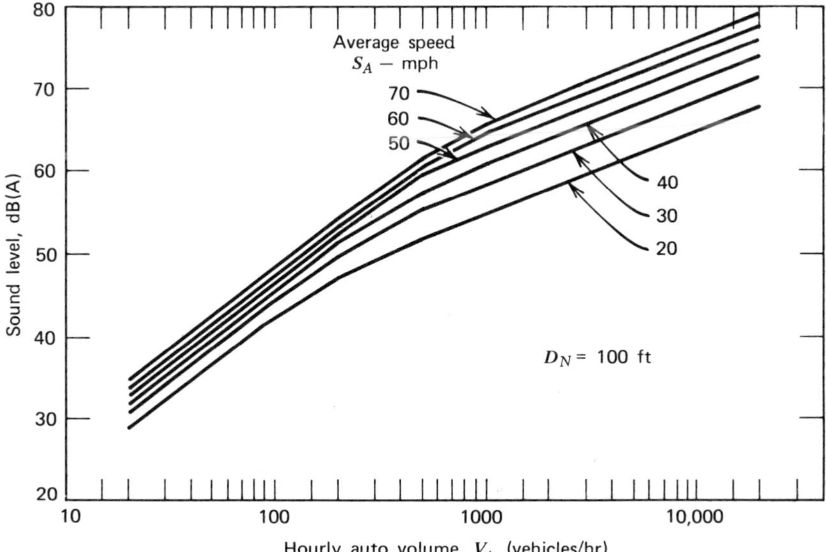

Figure 18 Plot of automobile noise, L_{50}, as a function of volume flow and average speed. (Ref. 15.)

The problem of a multiple lane highway is solved by using a "single-lane-equivalent" distance concept. It is assumed that the total traffic flow can be located in a single lane, whose lane-observer distance, D_E, is defined by

$$D_E = \sqrt{D_N D_F} \tag{7}$$

where D_N and D_F represent the nearest and farthest observer-lane distances, respectively.

A comparable noise-traffic density relationship for truck traffic is expressed by the equation:[16]

$$L_{50} = 10 \log V - 10 \log S - 15 \log D$$
$$+ 10 \log [\tanh (1.10 \times 10^{-3} \, VD/S] + 95 \text{ dB(A)} \tag{8}$$

This equation is plotted in Fig. 19 for a fixed distance of 100 ft from the traffic lane and for various values of truck speed, S_T. The apparent paradox that should be noted is that truck traffic noise, for a fixed volume flow, decreases with increasing vehicle speed. This is because truck traffic noise is assumed to be a function of vehicle density only* (distance being constant) and for a given volume flow, density decreases as vehicle speed increases.

A prediction of the overall traffic noise can now be made if data exists on the separate volume flows for automobiles and trucks, together with their average speeds. Decibel values can be added (as indicated in Chapter 2) that will yield total traffic noise levels for average roads on level terrain. Further corrections may be necessary as noted below.

EFFECTS OF ROAD SURFACE, GRADIENTS, AND INTERSECTIONS

There will be some deviations from predicted traffic noise levels because of road surface, road gradients, and the interruption of traffic flow at intersections.[17] At 50 ft from the edge of a highway an adjustment for road surface condition can be made by use of Table 3.

These adjustments are applicable to both automobiles and trucks at normal highway speeds.

An adjustment due to a road gradient applies primarily to heavy-duty trucks. An uphill grade will require a substantial increase in engine

---

* Except at high speeds where tire noise is often dominant. Trucks, having many gears, operate within a much narrower rpm range than automobiles, so that truck engine noise tends to be independent of vehicle speed.

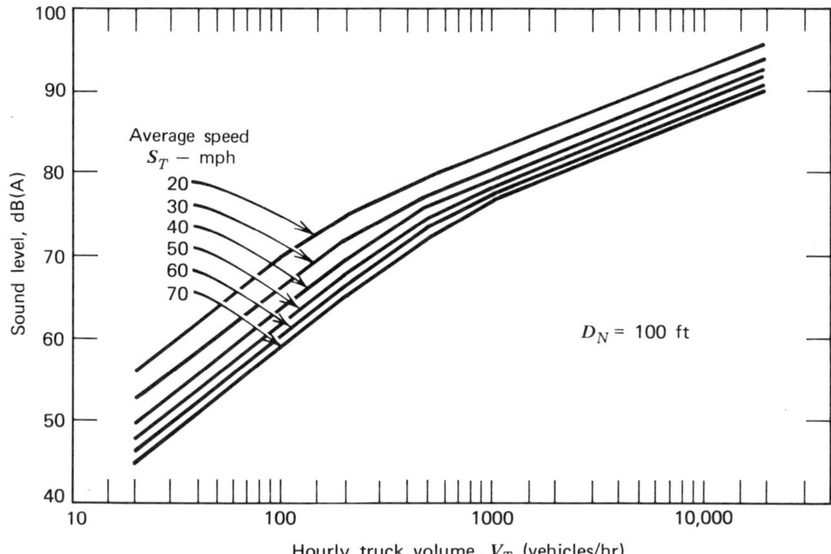

Figure 19 Plot of truck noise, L_{50}, as a function of volume flow and average speed. The graph includes assumptions that are not always valid in practice. (Ref. 16.)

Table 3 Effect of Road Surface on Noise Level[18]

Surface Type	Description	Adjustment, dB(A)
Smooth	Sealed-coated asphalt	−5
Normal	Moderately rough asphalt or concrete	0
Rough	Rough asphalt ($\frac{1}{2}$ in. voids) or grooved concrete	+5

power, and a recommended adjustment in noise level is given in Table 4.

When traffic is controlled at an intersection by a stop sign or a signal, there will also be an increase in traffic noise. Buses or trucks accelerating after a stop will contribute intrusive noise with 5 dB or more "peaks" over steady-state traffic noise levels, and there will be a lesser noise increment because of accelerating passenger cars. Thus residents in the immediate vicinity of hills or intersections will be exposed to higher traffic noise levels than for level-uninterrupted flow.

Table 4   Adjustment for Increased Noise Level of Trucks on Gradients[19]

Gradient	Adjustment
(%)	(dB)
≤ 2	0
3 to 4	+2
5 to 6	+3
≥ 7	+5

ATTENUATION BY BARRIERS

Traffic noise levels generated by a given traffic volume may be decreased by means of (1) a depressed roadway, (2) roadside barriers, (3) shielding by structures, and (4) landscaping.[20]

The depressed roadway configuration is costly to construct, but it provides an acoustically opaque mass of earth between the roadway and adjacent property. The acoustic "shadow," of course, is only a partial one because diffraction allows sound energy (primarily low frequencies) to "spill over" the embankment. The effective wavelength of traffic noise is about 2 to 3 ft so that the attenuation of traffic noise is close to the attenuation provided at a sound frequency of about 500 Hz. Figure 20 indicates the decrease in dB(A) that can be expected from a depressed roadway (of depth H), at an equivalent lane-observer distance D_E, and "curb"-observer distance D_C.[21] There is a practical attenuation limit of about 15 dB for automobiles and about a 10 dB limit for low speed trucks (because of their elevated exhaust stacks).

Roadside barriers are usually less effective than depressed roadways, primarily because there are practical limits to barrier heights. A depessed roadway may have an elevation difference of 20 ft, but it is usually not practical to build a barrier of this height over an extended portion of the noise-producing roadway. Barriers are thus employed in highly local situations. Figure 21 shows the noise decrement in terms of barrier height, H, equivalent-lane observer distance, D_E, and barrier-observer distance, D_B.

The effectiveness of shielding by structures will depend on structure type and spacing. A typical value of 3 to 5 dB per row of houses has been reported,[22] providing no line-of-sight contact exists between the roadway and the observer. However, even multiple rows of houses will

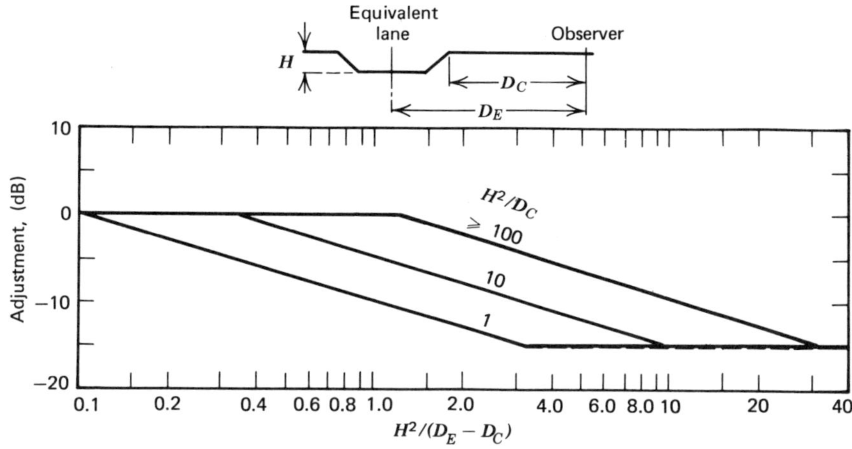

Figure 20   Adjustment in traffic noise levels for a depressed highway. (Ref. 21.)

not afford more than 10 dB attenuation. The presence of large multi-storied commercial buildings along a roadway will, of course, afford sub-stantial sound attenuation behind the buildings and at the same time increase the roadway noise level because of building reflections.

Landscaping provides little actual sound attenuation,[23] as noted in Chapter 3. Visual isolation from a roadway, however, may reduce sub-

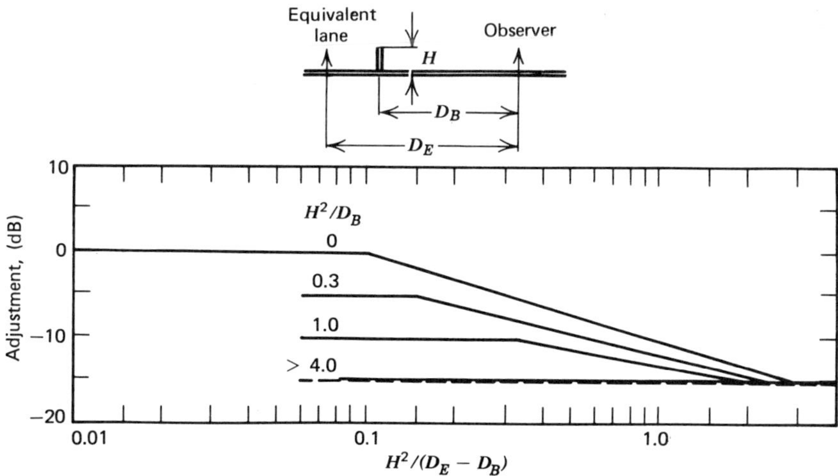

Figure 21   Adjustment in traffic noise levels because of barriers. (Ref. 21.)

stantially the subjective annoyance of a nearby highway. A design value of 5 dB noise attenuation for every 100 ft of planting depth has been suggested if the trees are at least 15 ft tall and if the density is such that no visual path exists between roadway and observer.

TRAFFIC NOISE INDEX (TNI)

A traffic noise index (TNI)* has been proposed as a method for summing, over a 24-hour period, exposures to traffic noise, without taking into explicit account the traffic volume or traffic mix. The traffic noise index is defined as:

$$TNI = 4(L_{10} - L_{90}) + L_{90} - 30 \tag{9}$$

where L_{10} and L_{90} are the noise levels, in dB(A), exceeded 10% of the time and 90% of the time, respectively.

Selected data from Griffiths and Langdon[24] are presented in Table 5 to show the correlation between observed noise levels and the traffic noise index in actual traffic flow surveys. The L_{50} values are included to indicate mean sound levels.

Table 5   Traffic Noise Index vs. Sound Levels in dB(A)[24]

TNI	L_{90}	L_{10}	L_{50}
98.5	54.5	73	64
91	49	67	57.5
89	61	75.5	68
85	51	67	58
82	52	67	59.5
79	49	64	56
72	48	61.5	53.5

Because the TNI tends to weight extreme values, it provides a scale that correlates fairly well with "dissatisfaction scores" in statistical sur-

* An index mostly used in England. Various computer programs have also been developed by the U.S. Department of Transportation, the Michigan Department of State Highways, and other groups that permit additional variables to be factored into calculations for predicting traffic noise; for example pavement elevation, barriers, surface types and grades.

veys.[25] A *TNI* of 74 has been suggested for planning purposes, in order to restrict the minimum distance that buildings should be located adjacent to highways. A *TNI* of 74 means that only one person in 40 is likely to be dissatisfied within a dwelling whose exterior wall is exposed to this traffic noise level. If, on the basis of a survey, a *TNI* at a distance of d_0 from a highway is greater than 74, then a new distance d_1 at which an acceptable *TNI* of 74 would be achieved can be computed by the formula:[26]

$$\text{Required reduction in } TNI = 45 \log_{10} \frac{d_1}{d_0} \tag{10}$$

RAIL TRANSIT SYSTEMS

Rail systems comprise long haul freight and passenger trains and a variety of subway, elevated, and surface vehicles. In the United States these vehicles operate on 334,000 miles of total trackage and along 205,000 route miles. An overall picture of the United States rail system, from 1950 to 1970 is shown in Figure 22,[27] together with typical noise levels within car interiors, and at a distance of 50 ft.

Long-haul railroad passenger traffic has declined to about 11 billion revenue passenger miles in 1970, or to about one-third that of 1950.[28] Freight hauling, however, has increased from about 590 to 776 billion freight-ton miles within the 1950 to 1970 period. Diesel-electric locomotives are used for 99% of the passenger-freight load. About 27,000 locomotives were in service in 1971; half of these were powered by 1800 horsepower engines or greater; lower powered vehicles are employed in switchyards and for short hauls.

Noise from these trains includes (1) diesel exhaust, (2) engine and air intake, (3) cooling fans, (4) wheel-rail interaction, (5) electric generator and electric traction motor noise, and (6) miscellaneous noises generated in freight and passenger cars primarily as a result of wheel-rail interaction. Other intermittent noises includes car impact sounds when trains are braking or accelerating, and the sound of sirens or horns that produces noise levels 10 to 20 dB greater than from other sources.

Rapid transit vehicles are all electric-powered. In 1971 there were 15 rail rapid transit systems in the United States; seven of these were subway and elevated, four furnished interurban surface transport. All of these systems use multiple-unit railcars, but there exists a wide variation in their noise-generating levels because of age and condition. Typical noise spectra 50 ft from these trains are shown in Fig. 23.[29]

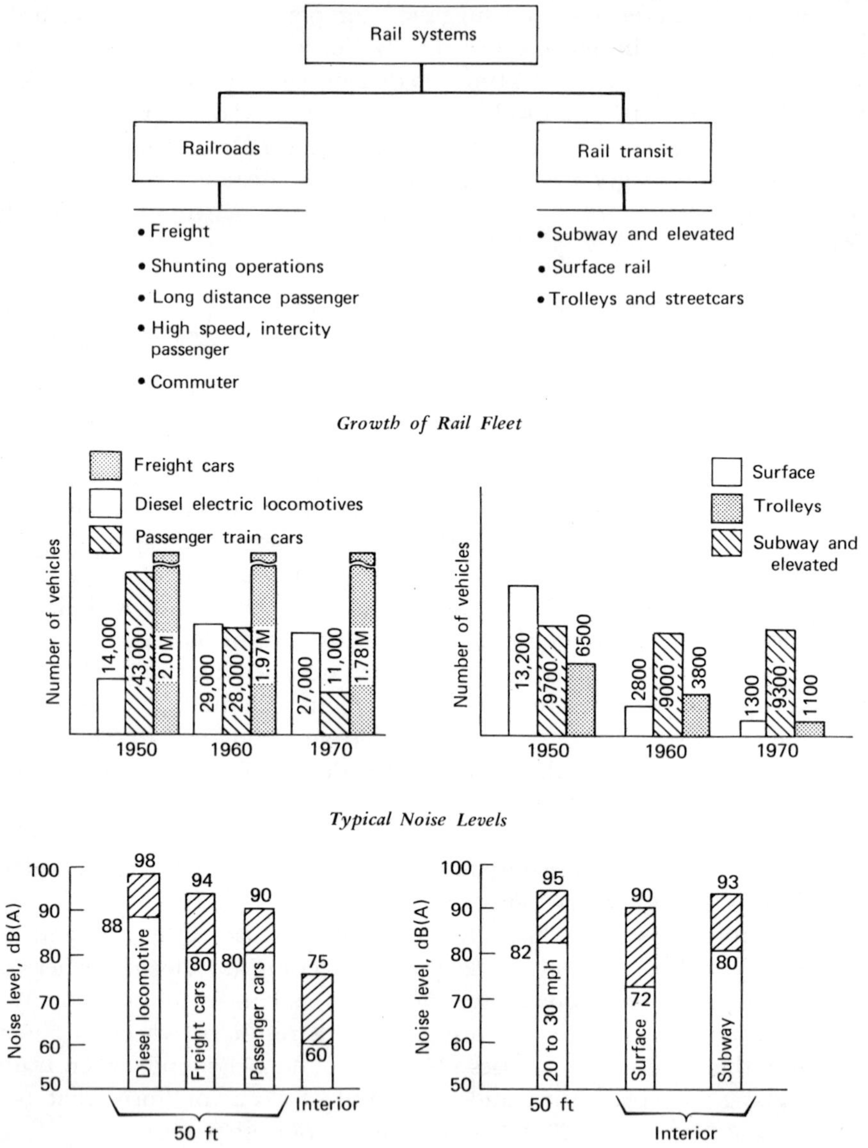

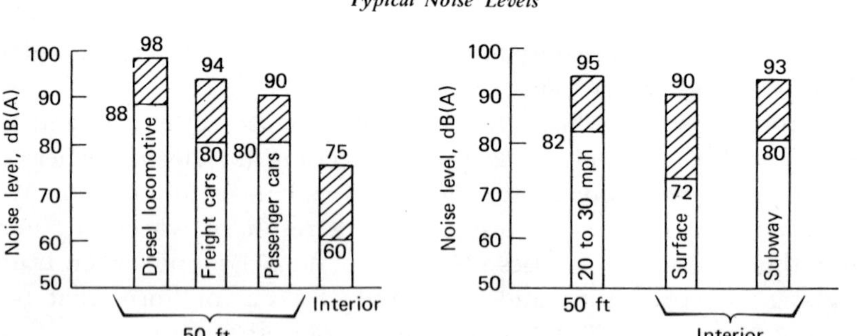

Figure 22 Approximate composition of the United States rail system for the period 1950 to 1970, together with typical noise levels within car interiors, and at a distance of 50 ft. (Courtesy of the U.S. Environmental Protection Agency, Ref. 27.)

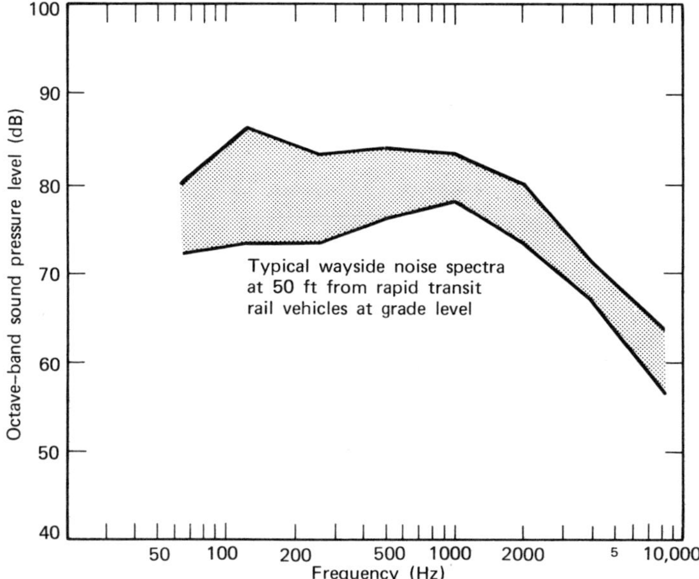

Figure 23   Typical octave-band noise spectra from rapid transit rail vehicles at a distance of 50 ft. (Courtesy of the U.S. Environmental Protection Agency, Ref. 29.)

The noise generated in communities having long-haul or rapid transit vehicles in their environs will depend primarily on track-observer distance, frequency, and train length. Rapid transit trains are short (~150 ft) so that their noise duration time is much less than that of a freight train whose length is often 3000 ft or greater. There exist no analytical formulas for relating train noise levels to speed, but Table 6 provides a rough range of dB(A) levels vs. distance for both railroad and rapid transit trains. Rapid transit vehicles provide the lower range of values listed; long-haul diesel-electric trains generate the higher noise levels.

HOVERCRAFT

Hovercraft, or air cushion-type vehicles, represent a relatively new type of transportation. They can be designed to be amphibious, although their predominant use throughout the world is as a high speed ferry over rivers and channels. Speeds greater than 40 mph are easily attainable; hence this type of vehicle is an attractive mode of transport for com-

Table 6   Noise Levels as a Function of Distance for Railroad and Rapid Transit Trains[30]

Distance (ft)	Noise Level, dB(A)
100	81–96
500	70–82
1000	61–76
2000	52–66
5000	39–54
10000	31–45

muters who wish to bypass the congested traffic of bridges and who work near a harbor.

Primary sources of noise are (1) the massive centrifugal "lift" fans that support the craft on an air cushion, (2) the driving propellers, and (3) the engines. Noise levels of these vehicles are comparable to those of single engine aircraft. Measurements have been reported[31] in the range of 84 dB(A) at a distance of 60 ft at engine idle, 90 to 95 dB(A) upon arrival, and peaks of 96 to 102 dB(A) when accelerating upon departure. Octave-band spectra of a commercial hovercraft, taken at different engine speeds and at a distance of 100 ft, are shown in Fig. 24.[32]

A terminal for a hovercraft should be located with regard to existing noise levels and public acceptability. If the craft is operating in a busy metropolitan harbor, such a craft will not add significantly to the general environmental noise. If in the future, however, these vehicles find increasing use near quiet suburban areas, then precautions should be taken to locate landing docks somewhat remote from residential dwellings.

RECREATIONAL VEHICLES

The dramatic increase in the number of vehicles designed primarily for recreational purposes has generated a new noise problem. Two decades ago, snowmobiles were nonexistent and the number of motorcycles operated in rural areas was very small. Today, there are large numbers of motorcycles, snowmobiles, and pleasure boats in operation. Not only do these vehicles produce high noise levels, but they inject intermittent and intrusive noise into rural areas and zones of relative quiet, so their

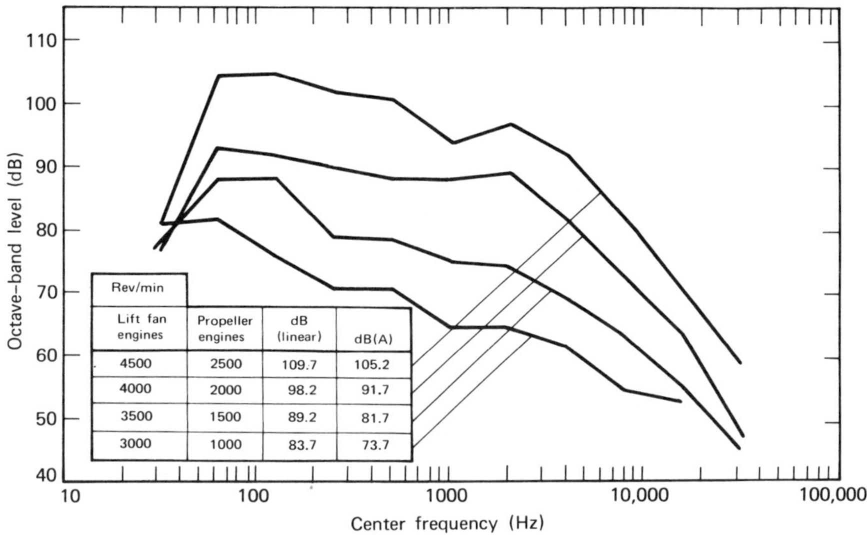

Rev/min			
Lift fan engines	Propeller engines	dB (linear)	dB(A)
4500	2500	109.7	105.2
4000	2000	98.2	91.7
3500	1500	89.2	81.7
3000	1000	83.7	73.7

Figure 24   Octave-band noise spectra from hovercraft, lift fan and propellers, for various engine speeds, at a distance of 100 ft. [From R. L. Trillo, *J. Sound Vib.*, **3**, 476 (1966).]

disturbance is therefore more noticeable. In many instances, of course, such vehicles are put to utilitarian use. Motorcycles are serving as an alternate for more costly automobiles, and their lower fuel consumption is making this vehicle attractive for commuter use. In the western United States, a growing number of farmers and ranchers are finding that snow-mobiles are the best means for rescuing cattle in snow storms, providing emergency feed, and having access to remote areas. And in inland lakes and waterways, small boats have been used to deliver the United States mail.

Figure 25 indicates (1) the approximate number of motorcycles, snow-mobiles, and pleasure boats in service in 1971, (2) the increase in the number of these vehicles from 1950 to 1970, (3) the range of typical noise levels in dB(A) at 50 ft, and (4) the range of noise levels encountered by a vehicle operator or passenger.[33]

Motorcycles

There is a substantial difference in the noise spectrum of a two-cycle and four-cycle motorcycle; the two-cycle exhibits a greater high frequency component than the four-cycle. All motorcycles will generate noise via

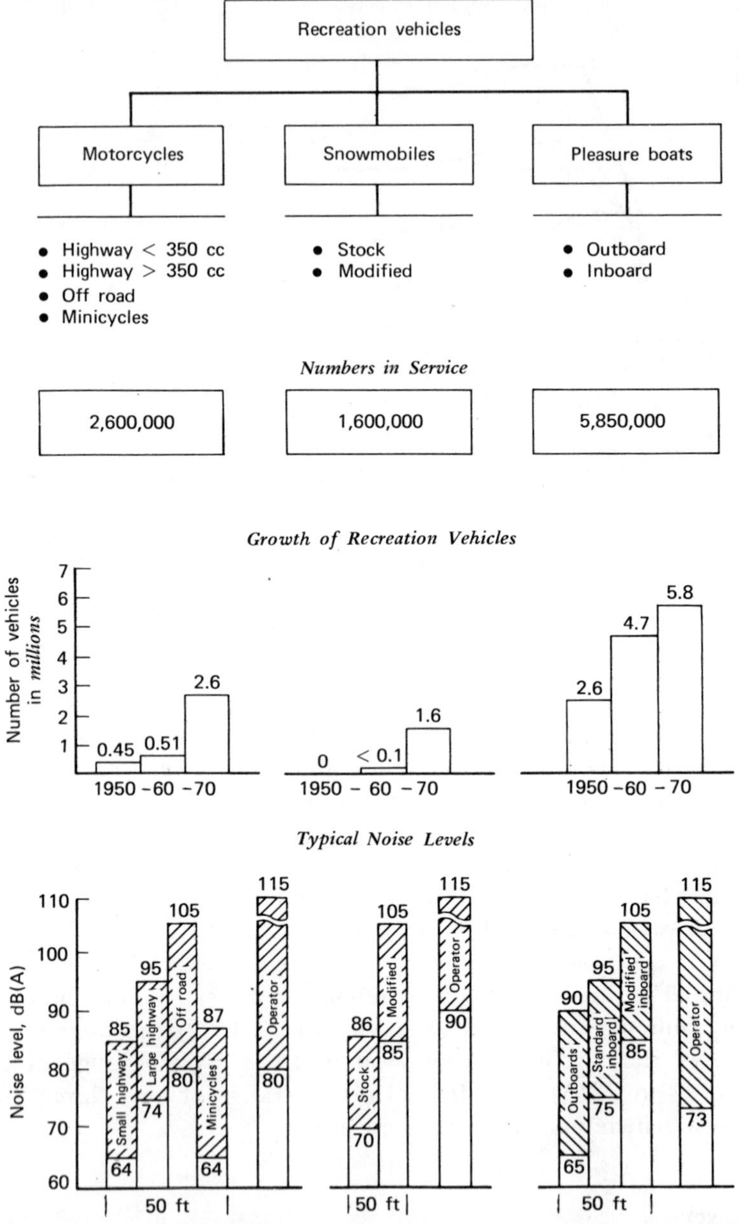

Figure 25 Growth in the number of recreational vehicles from 1950 to 1970, and the range of noise levels experienced by a vehicle operator. Typical noise levels at a distance of 50 ft are also indicated. (Courtesy of the U.S. Environmental Agency, Ref. 33.)

(1) the exhaust, (2) carburetor intake, (3) engine, and (4) tires. The relative noise contributions of these components in a typical four-cycle motocycle are, respectively, 86 dB(A), 82 dB(A), 78 dB(A), and 69 dB(A), at a distance of 50 ft, for a total noise level of 88 dB(A).[34] Exhaust noise is often sufficient to mask most of the other sources of sound. Until recently, many motorcycle enthusiasts have not even been interested in quiet vehicles. Also, since a large muffler volume is required to achieve effective muffling, there has been a general resistance to silencing the exhaust to the level of automobiles. The intake system is also noisy, because power is also lowered with restrictions upon air intake, and the aircooled engines are exposed and light in weight. Tire noise and also chain-drive noise are substantially lower than the noise from other sources.

An octave-band spectrum of the noise from a four-cycle vehicle, operating under a wide range of road conditions is shown in Fig. 26.[35] The upper bound shows the noise level for maximum acceleration; the lower limit is for cruising conditions at moderate speeds. The cycle measurement distance is 50 ft. As with a diesel truck, there is a substantial amount of sound energy emitted in the low frequency range. This means

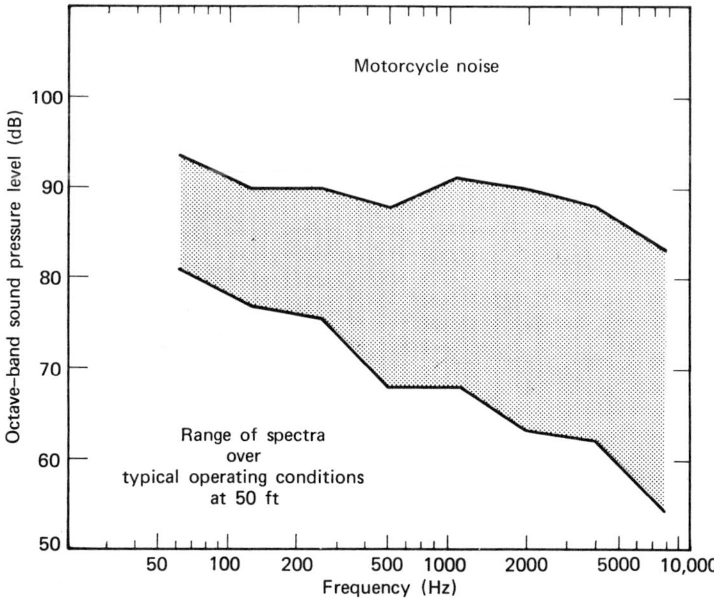

Figure 26 Octave-band spectral range of motorcycle noise at a distance of 50 ft. (Courtesy of the U.S. Environmental Protection Agency, Ref. 35.)

that motorcycle noise will intrude into nearby residential dwellings, regardless of their type of construction, and the long wavelength radiation will effectively diffuse over a broad area.

Snowmobiles

There is a great range in the noise outputs between older and more recently built machines. Old machines produced noise levels in excess of 95 dB(A) at a distance of 50 ft, and as high as 110 to 115 dB(A) at the operator position. Snowmobile powerplants of the lightweight two-cycle design are inherently noisy, and the customer demand for high horsepower has not prompted the design of quieter engines. Current legislation, however, is placing reasonable bounds on noise output, both to protect operator or passenger, and to reduce the environmental impact of these vehicles. State legislation is summarized in Table 7. The "effective date" refers to the date of manufacture to which the legislation applies; unless otherwise noted the snowmobile-measurement distance is 50 ft.

In addition to the environmental impact of snowmobiles, the snowmobile noise is potentially hazardous to the hearing of the operator.

A study was made of 10 current-model snowmobiles with engine displacements of 300 cc (~14 HP) to 775 cc (~55 HP).[36] One meter from the vehicles and at medium throttle, all machines generated a noise level of 100 dB(A) or greater (Fig. 27). The data suggests that until quieter snow-

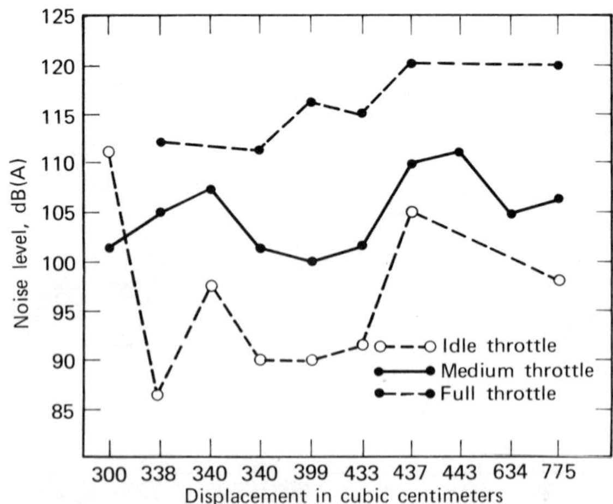

Figure 27   Sound levels for 10 snowmobiles measured at 1 meter from the back of the vehicles, under conditions of idle, medium, and full throttle. (From J. Curtis and R. C. Sauer, *Sound Vib.*, May 1973, p. 49.)

Table 7   Sound Level Limits for Snow-mobiles

State	Effective Date	Sound Level, dB(A)
Colorado	1/73	84
Connecticut	1/75	85
Iowa	7/73	82
Massachusetts	7/72	82
	7/74	73
Michigan	2/72	82
Minnesota	2/72	82
Montana	6/72	85[a]
New Hampshire	7/73	82
	7/78	73
	7/83	70
New Mexico	7/72	86
New York	6/72	82
	6/75	78
	6/78	73
Ohio	1/73	82
Oregon	1/73	82[b]
Rhode Island	6/72	82
	6/74	73
Utah	9/71	82
Vermont	9/72	82
Washington	1/73	82[b]
Wisconsin	7/72	82
	6/75	78

[a] 15 ft.
[b] 100 ft.

mobiles are built, either restricted exposure times or ear protectors should be considered. Eskimos who have used snowmobiles extensively have experienced substantial hearing losses at high frequencies.

Motorboats

Noise levels from inboard and outboard craft range from 65 to 95 dB(A) at 50 ft.[37] The lower limits relate to the noise levels of small outboard motors of low horsepower while the upper limits are reached by outboard craft of high horsepower with poor exhaust silencing systems. Most of the motorboat noise is from the exhaust. A substantial noise reduction has been achieved in modern outboard engines with underwater exhausts. Nevertheless, exhaust noise remains dominant as sound-deadening materials are being used in engine shrouds to attenuate engine noise. In general, modern craft of modest horsepower ratings do not present a serious noise hazard to the operator.

Noise from the hull—water impact can be detected by a distant observer, but it will usually be masked by exhaust noise unless one is specifically listening for this broad band type of noise. Inboard motorboats tend to be quieter than outboard engines of the same horsepower, although recent developments in outboard motor design promise even further sound reduction.

REFERENCES

1. U. S. Environmental Protection Agency, *Transportation Noise and Noise From Equipment Powered by Internal Combustion Engines*, Washington, D. C. 1971, p. 92.

2. J. Donovan and T. Ketcham, *Sound & Vib.*, 4, (October 1973).

3. U. S. Environmental Protection Agency, *Transportation Noise and Noise From Equipment Powered by Internal Combustion Engines*, Washington, D. C., 1971, *op. cit.*, p. 92.

4. *Ibid.*, p. 109.

5. R. F. Norris, *J. Acoust. Soc. Am.*, 8, 100 (1936).

6. C. R. Bragdon, *Noise Pollution*, University of Pennsylvania Press, Philadelphia, 1971, p. 48.

7. W. Tempest and M. E. Bryan, *Appl. Acoust.*, 5, 133 (1972).

8. General Motors Corp. *Proceedings of the Conference on Motor Vehicle Noise*, GM Proving Ground, April 1973, p. 13.

9. W. H. Close and T. Atkinson, *Sound & Vib.*, 28, (October 1973).

10. N. Olson, *J. Acoust. Soc. Am.*, 52, 129 (1972).

11. *Ibid.*

12. Study of Noise Pollution Aspects of Various Roadway Configurations, Prepared by the General Electric Company, for the New York State Department of Transportation, Philadelphia, Pa. (1971).

13. D. R. Johnson and E. G. Saunders, *J. Sound Vib.*, 7, 287 (1968).

14. Highway Noise: Report 117 of the National Cooperative Highway Research Program, National Academy of Science, Washington, D. C. (1971) p. 5.

15. *Ibid.*, p. 9.

16. *Ibid.*, p. 10.

17. T. E. H. Williams, *J. Sound Vib.*, **15**, 17 (1971).

18. Highway Report 117, *op. cit.*, p. 18.

19. Highway Report 117, *op. cit.*, p. 14.

20. W. E. Scholes, A. C. Salvedge, and T. W. Sargent, *Appl. Acoust.*, **5**, 295 (1972).

21. Highway Report 117, *op. cit.*, p. 17.

22. *Noise Environment of Urban and Suburban Areas,* Prepared by Bolt Beranek and Newman, Inc., for the U. S. Department of Urban Development, Washington, D. C. 1967.

23. *Effect of Highway Landscape Development on Nearby Property,* Report 75 of the National Cooperative Highway Research Program, National Academy of Science, Washington, D. C., 1969.

24. J. D. Griffiths and F. J. Langdon, *J. Sound Vib.*, **8**, 1 (1968).

25. C. G. Bottom and D. J. Croome, *Appl. Acoust.*, **2**, 279 (1969).

26. B. J. Smith, *Environmental Physics: Acoustics,* Elsevier, New York, 1970, p. 128.

27. U. S. Environmental Protection Agency, *op. cit.*, p. 134.

28. *Ibid.*, p. 133.

29. *Ibid.*, p. 143.

30. *Ibid.*, p. 146.

31. C. Duerden, *Noise Abatement,* Butterworths, London, 1970, p. 88.

32. R. L. Trillo, *J. Sound Vib.*, **3**, 476 (1966).

33. U. S. Environmental Protection Agency, *op. cit.*, p. 167.

34. *Ibid.*, p. 172.

35. *Ibid.*, p. 172.

36. J. Curtis and R. C. Sauer, *Sound & Vib.*, 49 (May 1973).

37. U. S. Environmental Protection Agency, *op. cit.*, p. 169.

10

AIRCRAFT NOISE

AIRCRAFT NOISE AS A SYSTEMS PROBLEM

Ever since the advent of aviation all aircraft have been noisy. Basically, all aircraft engines are heat engines that convert rapidly expanding gases into thrust. A small portion of this thrust energy, in some form, is converted to sound waves in the audible range. The early airplanes had rather low rotational velocity propellers, and unmuffled piston engines. The Wright brothers' craft in 1903 began with 12 brake horsepower. Propeller driven power plants increased up to 14,000 brake horsepower for the DC-7, that is, approximately a 30 dB power increment. As propeller tip speeds approached almost sonic velocity in the 1950s the propeller became the dominant noise source,[1] rather than the exhaust of the early unmuffled piston engines.

With the introduction of jet engines in the 1950s—turboprops followed by pure jets at the end of the decade—community noise exposure increased substantially. Several factors were involved. To the increased air traffic was added the louder noise levels of high thrust jets. Furthermore, pure jet aircraft could not maneuver easily at low speeds, and it became necessary to extend greatly the low-level approach glide path. Also, considerations of safety placed reasonable limits on takeoff and climb patterns. In 1960 the turbofan engine made its appearance, both for newly built aircraft and for replacement power plants. This turbofan increased the propulsion efficiency of the jet by extracting high-pressure energy from the combustion gas stream and employing it to drive a fan that

228

yielded additional thrust. From an engineering point of view the turbo-fan was a substantial step forward as it increased both the takeoff and cruise thrust; it also resulted, however, in the generation of the sirenlike *fan whine,* even though an overall reduction in emitted sound power was achieved.

In the mid-1960s a "quiet engine program" was begun that envisioned a reduction of engine noise of 15 to 20 dB below that of the DC-8 and the Boeing 707. Theoretical studies were then undertaken on fan whine, in order to minimize the production of these pure tones. In addition, the development of the so-called "high bypass ratio" engines led to engine configurations that are now used in the B-747, the DC-10, and the L-1011. These new engines, together with the use of sound-proofing materials, which were investigated by NASA and aircraft engine and airframe man-ufacturers, resulted in the development of aircraft that are noticeably quieter (by both objective and subjective measurements) than the early jets. Knowledgeable engineers believe we now may be approaching a technological plateau with respect to further noise reductions for con-ventional commercial aircraft. It is true that a further noise reduction is possible, in principle, through advanced engine design and the use of modified flight procedures.[2] But these designs and procedures must also satisfy economic and safety criteria. Also, some aerodynamic airframe noise will always exist regardless of the ultimate level of engine noise.

Over the long term, there are reasons for being neither too optimistic nor pessimistic about the aircraft noise problem. At present, airlines are transporting nearly 80% of all United States intercity passenger traffic traveling by common carrier. Figure 1 shows the dramatic growth of the United States commercial air fleet in terms of type, application, capacity, and noise levels.[3] The current annual passenger load of close to 200 million is expected to increase to 800 million by 1985, and there will be a substantial increase in air freight. Reducing the annoyance of aircraft noise with increased air service will require (1) the retrofitting or replace-ment of old jet aircraft,* (2) the introduction of new flight procedures with advanced instrument landing schemes, (3) a realistic "land use" planning program for the environs of all major airports, and (4) the development of vertical takeoff or landing (VTOL) and short takeoff or

---

* The JT8D engine, built by the Pratt and Whitney Aircraft Division of United Aircraft Corporation, powers a substantial portion of the nation's commercial air fleet, being used in the Douglas DC-9, Being 727 and Boeing 737 aircraft. Modified versions of this engine have been tested that reduce the "noise foot-print," in airport environs, by 75%.

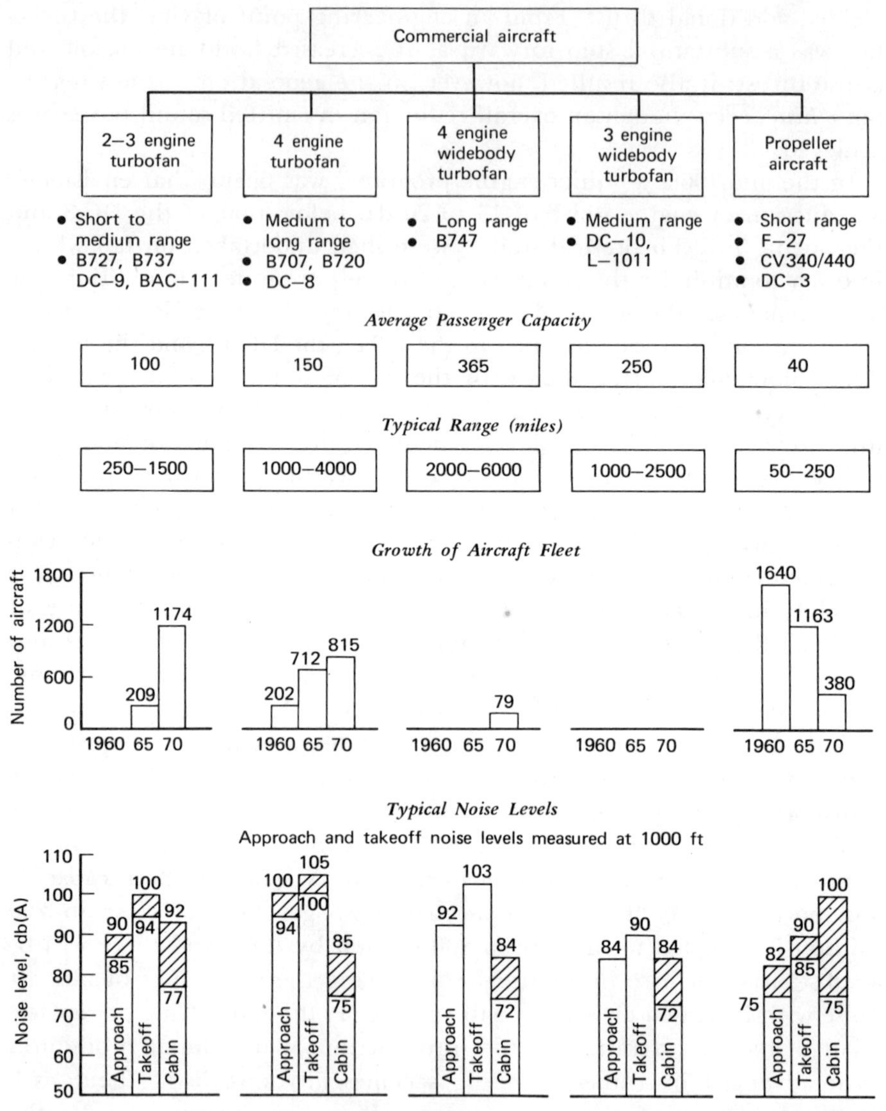

Figure 1   Noise characteristics of various commercial aircraft. (Courtesy of the Environmental Protection Agency, Ref. 3.)

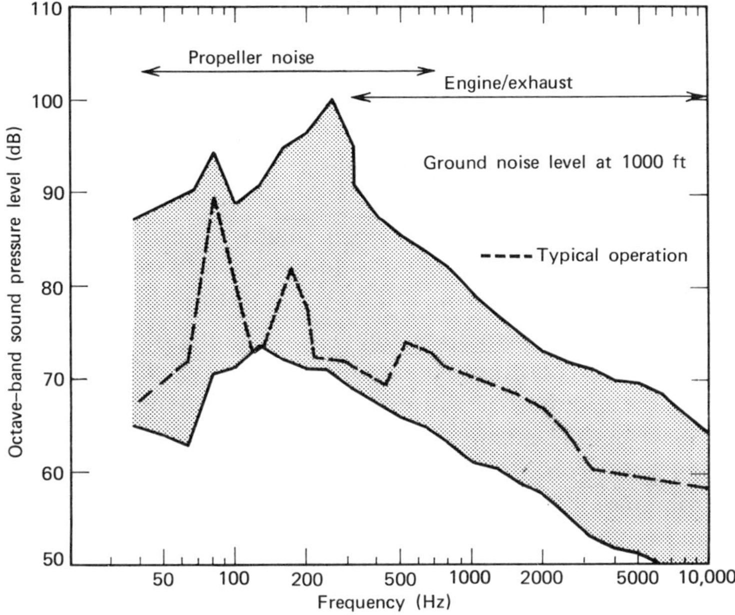

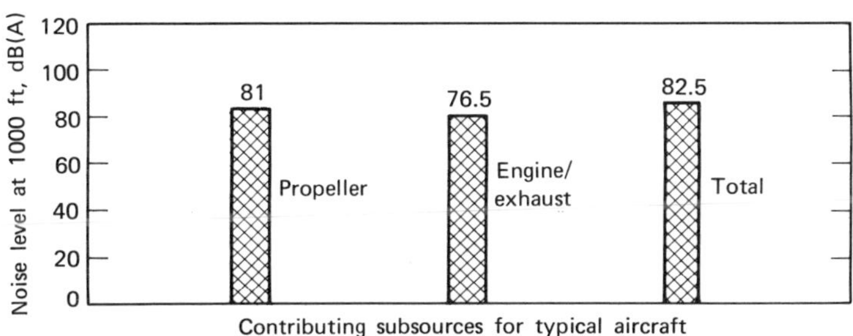

Figure 2   Noise levels and spectra of general aviation propeller aircraft. (Courtesy of the Environmental Protection Agency, Ref. 4.)

landing (STOL) aircraft. Short takeoff aircraft would be expected to operate from runways of 4000 ft.

The development of high speed ground transport would, of course, be highly desirable as it would reduce short-haul intercity air traffic. Implementation of this latter option, however, would require the establishment of new transportation priorities and a reappraisal of the nation's entire transportation system.

PROPELLER AIRCRAFT

Propeller aircraft comprise a small fraction of the commercial air fleet, as noted in Fig. 1, but they constitute the bulk of the "general aviation" aircraft (e.g., private, business, instructional, and corporate). There are approximately 150,000 of these airplanes distributed among some 11,000 airports throughout the United States. Most of these planes are of low horsepower rating, and their infrequent flights from local communities have posed no great environmental noise problem. At major airports, of course, noise from small private planes is masked by commercial aircraft. Noise levels and spectra of these general aviation propeller aircraft are shown in Fig. 2.[4] When these planes are flying at an altitude of 1000 ft, the noise at the ground below averages 81 dB(A) for propeller noise and 76.5 dB(A) for engine-exhaust noise for a total sound level of 82.5 dB(A).

Propeller noise consists of (1) discrete frequency or "rotational" noise arising from periodic disturbances of the air by the propeller and (2) broad band or "vortex" noise arising from random disturbances at the propeller.[5] The discrete frequency noise results from pressure waves being generated by the rotating propeller blades, the frequency of oscillation corresponding to the blade-passing frequency and harmonics. The actual magnitude and waveform of the oscillating pressure will depend on propeller design, rpm, thickness, and thrust or torque forces on a blade element. The propeller tip velocity is especially important. For subsonic tip speeds, the dominant pressure "peaks" will be limited to blade frequencies and harmonics below 1 kHz, but at supersonic tip speeds, harmonics up to 10 kHz will appear. Most propellers, however, are designed for 500 to 1000 ft/sec velocities,[6] so that virtually all periodic propeller noise is low frequency. The broad band or "vortex" noise is produced by air turbulence in the wake of the propellers and by complex fluctuating forces that are exerted by the propellers on the air stream.

HELICOPTER NOISE SPECTRA

The high radiated noise levels from helicopters have restricted public acceptance of these aircraft for the function they are best suited to perform; transportation to and from the center of cities. At moderate distances from a helicopter, the primary noise sources have been identified as (1) blade slap—when it occurs, (2) piston engine exhaust noise, (3) tail rotor rotational noise, (4) main rotor "vortex" noise, (5) main rotor rotational noise, (6) gear box noise, (7) turbine engine noise, and (8) miscellaneous aerodynamically and mechanically produced sounds. Since the piston engine is becoming obsolete as a helicopter power supply, studies of helicopter noise have focused on blade shape, rotational, and vortex noise, all of which are caused by the rotorblade-wake interactions. Essentially each large rotor blade functions like a wing in flight, and the lift on it generates a vortex wake behind it.[7] This vortex wake has a tendency to roll up into concentrated vortices with the result that each blade must pass over the concentrated vortex left by its predecessor. A severe aerodynamic and noisy interaction takes place under the name of "blade slap" that occurs under certain conditions of lift or descent. However, under virtually all flight conditions there will be a substantial amount of discrete frequency noise because of the very high amplitude pulsating sound pressures generated at the blade tip region. The result is a distinctive low frequency throbbing sound that propagates over great distances.

Figure 3 is a composite helicopter noise spectrum that identifies the high level main rotor discrete frequency noise, tail rotor noise, vortex noise, transmission noise and compressor noise from a gas turbine engine.[8,9] For gas turbine propelled helicopters and at an altitude of 1000 ft, an observer will hear noise dominated by main and tail rotor harmonics up to 1 KHz, with power plant noise contributing broad band sound (at lower sound pressure levels) from 1 to 20 kHz.

An appreciable noise reduction has been achieved in recent experimental helicopters. Figure 4 shows the flyover noise reduction of a modified helicopter Model HH-43B aircraft equipped with a Lycoming T53-

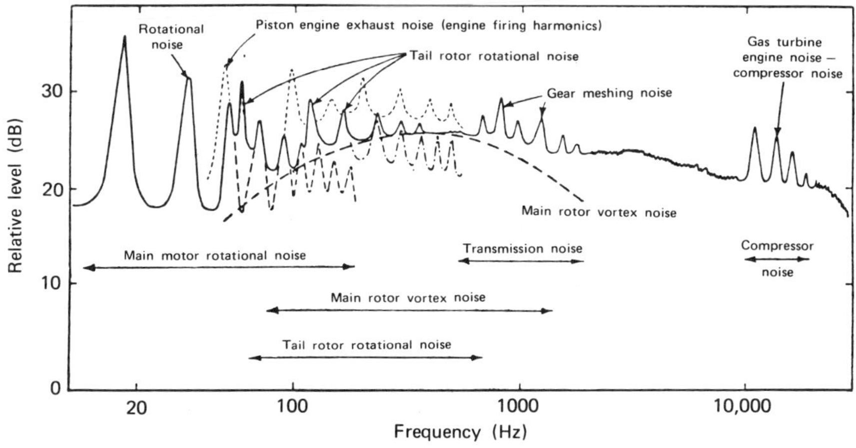

Figure 3 Narrow band spectrum of helicopter noise. (From E. J. Richards and D. J. Mead, *Noise and Acoustic Fatigue in Aeronautics*, Wiley, London, 1968.)

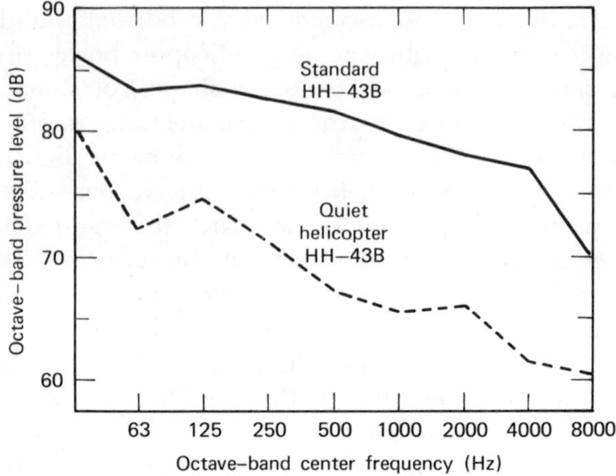

Figure 4   Helicopter noise reduction, in an experimental aircraft, during a 60-knot flyover at 200 ft altitude. [From M. A. Bowes, *J. Acoust. Soc. Am.*, **54**, 1214 (1973).]

L-1 turbine engine.[10] Tests were made during 60-knot flyovers at a 200 ft altitude. The noise reduction was made possible primarily through modifications to the engine and rotor system. Engine inlet noise was suppressed through the use of a large, reactively lined inlet duct; exhaust noise was reduced through the design of an exhaust system that allowed for a uniform expansion of the exciting exhaust flow to very low velocities. Engine-radiated noise was suppressed by vibration isolators between the engine and its supporting structure, and a high transmission loss enclosure was installed around the engine compartment. Rotor system noise suppression resulted from changes in rotor design and a reduction in rotor tip speed.

JET NOISE

A modern turbojet engine produces two kinds of noise. The first is the result of turbulence generated by the interaction of the high velocity jet with the stagnant atmosphere. The second is the high intensity whine caused by the high speed rotation of the engine's multibladed fan-compressor. Figure 5 is an engine schematic showing the general engine construction. On take off, the jet noise is dominant; during landing approach with reduced engine power, the whine from the fan-compressor inlet and discharge ducts tends to become the major noise source.

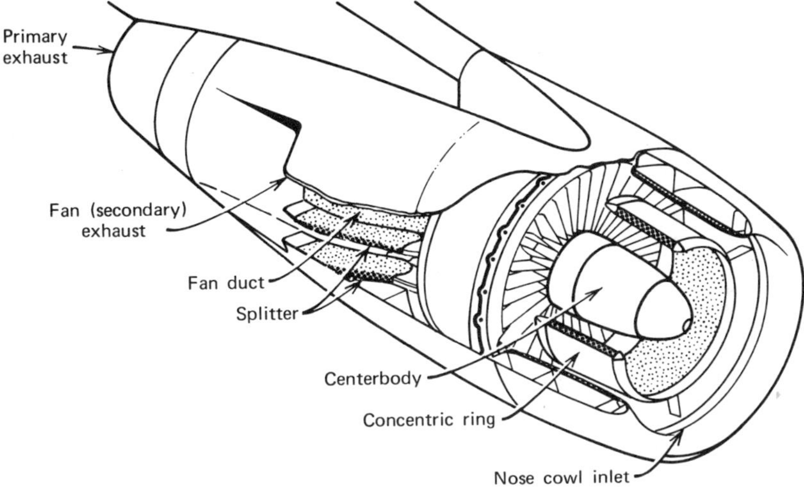

Figure 5   General construction features of a jet engine.

The jet exhaust noise will depend on gas exhaust velocity and many other engine design parameters. Classically, the radiated acoustic power is proportional to the eighth power of the jet velocity; therefore modern engines have been designed to produce jet streams of larger cross-sectional area and lower velocity to achieve lower noise generation. Close to the nozzle orifice, many small turbulent eddies will form, with increasingly larger eddies being formed downstream of the jet. The result is that the high frequency noise is produced close to the nozzle exit and lower frequency noise is produced in the large mixing region far downstream.

The major breakthrough in lower noise engines came with the "high bypass" ratio engine. In this scheme a large amount of the primary jet energy is supplied to the fan, which is caused to rotate at high velocities. The fan then generates a lower velocity air stream (bypass) that is concentric with the primary jet plume. Beyond the engine the fan-produced air shroud mixes with the primary exhaust gases to produce an air stream of large cross section and reduced velocity. Figure 6 is a schematic diagram indicating the mixing of the primary jet and fan exhausts. Bypass ratios ranging from 1:1 to 5:1 have led to substantially lower noise levels while maintaining, and sometimes increasing, engine thrust.[11] Figure 7 is a qualitative representation of the broad band jet noise and the higher frequency fan noise.

The generation of fan whine depends on fan design and rotational speed. The fan will generate some broad band noise plus discrete peaks

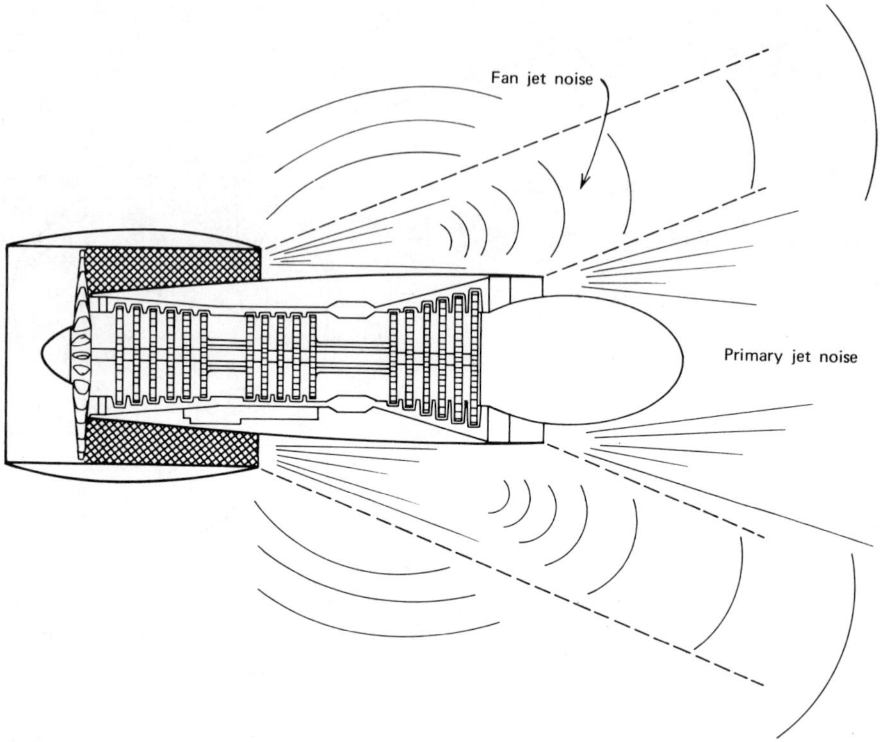

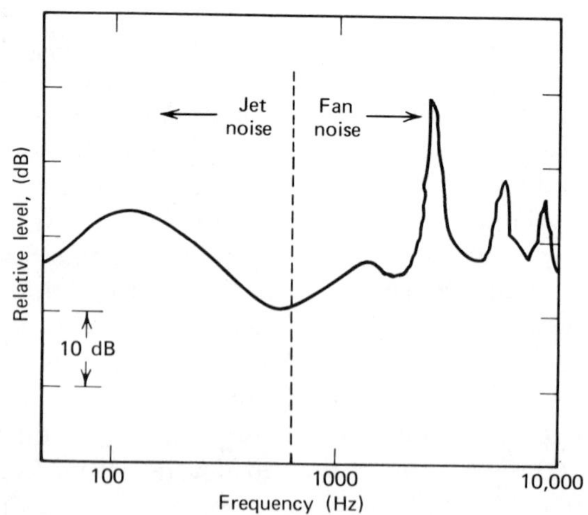

Figure 6   Schematic of fan-jet mixing for a high bypass engine.

Figure 7   Jet engine noise: broad band plus pure tone harmonics. [From R. A. Mangiarotty, *J. Acoust. Soc. Am.*, **48**, 783 (1970).]

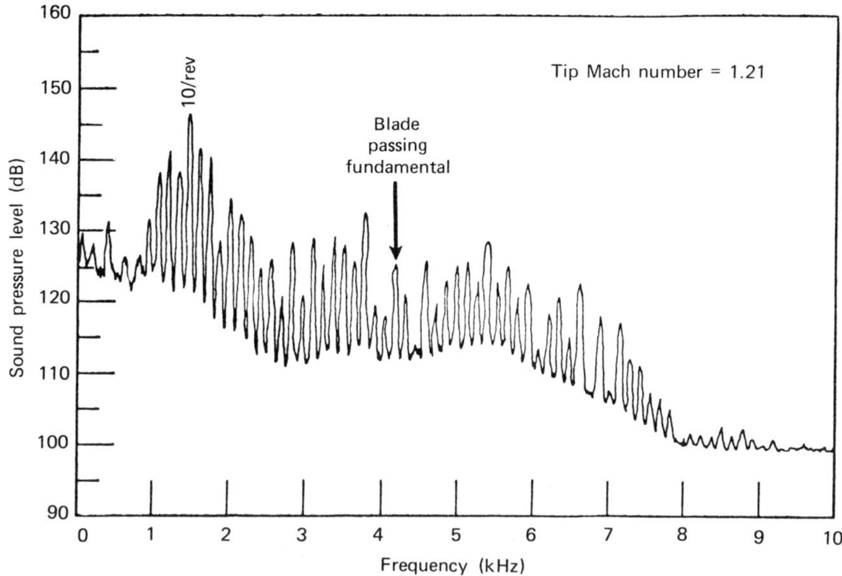

Figure 8   Narrow band spectral analysis of fan-produced noise, showing pure tone harmonics and subharmonics of the fan-blade fundamental. [From M. J. Benzakein, *J. Acoust. Soc. Am.*, **51**, 1428 (1972).]

of energy at frequencies that are related to the engine speed and to the number of blades and vanes in the various stages of the fan-compressor.[12] Recent research has shown that at transonic blade tip speeds, both harmonics and subharmonics* of the blade-passing frequency are produced, as shown in Fig. 8.[13] On most commercial turbofan engines, this fan whine is predominantly in the frequency range between 800 to 10,000 Hz. As a result, engineering efforts are being directed toward developing appropriate acoustic linings for engines to suppress fan noise at its source. The problem is a difficult one, however, because such lining materials must withstand environmental temperatures of $-85°$ to $700°F$, airflow velocities as high as 650 ft/sec. (in engine inlets and fan bypass ducts), and sound pressure levels as high as 170 dB—corresponding to fluctuating pressures of about 100 lb/ft².[14]

The basic types of linings that have been applied to jet engines to suppress fan noise have been (1) thick porous materials, (2) resonant or reactive linings, and (3) combination porous-resonant absorbers.[15] The

---

* Sometimes referred to as "MPT (multiple pure tone) noise" or "buzz-saw noise," because of its characteristic whine.

first type are constructed from such materials as special felts and glass fibers; this type will attenuate broad band noise, but it cannot attenuate selectively the large amplitude discrete frequencies. Resonant absorbers are fabricated with perforated plates with attached honeycomb-type cells. The cells have dimensions so that they act as resonant cavities for the dominant fan-generated frequencies. The combination broad band–resonant linings include the essential features of both types and, hence, are quite effective in attenuating fan whine.

Figure 9 is a photo of a modern jet engine, used in the DC-10, that has achieved reasonably low noise levels—via a high bypass ratio design plus acoustic linings.

SUBJECTIVE RESPONSE TO AIRCRAFT NOISE

Because aircraft noise has a unique frequency spectrum, various attempts have been made to relate this type of noise to the subjective annoyance experienced by persons living near airports. Specifically, rating schemes have been devised to account for (1) the distinctive noise spectrum of a single aircraft flyover and (2) the number of aircraft that are heard within a specified period.

Effective Perceived Noise Level (EPNL)

The principal components of noise from a single jet aircraft flyover are noted in Fig. 10. Prior to the aircraft reaching the observer, the high frequency fan whine emitted from the engine intake will be noted. When the aircraft is directly overhead, noise from the fan exit duct will become dominant, and as the plane recedes the lower frequency jet noise will be detected. It is clear that the perceived noise level (discussed in Chapter 6) will vary during the duration of the flyover and that it would be desirable to have a rating number that combines frequency, amplitude, and duration time. The effective perceived noise level (EPNL or EPNdB) is intended to provide such a rating number, and it embodies the results of many psychological studies relating to annoyance. It provides a correction for pure tone components as well as broad band noisiness. The procedure for obtaining an EPNdB level is quite complex as it requires summing 1/3 octave bands for the duration of the flyover in 0.5 second intervals. This number rating is important, however, because it is the present standard by which the Federal Aviation Administration certifies new aircraft.[16]

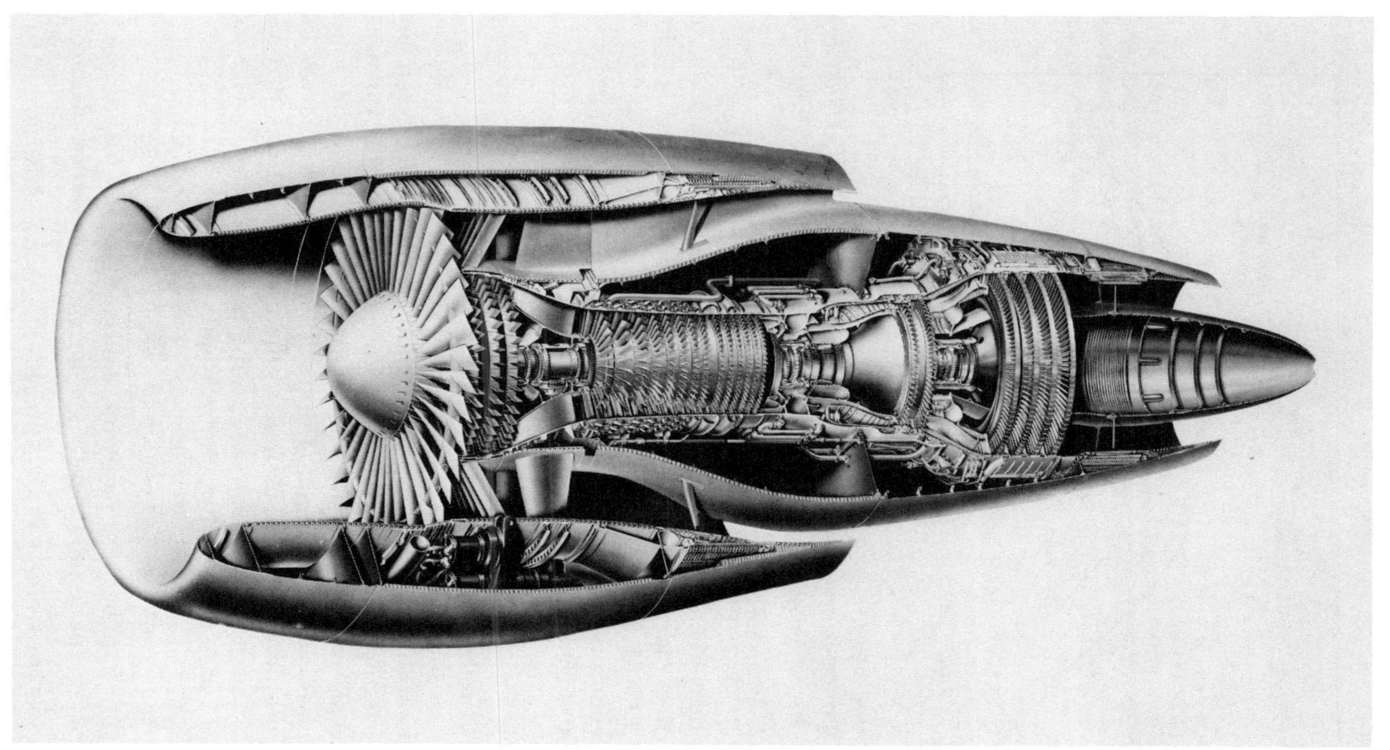

Figure 9 Photo of a high bypass ratio jet engine. (Courtesy of General Electric Co.)

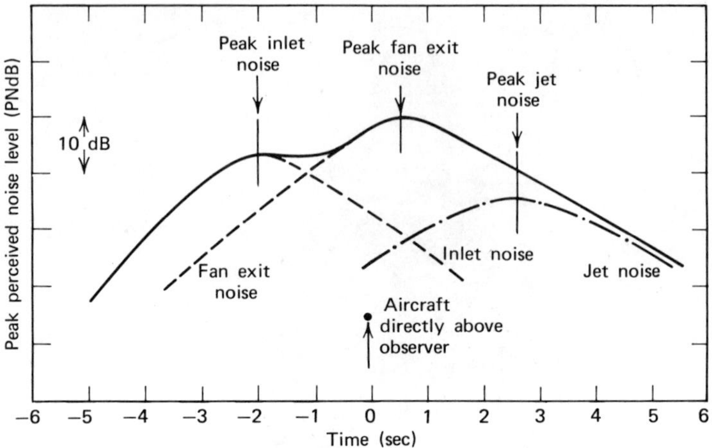

Figure 10   Principal components of peak perceived noise in a jet aircraft fly-over. [From R. A. Mangiarotty, *J. Sound Vib.*, **18**, 565 (1971).]

At airport takeoff and at a distance of 0.25 nautical miles from the runway centerline, older commercial aircraft will produce noise levels from 105 to 110 EPNdB.

Typical contours, in EPNdB, for an area surrounding a major airport, are shown in Fig. 11*a* and 11*b* for different type aircraft.

Noise and Number Index (NNI)

The noise and number index (NNI) is another subjective aircraft noise rating, and it applies to airport environs where a reasonable number of flights take place. It was developed in Great Britain based on a noise nuisance survey at the London (Heathrow) airport.[17] This survey indicated that people were more sensitive to the number of times they were disturbed by aircraft than they were to the peak noise level. The noise and number index is defined by the relation:

$$NNI = (\text{Average Peak Noise Level}) + 15\,(\log N) - 80 \qquad (1)$$

where N is the number of aircraft for a specified period (day or night), and the average peak noise level is a logarithmic average defined by $10 \log (1/N) \sum_{1}^{N} 10^{L/10}$, and L is in PNdB. The value 80 is subtracted to

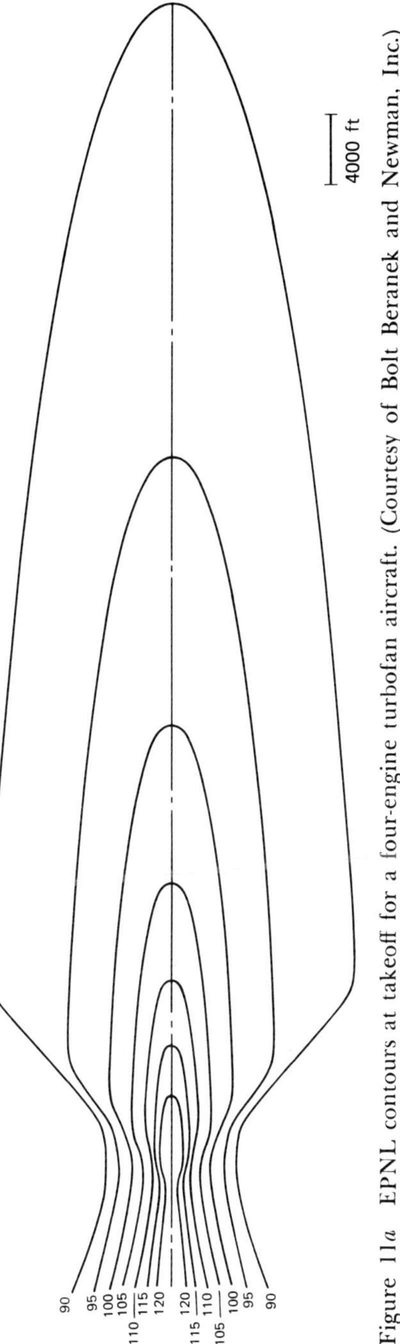

Figure 11*a* EPNL contours at takeoff for a four-engine turbofan aircraft. (Courtesy of Bolt Beranek and Newman, Inc.)

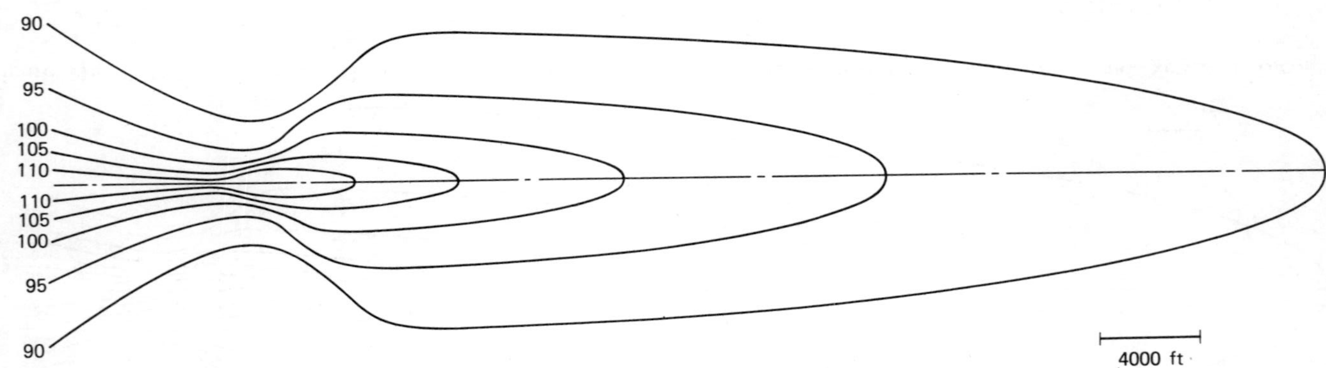

Figure 11*b*   EPNL contours at takeoff for a high bypass ratio "quiet" four-engine turbofan aircraft. (Courtesy of Bolt Beranek and Newman, Inc.)

normalize the index to 0 for conditions of annoyance. If the perceived noise level is approximated by the use of A-weighted sound levels, then the number and noise index is:[18]

$$NNI \simeq (\text{Average Peak A-Level}) + 15 \log N - 67 \qquad (2)$$

An NNI of 20 to 30 is considered intrusive, 40 is annoying, 50 very annoying, and a rating of 60 is classified as intolerable.

Composite Noise Rating (CNR)

A composite noise rating has been defined in terms of the effective perceived noise level and the number of day and night flights:

$$CNR = EPNL + 10 \log (N_d + 10 \, N_n) - 12 \qquad (3)$$

where $EPNL$ = effective perceived noise level
$\quad N_d$ = number of daytime flights
$\quad N_n$ = number of night flights

The scale has been used by the FAA and the military, but it has been superceded by the *noise exposure forecast.*

Noise Exposure Forecast (NEF)

Another rating scheme, that is now widely used in the United States, is termed the *noise exposure forecast* (NEF). It is designed to assess the impact of aircraft noise on a community, taking into account the total mix of aircraft, the number of operations in day and night periods, and the subjective noise levels generated by each type of aircraft. In practice, computer calculations are required to provide NEF ratings and to determine contours surrounding airports, but the parameters and relationships entering the computations are as follows:[19]

For a specific class of aircraft, i, on a flight path, j, the $NEF_{(ij)}$ is

$$NEF_{(ij)} = EPNL_{(ij)} + 10 \log \left[\frac{n_{D(ij)}}{K_D} + \frac{n_{N(ij)}}{K_N} \right] - C \qquad (4)$$

where $EPNL_{(ij)}$ = effective perceived noise level at a point in the airport environs resulting from a particular class of aircraft following a particular flight path.

$n_{D(ij)}$ = number of daytime (0700–2200) operations, of respective aircraft class i, and flight path j, where each landing and each takeoff is counted as a single operation.

$n_{N(ij)}$ = number of nighttime (2200–0700) operations.

$$K_D = 20, \qquad K_N = 1.2, \qquad \text{and } C = 75$$

Values of the constants K_D and K_N have been assigned so that one nighttime flight contributes as much to the *NEF* as 17 daytime flights, and the value of the constant C has been designated so that *NEF* numbers typically lie in the range of 20 to 40 to avoid confusion with other composite noise rating numbers.

For a specific ground location, the total *NEF* number will be the sum of all the $NEF_{(ij)}$ values, for all aircraft and all flight patterns, computed by the relation:

$$NEF = 10 \log \sum_i \sum_j \text{antilog} \left[\frac{NEF_{(ij)}}{10} \right] \tag{5}$$

Analytical models have been developed to forecast *NEF* values, for the year 1985, at many of the major United States airports. Reasonable assumptions have been made with respect to (1) aircraft fleet composition, (2) flight procedures, (3) mix of aircraft within the fleet, (4) ratio of day to night operations, and (5) total number of operations. Contour maps have then been generated by computer. One such map,[20] for the Stapleton International Airport at Denver, Colorado, is shown in Fig. 12. The projected 1985 air carrier operations is estimated at 282,000, and the corresponding passenger load is estimated to be 22,000,000 to 28,000,000. *NEF* contours of 30, 35, 40, and 45 circumscribe the north-south and east-west runways.

Areas encompassed by such *NEF* contours for the year 1985 at 12 major airports, are shown in Table 1. The "total" area, in square miles, is the entire area enveloped by the designated *NEF* contour. Where the term "less water" is used, the area over water has been excluded. The "net" area excludes both the water areas and the land occupied by the airport, itself.

Present federal guidelines for planning recommend that new residential construction should not be undertaken in airport environs exposed to values of $NEF = 30$ and higher.

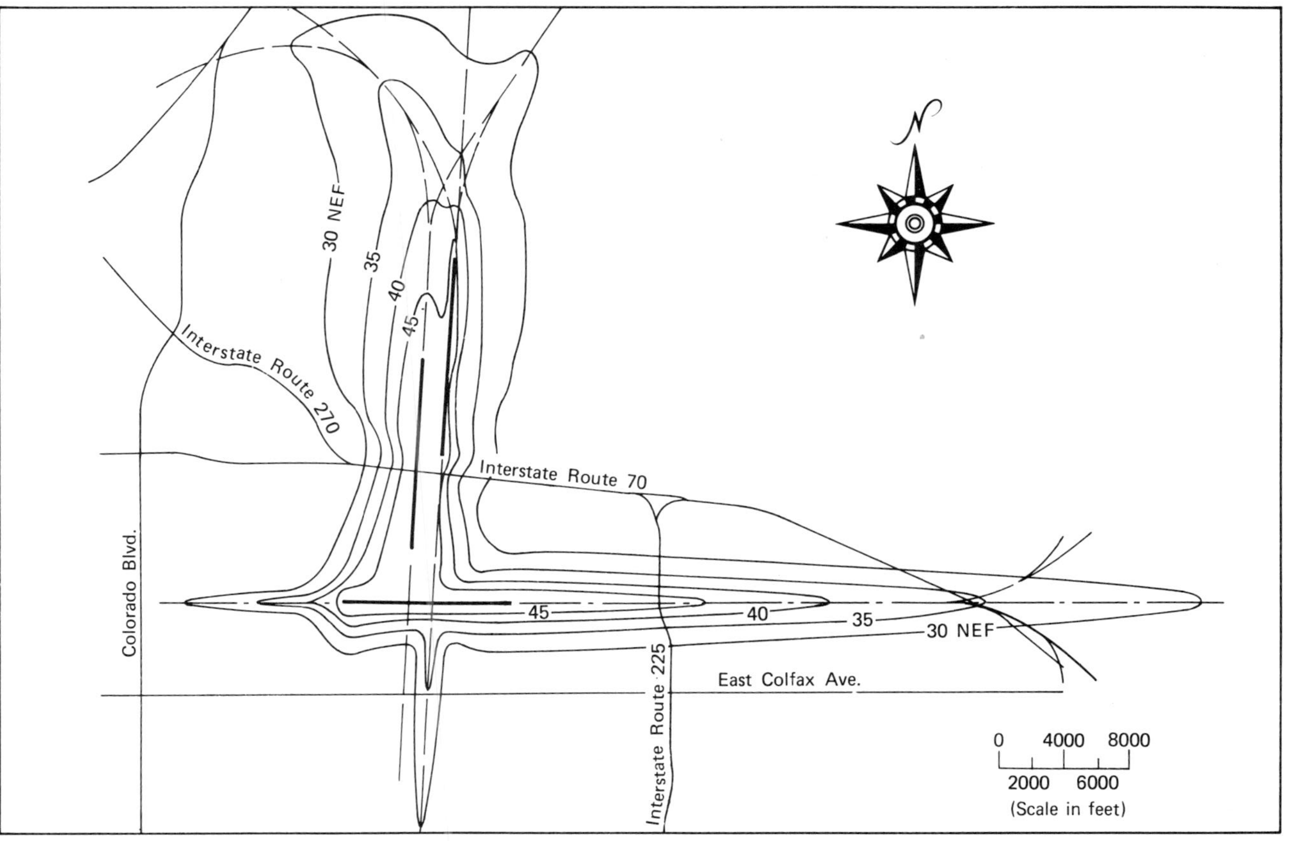

Figure 12   Noise Exposure Forecast contours for 1985 civil air carrier operations at the Stapleton (Denver) International Airport. (Ref. 20.)

Airport		NEF Values			
		30	35	40	45
Atlanta	Total	37.7	17.9	9.8	4.2
	Net	33.7	14.1	6.2	1.7
Chicago	Total	55.3	25.5	12.9	8.9
	Net	45.2	16.4	5.3	2.8
Denver	Total	21.1	10.4	5.5	3.0
	Net	14.3	4.9	1.5	0.3
Detroit	Total	29.0	14.2	7.3	3.6
	Net	24.0	9.8	3.3	0.7
Indianapolis	Total	12.2	5.9	2.7	1.2
	Net	9.3	3.7	1.2	0.4
Kansas City	Total	21.6	10.4	5.0	2.3
	Net	13.6	4.3	1.1	0.1
JFK	Total	40.4	24.4	14.3	6.1
	Less water	29.7	17.8	10.9	4.1
	Net	21.8	10.2	3.8	0.4
LaGuardia	Total	23.4	9.8	4.3	1.9
	Less water	17.6	6.0	2.2	0.9
	Net	16.4	4.8	1.2	0.2
Portland	Total	23.7	11.8	6.0	2.9
	Less water	16.9	8.1	4.3	2.3
	Net	14.1	5.3	1.8	0.7
San Diego	Total	13.4	6.1	2.9	1.2
	Less water	10.7	5.7	2.7	1.2
	Net	9.9	4.9	2.1	0.7
Seattle-Tacoma	Total	27.2	12.5	6.2	2.7
	Net	25.7	10.5	4.5	1.3
Tampa	Total	24.6	12.8	5.3	3.1
	Less water	20.2	10.7	5.3	3.0
	Net	16.0	6.8	2.0	0.7

Community Noise Equivalent Level (CNEL)

Another subjective rating scale has been adopted in the Noise Standards for California Airports.[21] Termed the Community Noise Equivalent Level, it can be computed by the expression

$$CNEL = NL - 35 + 10 \log N \tag{6}$$

where NL is the aircraft flyover noise level in dB(A) and N is the total number of flights per day. Equation 6 is appropriate with these qualifications and assumptions:

1. Air traffic is assumed to consist of 70% daytime flights (within 12 hrs), 23% evening flights (within 3 hours), and 7% nighttime flights (within 9 hrs).
2. The flyover noise levels, NL, are at least 20 dB above the ambient noise.
3. Mean duration of a flyover is 15 sec.
4. All aircraft flyover noise levels are measured at a given location.

An alternative and more general expression is:

$$CNEL = NL + 10 \log (N_d + 3N_e + 10N_n) - 38.4 \tag{7}$$

where N_d, N_e, and N_n are the number of day, evening, and night flyovers, respectively. In terms of this $CNEL$ rating, the Department of Aeronautics of the State of California has proposed the exposure limits for its airports listed in Table 2.

Table 2 Proposed *CNEL* Limits for California Airports

Date	CNEL
Until 12/31/75	80
1/1/76–12/31/80	75
1/1/81–12/31/85	70
1/1/86–	65

NOISE CONTOURS NEAR AIRPORTS

Takeoff and Landing Patterns

As indicated in the previous section, the noise profiles near airports will be affected by the specific patterns of takeoff and landing, as well as aircraft traffic density. A substantial additional amount of noise is also due to ground operations and engine "run-up." The weather will also cause large perturbations in noise levels. The wind direction and velocity can extend or shorten the noise contours, and the pressure, temperature, and wind speed substantially can affect the needed takeoff distance. The difference in lift-thrust due to fluctuations in air density alone can result in a difference of almost 2000 ft in takeoff roll for a heavily loaded jet. However, for equally favorable weather conditions, the noise level under a takeoff path (for a given type of aircraft) will be dependent upon (1) specific flight path and (2) the rate-of-climb programming. Figure 13 illustrates the general concept of using preferential routes to avoid maximum noise exposure to builtup areas. At distances of three miles from takeoff, the ground exposure noise reduction can be substantial. The programmed rate-of-climb is sketched in Fig. 14. All aircraft take off initially with maximum thrust, but a reduction in power to 50 to 75% of maximum thrust is acceptable at a later stage in the climb. Lower power can easily decrease a ground sound pressure level by 10 dB, but then the lower climb rates tend to have the noise affect a greater ground area. Specific climb regimes depend on particular airport environs and considerations of aircraft safety. New climb procedures suggest that planes climb to 1500 ft, reduce power until an altitude of 3000 ft is reached, after which a normal climb rate is resumed.

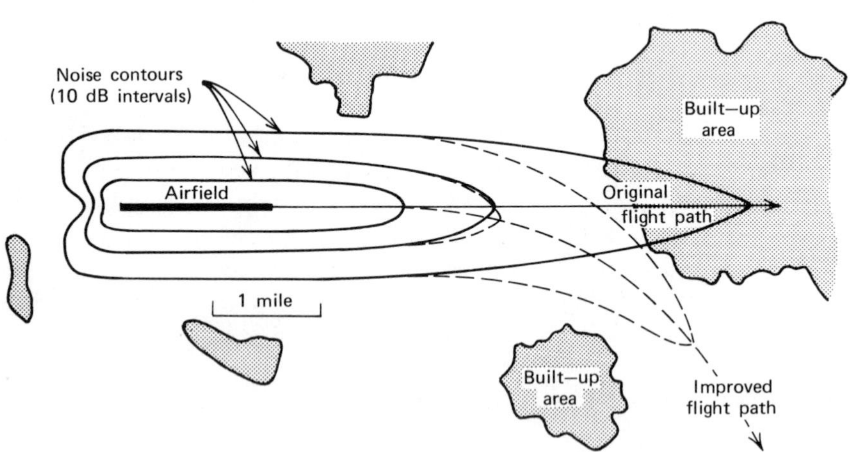

Figure 13   Use of preferential air routes to lessen noise exposures in built-up areas. (From E. J. Richards and D. J. Mead, *Noise and Acoustic Fatigue in Aeronautics*, Wiley, 1968.)

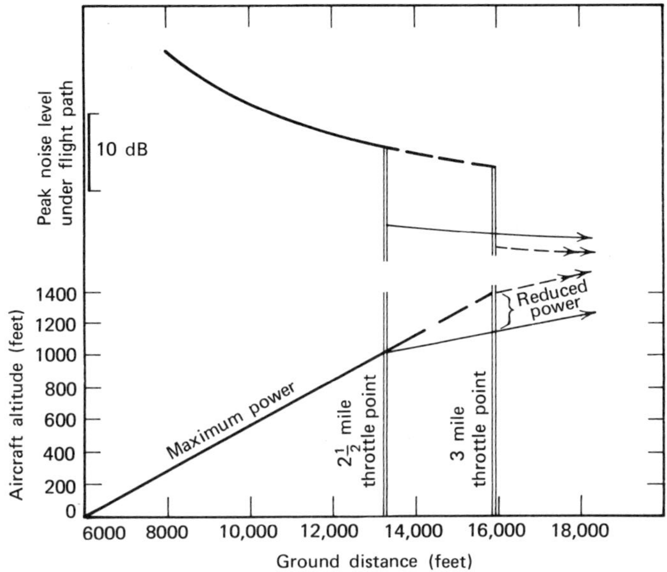

Figure 14   Programmed rate-of-climb to minimize exposures at ground level. (From E. J. Richards and D. J. Mead, *Noise and Acoustic Fatigue in Aeronautics*, Wiley, 1968.)

The landing pattern is set by airport instrumentation and approach angles determined by the FAA.[22] At present, aircraft use 3° instrument landing system (ILS) glide paths, descending at a rate of about 600 ft/ min. Some recent NASA-sponsored tests have indicated that a much steeper approach angle of 6° at a descent rate of 1200 ft/min is feasible, with a shift to a standard ILS approach at an altitude of about 800 ft. This two-segment procedure is reported to reduce the noise "footprint" of 90 EPNdB for a B-727 aircraft from 5.5 mile to 2.0 mile. The greatest improvement would occur from 3 to 8 miles from the landing runway along the ILS approach path, where a 5 to 15 dB decrement might be realized.

Noise Under Air Route Corridors

A serious problem in the aircraft industry is that noise cannot be limited to an airport environs alone, but it accompanies very extensive approach and departure routes. These approach and departure corridors are very long, and in some instances aircraft are flying below a 5000 ft altitude for a distance of 30 miles. As a result, many communities are exposed to noise levels 20 dB(A) or more above the outdoor ambient. The extent of the air-corridor problem has been studied extensively[23] for the area surrounding the Los Angeles International Airport.

In this area there are some 500 or more flights per day, over 300 from eastern United States, and somewhat less than 100 each from north and south. Approach routes from north and south are sometimes over the ocean, but the eastern traffic requires flight paths over Beverly Hills, Santa Monica, Redondo Beach, and about five other major municipalities. The four major arrival and departure routes over the Los Angeles basin are sketched in Fig. 15. Most flights are two-, three- and four-engine jet aircraft ranging in size from the DC-9 to the B-747, and the flight corridors extend some 15 to 50 miles over high density residential areas. Directly under the northeast departure route, noise levels from Santa Monica are in the 70 to 77 dB(A) range, and it is estimated that about

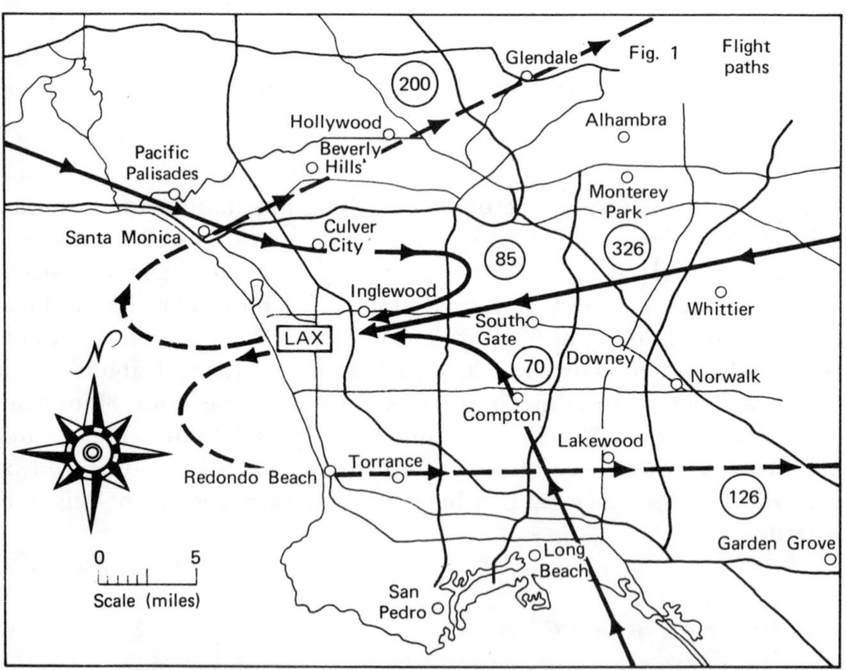

Figure 15   Major arrival and departure routes over the Los Angeles area. [From P. Hurdle, S. R. Lane, and W. C. Meecham, *J. Acoust. Soc. Am.*, **50,** 32 (1971).]

500,000 people are subject to significant noise intrusion levels along this path. Similar conditions apply to the other arrival corridors, so that it is estimated that more than one million people are exposed to 15 to 20 dB(A) noise intrusion levels from 100 to 300 times daily, exclusive of communities immediately adjacent to the airport. In the immediate airport area (10 square miles containing a population of 70,000) exposure levels reach 100 PNdB or 90 dB(A).

Similar situations exist around other major airports, where frequent jet flyovers generate high noise levels in otherwise quiet communities.

CERTIFICATION OF NEW AIRCRAFT

Because the environmental impact of aircraft extends over such extensive areas, a major technological objective has been to reduce aircraft noise, per se, because there are limited options commensurate with safety that can be exercised in takeoff and landing procedures and routes. For jet aircraft requiring certification by the Federal Aviation Administration after 1969, specific noise limits have been imposed. Three measurement locations are required in certification testing, as indicated in Fig. 16.

1. *Landing.* 1 nautical mile from landing threshold, directly under the aircraft path.
2. *Takeoff.* 3.5 nautical miles from the start of the takeoff roll and on the extension of the runway centerline.
3. *Sideline.* At the location of maximum noise along a line parallel to and at a distance of 0.35 nautical miles from the runway centerline, for aircraft that have four or more engines; and 0.25 nautical miles from the runway centerline, for aircraft which have three or fewer engines.

The maximum permissible noise limits at the three measurement positions are given in terms of the aircraft's maximum certified gross weight. This gross weight dependence simply takes into account that an increased gross weight demands more engine thrust—with an accompanying higher noise level. Figure 17 indicates the FAA noise limits for takeoff in terms of effective perceived noise levels (EPNdB) vs. gross weight at the specified 3.5 nautical mile location. Figure 17 also indicates noise levels produced from a number of existing aircraft.[24]

It should be noted, however, that there are nearly 1800 existing large turbojet airplanes, having about 4,000,000 operations per year in the United States, that are not covered by the new aircraft certification standards. These older aircraft have the greatest noise impact in the vicinity of most commercial airports.[25]

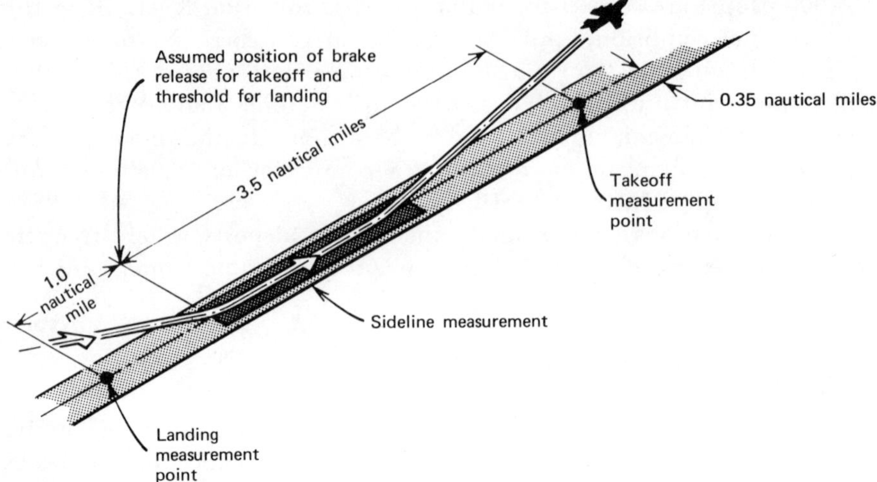

Figure 16   Measurement locations for determining noise levels in aircraft certification testing.

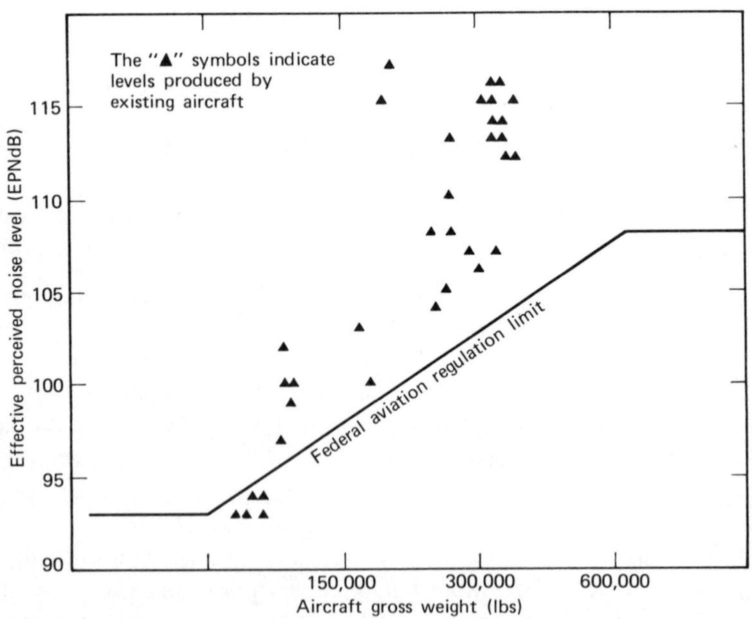

Figure 17   Noise limits (EPNdB) vs. aircraft gross weight at 3.5 nautical miles (Ref. 24.)

COMMUNITY REACTION TO AIRPORT NOISE

While the reduction of aircraft noise is an engineering problem, the public acceptance of noise surrounding airports involves other factors, and several investigators have attempted to identify these factors.[26,27] One of the most comprehensive sociological studies was conducted for the airports serving Atlanta, Dallas, Denver, and Los Angeles.[28] As a result of more than 4000 interviews, the seven major social-psychological predictors of *annoyance* were found to be (1) fear of aircraft crashing in the neighborhood, (2) distance from the airport, (3) susceptibility to noise, (4) noise adaptability, (5) city of residence, (6) belief in the misfeasance in the aircraft or airport industries, and (7) the extent to which the airport was deemed to be important to the local economy. Additionally, data was obtained to determine the specific daily activity that was most interfered with, among residents who indicated they were "extremely disturbed" or "highly bothered" by aircraft noise. The activities most interfered with are listed in Table 3.

Table 3   Interference of Daily Activities by Aircraft Noise[28]

Activity	Percent Highly Annoyed
TV and radio reception	20.6
Conversation	14.5
Telephone	13.8
Relaxing outside	12.5
Relaxing inside	10.7
Listening to records and tapes	9.1
Sleeping	7.7
Reading	6.3
Eating	3.5

There are no simple solutions to the airport problem because of the vast amounts of land acreage and airspace involved. To accommodate a large number of aircraft with safety, a major airport requires two parallel runways some $2\frac{1}{2}$ miles long and at least 6000 ft apart. Reasonably high noise levels must be expected over 10 square miles regardless of landing and takeoff procedures for the present air fleet. Hence, the appropriate *land use* of airport environs, as discussed in Chapter 13, is one of the few practical options.

SONIC BOOMS

Nature of the Sonic Boom

A sonic boom is a short duration pressure transient. It occurs during the passage of an aircraft whose speed is greater than the local speed of sound in the atmosphere.* At low altitudes, this speed is in the range of 770 mph. At ground level the boom from an aircraft flying at 40,000 ft, may be discerned as a sequence of two sharp bangs that are separated by a short time interval. A qualitative explanation of the sonic boom is that this is simply a shock wave that is generated by the pressure buildup at the surfaces of the aircraft in the same fashion that a "bow" wave is produced by a boat in the water. At a considerable distance from an aircraft, not only is a bow wave detected, but a trailing shock is also formed. At a point on the ground the bow shock is detected as an abrupt *overpressure* in comparison to the ambient atmosphere. The sound pressure then decreases almost linearly to a value below the ambient; this pressure is termed the *underpressure*. An abrupt rise in pressure back to the ambient level occurs as the trailing shock passes. Figure 18 indicates the shock wave reaching the earth and being reflected, and the corresponding over- and underpressure pattern that is termed the *boom signature*. This boom signature is also described as an "N"-wave of peak pressure, ΔP, and duration time, ΔT.

The magnitude of ΔP at ground level may be in the range of 1 to 10 lb/ft^2 above atmospheric pressure ($\sim$2116 lb/ft^2); ΔT will be in the range of 50 to 300 msec, depending primarily on the size of the aircraft and its operating altitude. The ΔT for a fighter aircraft is of the order of 0.1 sec; whereas the duration time of the British-French *Concorde,* is about 250 to 275 msec at an altitude of 40,000 ft. At this altitude the characteristic overpressure of the *Concorde* is about 2.4 lb/ft^2.[29]

Ground Exposure Pattern

This *boom signature* is first generated by an aircraft as soon as it reaches a supersonic velocity, and continues for the duration of its supersonic

---

* The first transonic flight took place in 1947 when Captain Charles E. Yeager, in an X-1 research plane, crossed what was then known as the sound barrier. By September 1, 1974, a cruising speed of 1817 mph had been attained in an Air Force SR-71 transatlantic crossing, when Major James V. Sullivan covered 3470 statute miles from New York to London in 1 hr, 55 min, at an average altitude of over 80,000 ft.

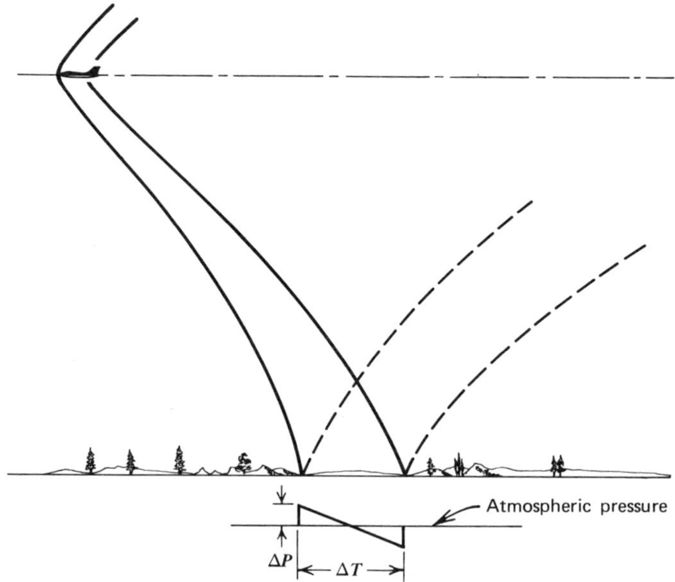

Figure 18   Nature of the sonic boom.

flight. The ground corridor directly under the aircraft's flight is thus termed the *boom carpet,* a zone which may extend for thousands of miles. The actual intensity of the sonic boom and the width of the boom carpet will depend upon the aircraft's altitude, Mach number,* volume, length, and aerodynamic design. To some degree the width of the boom carpet is also dependent upon atmospheric conditions.

Figure 19 illustrates the generation of the sonic-boom pressure field that results in a conical volume of air being disturbed behind the aircraft nose. The figure also depicts the ground "footprint" (the shaded area) that is bounded by the hyperbolic line marking the intersection of the cone with the ground plane. N-wave signatures will be characterized by decreasing overpressures and increasing signature times as the distances from the ground centerline increases. Furthermore, the apex angle of the cone decreases with increasing Mach number, and increases as aircraft speed is decreased, so that at lower speeds the ground exposure region (for a given altitude) is broader. Hence, the ground area affected

* *Mach number* is the ratio of the speed of an object to the speed of sound in the surrounding atmosphere. The term is in honor of the Austrian physicist and philosopher, Ernst Mach (1838-1916).

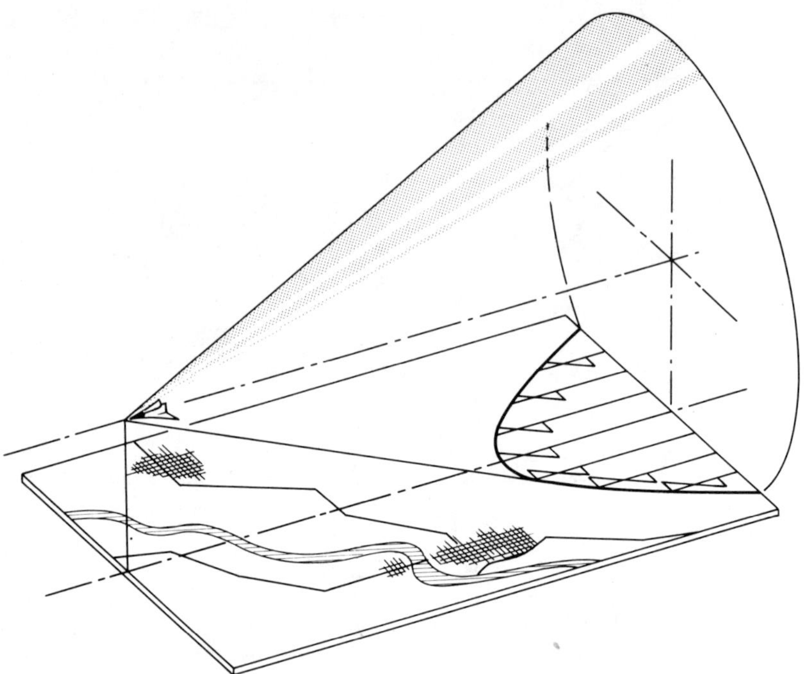

Figure 19   Schematic of the sonic boom pressure field.

by a sonic boom is finite in width. The "boom carpet" that follows a supersonic jet may be as wide as 60 miles for a supersonic transport at a normal cruise altitude of 60,000 ft; a rough rule of thumb is that the lateral extent in miles is equal to the aircraft's altitude, in thousands of feet.[30]

Figure 20 shows the general extent of the boom carpet for proposed supersonic flights between airports (a) and (b). Sonic booms will begin at a point about 100 miles from takeoff and cease at approximately the same distance from the destination. Hence sonic booms will not be experienced at the airports or in their immediate vicinity,[31] even though the overall boom carpet may extend for thousands of miles. Maximum overpressures will be generated on the ground when the aircraft is accelerating in the transonic speed range, and overpressures directly under the aircraft flight path will be greater than at the edges of the "carpet."

Because most supersonic flights are anticipated over oceans, a study has been made of air-water interactions from these pressure transients. For

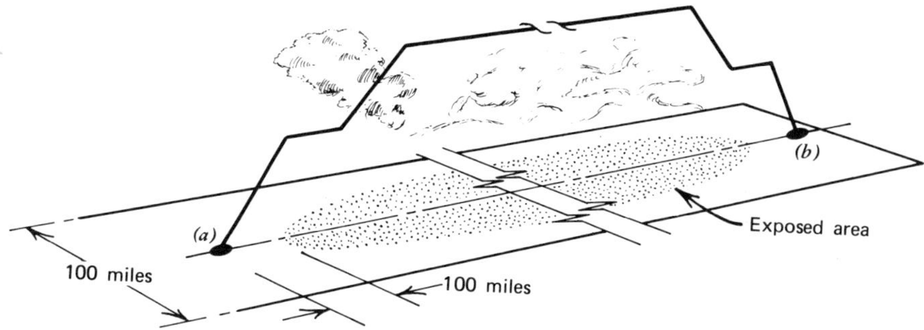

Figure 20   Extent of the sonic boom carpet.

present and planned supersonic speeds (<1000 m/sec) the N-wave from
an aircraft in level flight has an angle of incidence on the water surface
of greater than 19°. This minimum angle of incidence exceeds the critical angle for the passage of a sound wave from air to water, which is
about 14°. The net result is that each sinusoidal component of an N-
wave is totally reflected, and a pressure transient is generated in the
water, whose amplitude decreases exponentially with depth below the
surface.[32] Such pressures, however, are substantially less than pressures
known to be harmful to marine life.

Frequency Spectrum of N-Waves

The frequency spectrum of an N-wave is shown in Fig. 21.[33] The high
frequency envelope of this spectrum decreases at a rate of roughly 6 dB
per octave. The higher frequencies are significant because of loudness
and the lower frequencies are important because of the effect of sonic
booms on structures. Since much of the sonic boom energy is seen to exist
in the infrasonic range, many studies have been made on the structural
damage to buildings.[34,35,36] In many practical situations, of course, a portion of this infrasonic energy will be converted to audible sound by
means of harmonic generation, that is, as windows or structures give rise
to various modes of vibration. Nevertheless, most of the energy content
of a boom is inaudible, even though the boom is heard as two sharp
bangs. It has been found that the energy maximum is near a frequency
that is equal to $1/\Delta T$, where ΔT is the duration of the "N"-wave (in
seconds).

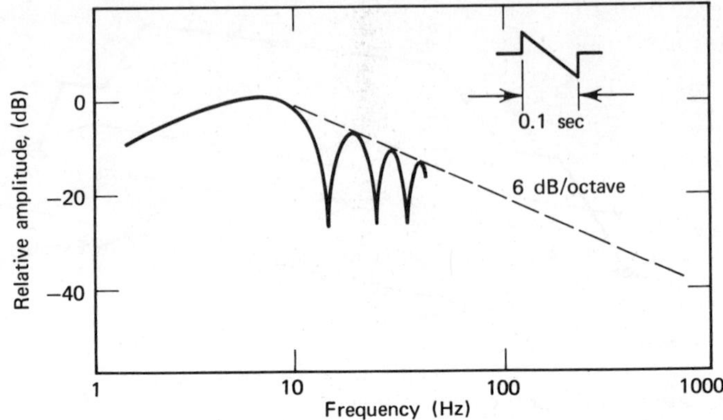

Figure 21   Characteristic frequency spectrum of N-waves. [From A. D. Pierce and D. J. Maglieri, *J. Acoust. Soc. Am.*, **51**, 702 (1972).]

NEAR-SONIC TRANSPORTS

Because of the sonic boom problem, as well as other potential environmental problems, further development of the proposed United States supersonic transport (SST) has been deferred. If used for transcontinental service over major existing routes, about one-third to one-half of the United States population would be exposed to several sonic booms per day. Therefore, a substantial research and engineering effort is being focused on the design of an aircraft that would have a cruising speed just under the velocity of sound. It has been shown that a Mach 0.98 aircraft (as opposed to current Mach 0.84 jets) could be designed that would not cause a sonic boom regardless of wind and temperature conditions.[37] Based on wind tunnel tests and engine thrust calculations, this aircraft would also meet the FAR-36* noise level limits for takeoff and landing. An artists's conception of this new aircraft (tentatively designated as the Boeing-767) is depicted in Fig. 22.

Such an aircraft would meet, in part, the design goals of the air-frame and jet engine manufacturers for a faster aircraft, and the public would benefit from a plane that would be substantially less noisy than most of those in the present air fleet. The aircraft is designed for both transcontinental and overseas routes. It would reduce San Francisco-New York flight times by one hour, and it has a range that would allow it to travel the 5869 nautical mile New York-Tokyo great circle route, nonstop.

* Federal Aviation Regulation, Part 36, published in the Federal Register, Nov. 21, 1969.

Figure 22 Artist's conception of a near-sonic Boeing 767. (Courtesy of the Boeing Co.)

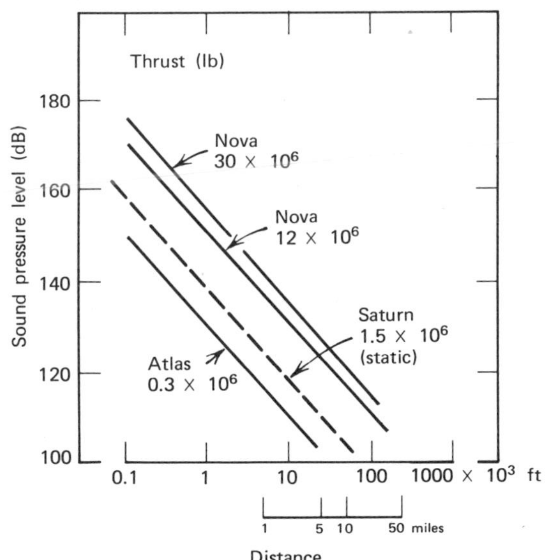

Figure 23 Sound pressure levels generated by large rocket-powered vehicles at launch. (From Regier, Mayes, and Edge, Ref. 39; courtesy of the National Aeronautics and Space Administration.)

ROCKET-POWERED VEHICLES

Rocket-powered vehicles have little general environmental impact because launching sites are remote from communities. They represent, however, the most powerful sound sources in existence. A Saturn rocket generates approximately 2.5×10^7 W of sound energy with a maximum sound pressure level occurring in the frequency range of 60 Hz.[38] The overall sound pressure level of this and several other rocket-powered vehicles as a function of distance is shown in Fig. 23.[39]

REFERENCES

1. F. W. Kolk, *Astronautics and Aeronautics,* 20 (October 1972).

2. H. B. Safeer, *Sound & Vib.,* 22 (October 1973).

3. U. S. Environmental Protection Agency, *Transportation Noise and Noise From Equipment Powered by Internal Combustion Engines,* Washington, D. C., 1971, p. 10.

4. *Ibid.,* p. 77.

5. E. J. Richards and D. J. Mead, *Noise and Acoustic Fatigue in Aeronautics,* Wiley, London, 1968, p. 182.

6. H. H. Hubbard, D. L. Lansing, and H. L. Runyan, *J. Sound Vib.,* **19,** 227 (1971).

7. M. V. Lowson and J. B. Ollerhead, *J. Sound Vib.,* **9,** 197 (1969).

8. Richards and Mead, *op. cit.,* p. 201.

9. Lowson and Ollerhead, *op. cit.,* p. 201.

10. M. A. Bowes, *J. Acoust. Soc. Am.,* **54,** 1214 (1973).

11. K. W. Bushell, *J. Sound Vib.,* **17,** 271 (1971).

12. J. W. R. Griffiths, *J. Sound Vib.,* **1,** 127 (1964).

13. M. J. Benzakein, *J. Acoust. Soc. Am.,* **51,** 1428 (1972).

14. R. A. Mangiarotty, *J. Acoust. Soc. Am.,* **48,** 783 (1970).

15. R. A. Mangiarotty, *J. Sound Vib.,* **18,** 565 (1971).

16. Federal Aviation Regulations, Part 36: Noise Standards, Aircraft Type Certification. Federal Register, Vol. 34, No. 226, November 25, 1969.

17. Richards and Mead, *op. cit.,* p. 147.

18. A. P. G. Peterson and E. E. Gross, Jr., *Handbook of Noise Measurement,* General Radio Co., Concord, Mass., 1972, p. 35.

19. L. L. Beranek, *Noise and Vibration Control,* McGraw-Hill, New York, 1971, p. 581.

20. "Aircraft Noise Analysis for the Existing Air Carrier System," a report prepared for the Aviation Advisory Commission by Bolt Beranek and Newman, Inc., September 1, 1972.

21. M. Rettinger, *Acoustic Design and Noise Control*, Chemical, New York, 1973, p. 256.

22. Safeer, *op. cit.*, p. 23.

23. P. Hurdle, S. R. Lane, and W. C. Meecham, *J. Acoust. Soc. Am.*, **50**, 32 (1971).

24. U. S. Environmental-Protection Agency, *op. cit.*, p. 36.

25. U. S. Environmental Protection Agency, *Laws and Regulatory Schemes for Noise Abatement*, Washington, D C., December 1971, pp. 1-18.

26. F. L. Sawyer, *J. Sound Vib.*, **5**, 335 (1967).

27. C. B. Pattarini, *J. Sound Vib.*, **5**, 370 (1967).

28. W. R. Hazard, *J. Sound Vib.*, **15**, 425 (1971).

29. C. H. E. Warren, *J. Acoust. Soc. Am.*, **51**, 783 (1972).

30. H. W. Carlson and D. J. Maglieri, *J. Acoust. Soc. Am.*, **51**, 675 (1972).

31. H. H. Hubbard, *Proceedings of the Sonic Boom Symposium*, St. Louis (1965).

32. J. C. Cook, T. Gaforth, and R. K. Cook, *J. Acoust. Soc. Am.*, **51**, 729 (1972).

33. A. D. Pierce and D. J. Maglieri, *J. Acoust. Soc. Am.*, **51**, 702 (1972).

34. B. L. Clarkson and W. H. Mayes, *J. Acoust. Soc. Am.*, **51**, 742 (1972).

35. U. S. Environmental Protection Agency, *Effects of Sonic Boom on Structures*, Washington, D. C., 1971.

36. National Aeronautics and Space Adm. Report CR-227, *Studies of Sonic Boom Induced Damage*, Washington, D. C., May 1965.

37. L. T. Goodmanson, *Astronautics and Aeronautics*, 46 (November 1971).

38. R. N. Tedrick, *J. Acoust. Soc. Am.*, **36**, 2027 (1964).

39. A. A. Regier, W. H. Mayes, and P. M. Edge, Jr., *Sound*, 1, 7, (November–December 1962).

11

INDUSTRIAL NOISE CONTROL

PLANNING FOR NOISE CONTROL

The basis for any viable noise control program is an appreciation of its importance and a commitment to undertake a *continuing* noise abatement effort. In industry this means that management must delegate responsibility to a knowledgeable staff engineer or consultant and obtain cooperation from its entire planning and production groups. If new production facilities are to be built, the site location, architectural design of the plant, and the acquisition of production facilities must be given careful acoustical consideration. It is a substantially simpler problem to design an industrial plant with moderate noise levels both within the plant and at the property line, than it is to do so after a facility is operational. Building design and plant layout can provide for an isolation of the noisiest operations, and purchase orders for new machine tools or processing equipment can include noise levels as a part of the specifications. There are practical limits to reducing noise, of course, for high speed machinery often generates high noise levels, and no manufacturer can afford to invest in production facilities that will markedly increase the unit cost of products.

Nevertheless, several options are usually available in the planning stage. Excessively noisy automatic equipment can sometimes be monitored remotely by closed-circuit television. Without excessive costs, rooms can occasionally be designed with heavy glass partitions, so that visual contact can be maintained with equipment that requires only infrequent servicing. Also, the acoustical treatment of walls and ceilings is relatively inexpensive while a plant is under construction.

It is not easy to place a dollar value on the benefits of an industrial noise control program. Since the government has drawn up regulations requiring conformity to certain standards (see Chapter 12), an occupational noise control program is mandatory to some extent. In addition, industrial managements can realize a return on its investment through improved employee morale, greater employee freedom from fatigue and annoyance, clearer communication, safety, and—occasionally—increased productivity. A reasonably acceptable aural environment is also a worthy goal in itself.

AUDIOMETRIC TESTING AND PERSONNEL PROTECTION

In a large industrial plant the responsibility of personnel protection is shared by the medical director and the safety engineer. Initial audiometric testing is now becoming an integral part of the preemployment physical examination, and such testing is important to both employer and employee. For many employees these tests will be the first definitive measurement of their hearing sensitivity; for employers the tests provide the data against which future changes in hearing acuity may be compared. Subsequent tests, of course, will indicate changes due to age and noise-induced shifts resulting from both occupational and nonoccupational exposures (see Chapter 17). In many instances, the latter will be important. Persons whose hobbies include hunting, small arms target practice, power boating, rock and roll music, flying private aircraft, and so on, may have a negligible portion of their hearing loss attributed to occupational causes. Only in very few instances is it possible to clearly differentiate between occupational and nonoccupational shifts in the hearing threshold. One such instance is the case of telephone linemen who in field work use a phone receiver for testing and communication. Invariably these men use a single ear (right or left but not alternately), so that a single ear is exposed to line and other noises.* Thus, a long-term comparison of right and left hearing sensitivity is meaningful: one ear is exposed to occupational sounds while the other ear is exposed to all other nonoccupational noise and serves as a reference.[1]

The modern professional safety engineer will sometimes be charged with the supervisory responsibility for all aspects of in-plant safety, for

---

* Truck drivers also have somewhat greater occupational exposure to their left than right ear because of traffic noise.

example, mechanical, electrical, chemical, radiation protection, and noise. There is some merit in having all safety responsibility centralized, for in practice these several areas are interrelated in the overall operation of a manufacturing or processing plant. Furthermore, both industrial management and government regulations require complete records on employee safety. In the near future, the use of ear protective devices (e.g., ear plugs and ear muffs) will be as common as safety glasses, and the safety engineer is best equipped to keep up to date with the overall aspects of plant safety.

In industrial operations where noise levels approach the limits specified in the Walsh-Healey and Occupational Safety & Health Act (OSHA) (Chapter 12), employee testing may be frequent, and some employees will be required to wear the individual audio monitors described in Chapter 7. Exposure records should include, when feasible, the octave-band analysis of the noise to which the employee is exposed, duration times, and the type of ear protective device that is used. Because there are large individual physiological differences in sensitivity to noise exposure, a fairly complete record should be maintained for employees exposed to high sound levels. Furthermore, such records provide an invaluable clue for transferring employees to alternative assignments where noise is a negligible health hazard.

The American Academy of Opthalmology and Otolaryngology has been active in providing specific guidelines for hearing conservation programs in industry.[2]

CONSTRUCTION NOISE

Noise from construction and demolition affects a substantial number of persons in major urban centers. Relatively few complaints are registered during the demolition activities, because such work is usually of short duration. Building construction, and the repair of water, sewer, gas, electric services and accompanying street repairs, however, often takes place over an extended period. The primary source of construction noise is from equipment; for example, gasoline-powered and diesel engines, air compressors, mechanical and hydraulic systems, pneumatically powered drills and hammers, and steam pile drivers. The substitution of non-impact (hydraulic or electric) tools for some impact machinery has allowed a significant noise reduction for equipment, although their utilization depends on the specific work to be undertaken. The replacement of welding for riveting is another noise reduction option.

The typical noise levels in the environs of various construction projects have been measured in a recent study, and the results are presented in Table 1. Noise levels at the construction site boundary are given in dB(A), and in terms of a Noise Pollution Level (NPL)* that is intended to take into account the effect of the number and duty cycle of all equipment that is utilized at the site.

Table 1   Typical Noise Levels at Construction Sites[3]

Type of Site	Domestic Housing		Office Buildings, Etc.		Sewers, Roads, Etc.	
	dB(A)	NPL	dB(A)	NPL	dB(A)	NPL
Ground clearing	84	100	84	99	84	100
Excavation	88	106	89	104	89	105
Foundation	81	99	78	85	88	108
Erection	82	97	85	97	79	88
Finishing	88	106	89	104	84	100

Because of its temporary character, construction noise is usually considered to be an environmental problem of lesser impact than traffic or aircraft noise. Nevertheless, the amount of annual construction activity is substantial. In 1970 the total number of projects in the United States was approximately (1) residential—514,000; (2) nonresidential—63,000; (3) municipal streets—336,000; and (4) public works—485,000.

For high-rise structures construction activity may last for two years or more, so that attaining minimum noise levels becomes very important. If the construction site is extensive, some noise reduction can be realized by locating the noisy equipment as far as possible from the site boundary and providing enclosures for stationary equipment. Both noise and cost reductions are also being achieved by the extensive use of prefabricated structures, which reduce on-site activity and shorten construction time.

In an effort to encourage the manufacture of substantially quieter construction equipment, the General Services Administration has imposed noise specifications on all equipment to be used on federal construction

---

* The NPL, [dB(A)] is defined as the sum of the A-weighted average sound pressure level and 2.56 times the standard deviation of the A-weighted sound pressure level; thus the NPL is intended to account for the effect of steady noise plus the annoyance due to fluctuations.

projects. These limits for various types of construction equipment are listed in Table 2.

NOISE CONTROL IN MANUFACTURING PLANTS

There are some 311,000 industrial plants within the United States, most of which are engaged in some form of manufacturing activity. A majority

Table 2   Noise Specifications for Construction Equipment at a Distance of 50 Feet [dB(A)][4]

Equipment	Effective July 1, 1973	Effective January 1, 1975
Earthmoving		
Front loader	79	75
Backhoes	85	75
Dozers	80	75
Tractors	80	75
Scrapers	88	80
Graders	85	75
Trucks	91	75
Pavers	89	80
Materials handling		
Concrete mixer	85	75
Concrete pump	82	75
Crane	83	75
Derrick	88	75
Stationary		
Pumps	76	75
Generators	78	75
Compressors	81	75
Impact		
Pile drivers	101	95
Jackhammers	88	75
Rock drills	98	80
Pneumatic tools	86	80
Other		
Saws	78	75
Vibrators	76	75

of the plants are located within large metropolitan areas, but an increasing number are now being built in suburban and rural communities. Major industries include aircraft, apparel, automotive, chemical, cement, ceramics and glass, containers and packaging, electrical products, lumber, leather, machines and machine tools, metals and metal products, paper, printing and publishing, petrochemicals, pharmaceuticals, plastics, and textiles.

In-Plant Noise Sources

Sources of noise in a manufacturing plant will include fabrication and assembly machines, plus a wide variety of facilities for material transport and general plant services.

In a large plant, a rule of thumb is that the noise produced by a number of machines of the same type can be estimated by adding $10 \log_{10} N$ to the sound pressure level generated from one machine alone,[5] that is,

$$L_p(N) = L_p + 10 \log_{10} N \tag{1}$$

where $L_p(N)$ = sound pressure level of N machines
$\quad\quad L_p$ = sound pressure level of a single machine
$\quad\quad N$ = number of machines of the same type

A selected list of machines and facilities appears in Table 3.

Many of these devices generate noise levels of 90 to 115 dB(A), within 2 to 5 ft of the apparatus. Impact noise from riveting or punch presses is usually very intense. An octave-band sound pressure level measurement of a punch press, operating at 720 impacts/min in a can manufacturing plant is shown in Fig. 1.

The high sound pressure level readings result from the high "peak" levels generated by impact and the high repetition rate (12/sec). The short term transient associated with a single impact must be observed by an oscilloscope. The recording of the transient associated with a single impact of a 1800 lb drop hammer in a forming operation is shown in Fig. 2. The transient is characterized by a rapid rise to 120 dB (at 20 ft), followed by a low frequency oscillation that is caused by the "ringing" of the hammer.

High sound levels are also generated from grinders as shown in Fig. 3. This 1/3 octave-band spectrum shows that most of the sound energy from a grinder is in the high frequency range, so that speech communication in the immediate vicinity of a grinder is virtually impossible. The spectrum corresponds to an A-weighted level of almost 110 dB.

Table 3 Noise Sources in Manufacturing Plants

Air hammers	Polishers
Air hoists	Presses, pneumatic
Boring machines	Presses, punch
Chippers	Pumps
Combustion	Reciprocating devices
Compressors	Riveting devices
Conveyor belts	Sand blasting
Cranes, overhead	Saws
Cutting machines	Screw machines
Drop forges	Shakers
Drop hammers	Shot blasting
Fans	Spraying
Fork trucks	Tamping machines
Furnaces	Transformers
Gas and fluid flow	Tumblers
Grinders	Turbines
Hammers	Turntables
Injection molding	Vibrators
Lathes	Welding machines
Motors	Wire drawing
Planers	Wrenches, pneumatic

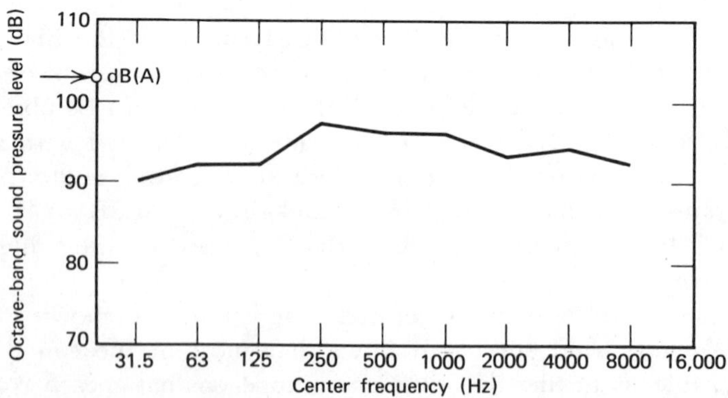

Figure 1 Octave-band sound pressure levels produced by a punch press (720 strokes/min) in a can-manufacturing plant. (Courtesy of the U.S. Environmental Protection Agency.)

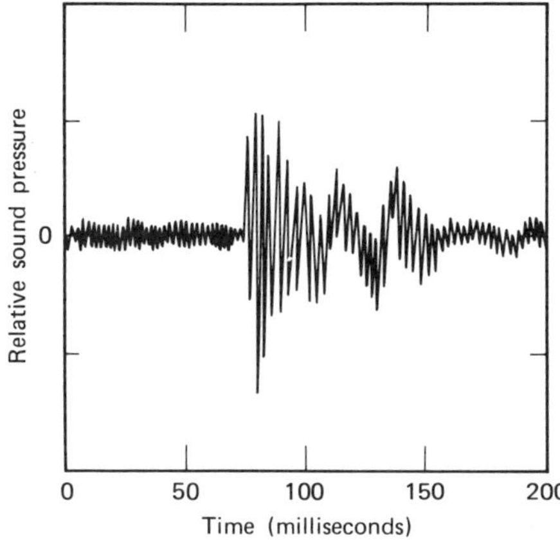

Figure 2   Oscillographic recording of impact noise generated by a drop hammer. (From A. P. G. Peterson in *Handbook of Noise Control*, ed. by C. M. Harris, Copyright McGraw-Hill 1957, used with permission of McGraw-Hill Book Co.)

Other important in-plant noise sources include piping systems and valves. At times, the noise generated from these elements will exceed that of machinery. High velocity flow of air, steam, or other fluids that undergo an abrupt change in pipe diameter will give rise to turbulence and resultant noise. Thus, sharp bends and constrictions should be minimized; a rough rule of thumb is that the radius of a piping bend should be at least five times the pipe diameter.

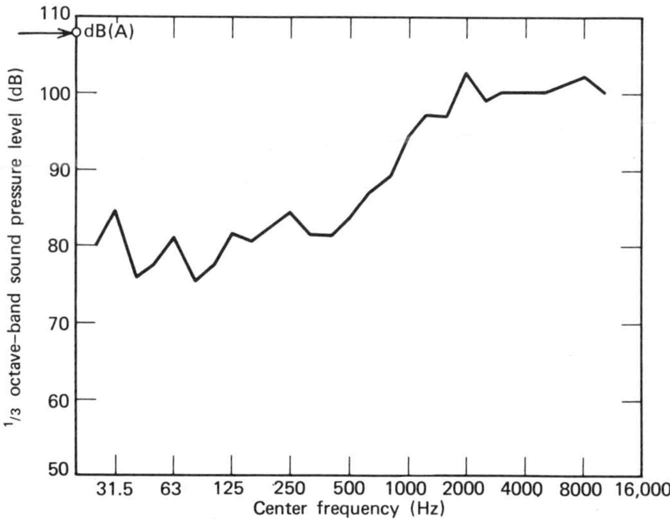

Figure 3   One-third octave-band sound pressure levels measured near a rough grinding operation in an automobile assembly plant. (Courtesy of the U.S. Environmental Protection Agency.)

Directivity of Sound Pattern

In some instances there will be a directional asymmetry to the sound radiated from a machine. Figure 4 indicates the magnitude of such an asymmetry as a function of frequency for a hand grinder running under no load at 22,000 rpm. (The microphone was rotated around the grinder at a distance of 30 in. in a plane containing the axis of the grinder.) The low frequency sound is uniformly radiated, as discussed in Chapter 2, but the high frequency sound is "beamed" in directions that depend upon grinder construction. At 10,000 Hz, there is a maximum directional differential of 10 dB. This suggests that in some instances it is

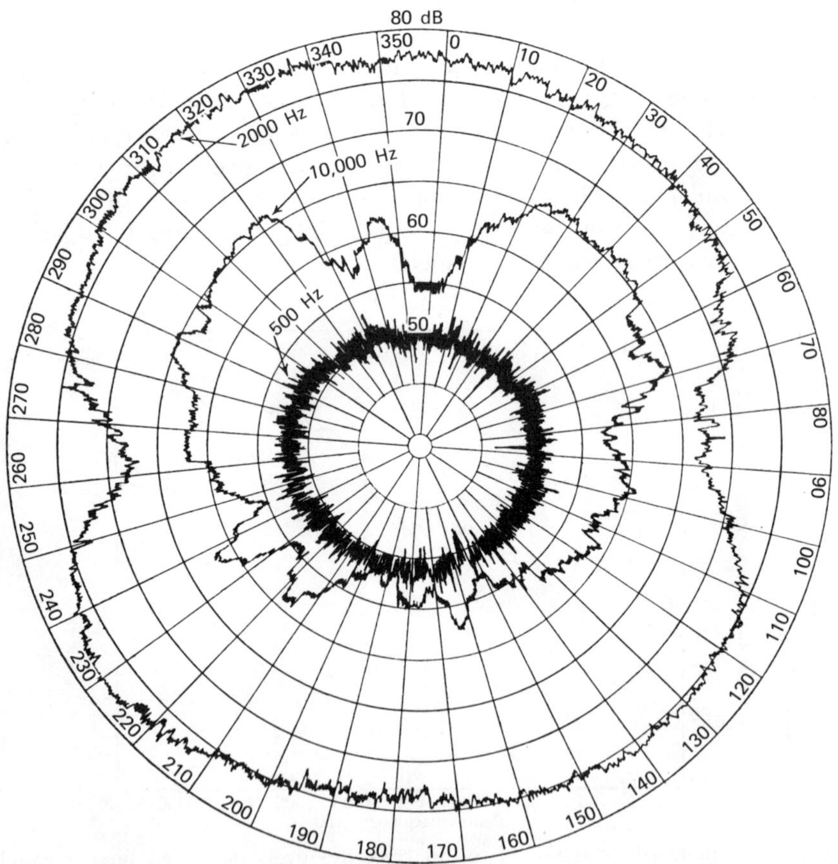

Figure 4   Directional sound pattern produced by a high-speed electric hand grinder running under no load at 22,000 rpm. (Courtesy of General Electric Co.)

possible to orient stationary machinery within a plant to take advantage of sound level minima for the benefit of operators or persons working in close proximity to such equipment. In this case, however, the plant must be acoustically treated so that there will be a minimum of reverberant sound.

Operator and Machine Enclosures

While the ideal solution to the in-plant noise problem is the procurement of quiet machinery, in practice a substantial amount of production equipment will remain noisy. High production rates require high speed machinery and, in general, high noise levels accompany high rates of impact, grinding, polishing, forging, and so on.

When the plant is a large one in which the overall noise level results from many machines, enclosures of the type shown in Fig. 5 may be used.

Figure 5   Prefabricated acoustic enclosure for personnel protection. (Courtesy of Industrial Acoustics Co., Inc.)

The enclosures are available in many models and sizes. They may be used either for supervisory personnel or for operators who may be monitoring one or more automatic machines. They permit visual contact to be maintained with large and expensive machinery, and provide noise reductions of 30 dB(A). In some instances such installations provide more personnel protection than any other means.

When only one or two machines are the dominant noise sources, it may be feasible to isolate the noisy equipment in a small area enclosure of the type shown in Fig. 6.

Usually these partition enclosures are made of fiberglass cloth and lead or lead-impregnated plastics, so as to provide a limp, dense, sound barrier. These curtains can be opened to allow for the easy transfer of materials, and they provide lower sound levels in adjacent areas, so that only a single operator (within the enclosure) is required to wear ear protection.

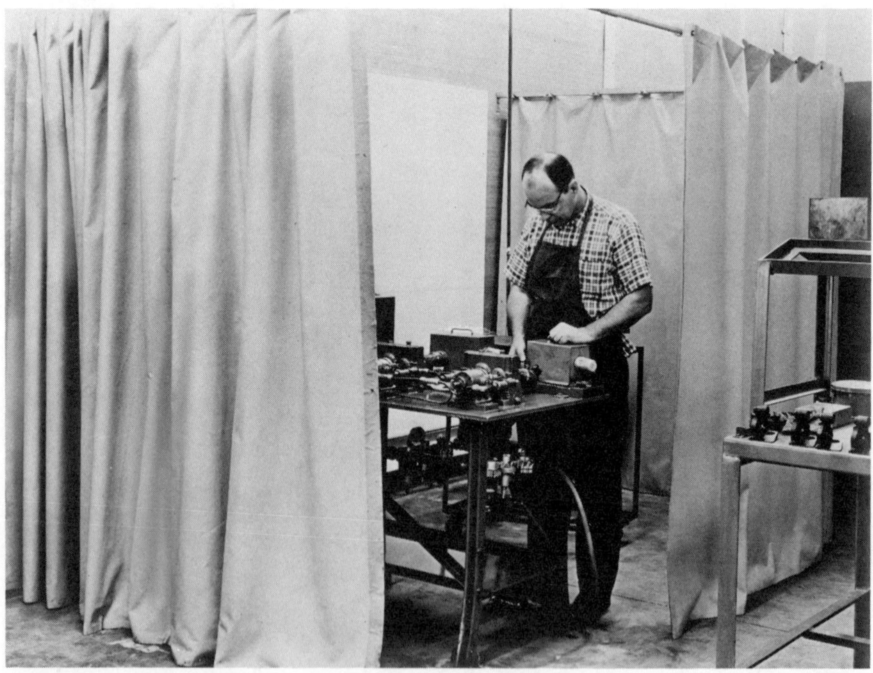

Figure 6   A floor-to-ceiling enclosure with curtains made of fiberglass cloth and lead or leaded vinyl can be used to isolate noisy machinery within an industrial plant. (Courtesy of Singer Partitions, Inc.)

A third type of enclosure applies to equipment that generates high noise levels, but only infrequently requires inspecting or servicing; for example, compressors, vacuum pumps, and motors. Since such equipment often radiates noise by vibrations through a building structure, vibration isolation is of paramount importance. Figure 7 is a schematic diagram of such an enclosure in which the apparatus is mounted on vibration isolators and ventilation for the unit is provided via a lined duct.

In new plant layouts or where major modifications are being made, extended glass partitions can be installed to separate machines and operators. Because an increasing number of production machines are operated automatically, it is likely that large transparent partitions will be one alternative for reducing noise in a factory. A single leaf glass wall weighing 6 lb/ft^2 will provide a midfrequency noise insulation of about 35 dB. If a greater sound reduction is required, a double glass wall may be used. Transmission loss curves for various glass thicknesses as a function of frequency can be obtained from glass manufacturers.

Vibration Isolation

The general problem of vibration isolation is especially important, because in some instances a remote part of a building may be a more efficient radiator of sound than the primary source (e.g., drill and lathe). In

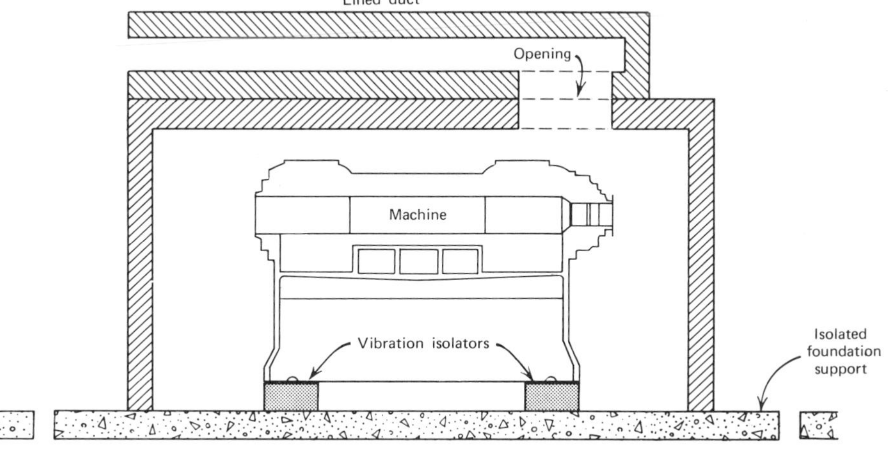

Figure 7   Enclosure for machinery with vibration isolators and liner air cooling ducts.

this case, the secondary radiator acts as a sound amplifier. In an effective radiator, the vibrating element will have dimensions that are equal to or greater than the wavelength of the frequency of the noise. Thus, an efficient radiator at 100 Hz is likely to have dimensions of over 11 ft; at 1000 Hz it may be only about 13 to 14 in.; at 5000 Hz the wavelength of sound is only 2.6 in. Hence, we would expect the higher frequency sound to be emitted from the primary machine elements and low frequency sounds to be transmitted and amplified through panels, protective housings of large areas, surface plates, floors, and walls. Also, mechanically generated noise is most effectively transmitted as airborne sound when the radiator is excited to one of its resonant frequencies or modes. A portable drill, for example, will almost always sound louder when clamped to a table or some metallic surface, unless the support is very massive. It is for this reason that some industrial equipment meeting noise standards at the factory will exceed noise specifications after installation.

To minimize this transfer of energy between a sound source and a secondary radiating surface, it is therefore necessary to provide some means of vibration isolation. In the general case we seek a large "impedance mismatch" between the vibrating source and its mechanical support or interconnections. As defined in Chapter 2, the acoustical impedance is the product of the density of the material and the velocity of sound in the material.

A high impedance material permits large amounts of energy to be transmitted by vibrational excitation, whereas a low impedance material is an inefficient medium for transferring energy. Table 4 lists the acoustical impedance of several common materials, and it indicates why rubber and cork are often chosen for vibration mounts.

If the equipment is massive, vibration isolation can be achieved by mounting the equipment on its own separate foundation (if at ground level). It has been shown that a 5 to 10 dB(A) sound reduction can often be realized from vibration isolation alone.

Noise Reduction With Acoustic Materials

In Chapter 4 we noted that the reverberation time, T, could be approximated by the expression $T = 0.05 \ V/A$ where V is the room volume (ft^3) and A is the total absorption (in sabins). Because industrial plants often have exceedingly large volumes and the structural walls and ceilings are highly reflective, the reverberation time may be 3 sec or greater.

Table 4 Acoustical Impedances of Some Common Materials[6]

Material	Acoustical Impedance, ρc (in.-lb/in.^{3}-sec)
Soft rubber	100
Cork	165
Pine	1,900
Water	2,000
Concrete	14,000
Glass	20,000
Lead	20,500
Cast iron	39,000
Copper	45,000
Steel	59,500

As a result, the entire plant is filled with reverberant sound so that throughout it there is little difference in the noise level. Even a single noisy machine can make an entire building noisy, for the 6 dB decrease in sound level with a doubling of distance applies only when there are no reflections. A substantial decrease in reverberation time and overall noise level can be achieved through the use of

(a) Sprayed-on materials
(b) Acoustic tile
(c) Resonant absorbers
(d) Diaphragmatic absorbers
(e) Space absorbers

(a) Sprayed-On Materials

A number of spray-on materials that consist of cellulose or foamed elastomers are available commercially. They are specifically designed as a sound control material for manufacturing plants, airports, gymnasiums, and large volume structures. These materials can be applied directly to concrete walls or roof structures up to thicknesses of 1-in; they are fire-resistant and can be applied easily over areas in which there are electrical conduits and other protruding service lines. Offering an aesthetic improvement as well as reverberation control, they can sometimes reduce plant noise from 8 to 12 dB(A), when applied to a building with highly reflecting walls and ceilings.

(b) Acoustic Tile

Acoustic tile is widely applied in both existing and new plants. In new building construction, it is common practice to use a suspended ceiling so as to leave a shallow space between tile and the structural ceiling for accommodating ducts, electrical services, and steam lines, for instance. Adequate clearance, however, must be allowed for equipment such as overhead cranes and hoists. Tiles have excellent sound-absorbing characteristics at high frequencies and an application of tile to ceilings alone will usually provide a noise reduction coefficient (NRC) of 0.60 for the ceiling area. For building spaces with heights of 9 to 15 ft, such a ceiling treatment will bring the overall room absorption coefficient well above the recommended lower limit of 0.20.[7]

The addition of tile on the upper wall area of a building can provide an even further reduction of reflected sound. A practical upper limit for sound conditioning any large space is an average (overall) absorption coefficient of 0.50. With this value the audibility of reflected or reverberant sound in relation to direct sound is negligible, and any further use of sound-absorbing materials has little effect.[8]

(c) Resonant Absorbers

The absorption of low frequency sound of high intensity is sometimes achieved by the use of specially designed units, fabricated in the general shape of building blocks (Fig. 8). Usually termed *Helmholtz* resonators, they derive their sound-absorption qualities from being a "tuned" or resonant element, as in certain automobile mufflers. They can be used as a structural material for either indoor or exterior walls, and they have been successfully applied in diesel engine and other plants where much of the sound is low frequency. The resonant frequency, f, of such a Helmholtz resonator has been shown to be[9]

$$f = \frac{c}{2\pi} \frac{A}{\sqrt{vV}} \text{ Hz} \tag{2}$$

where c is the velocity of sound, V is the resonator chamber volume, A is the cross section of the resonator slit or aperture, and v is the volume of the aperture. The air in the aperture vibrates somewhat like a single mass and the larger volume of air in the chamber acts as a spring and provides absorption of the sound energy.

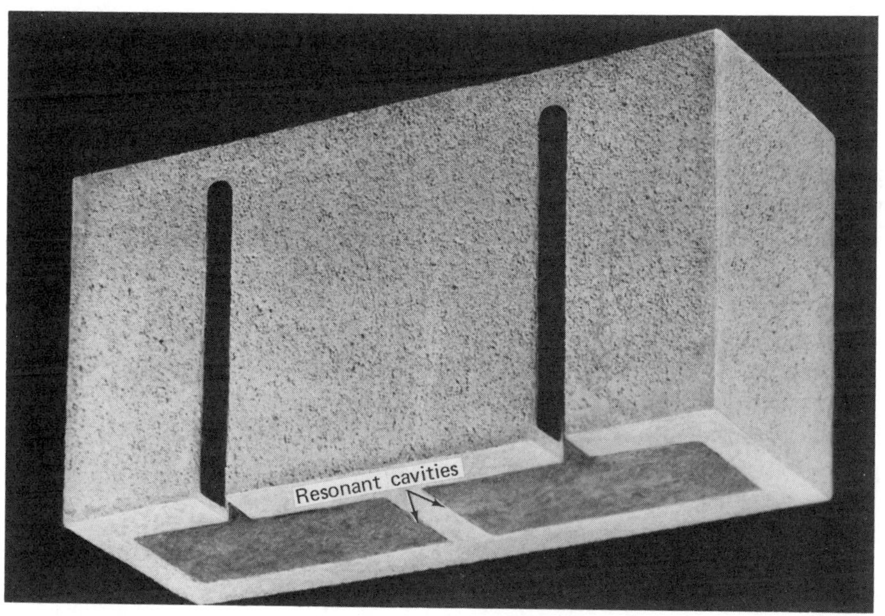

Figure 8   A diesel engine test facility at General Motors. Walls are constructed of 85,000 masonry resonant absorbers. (Courtesy of Proudfoot Co., Inc.)

(d) Diaphragmatic Absorbers

Diaphragmatic absorbers are usually in the form of large area panels which are also effective for the low frequency absorption of sound. If a panel is massive and stiff, most sound energy will be reflected, but if a panel is light and flexible, it can absorb sound by being set into vibration. The rate of sound absorption of such a panel is proportional to the product of the amplitude and frequency of vibration, to its internal coefficient of damping, and to the frictional losses at the edges of its mounting.[10] Because large panels have many vibrational modes, these diaphragmatic absorbers are effective over a broad frequency range. Commercial fiberboards have absorption coefficients of 0.2 to 0.3 in the range of 1 to 2 kHz, if mounted to wood strips so that they can undergo vibration rather than being cemented rigidly to massive walls. Such panels have also been employed effectively to control the frequency response of sound-recording studios and auditoria.

(e) Space Absorbers

Space or suspended absorbers refer to the general class of sound absorbers that are suspended as individual elements from a ceiling, instead of being installed as a continuous ceiling or wall surface. They consist of various acoustical materials, fabricated with suitable binders, in hollow pyramidical, spherical, cubical, conical, or other geometric shapes. In shops in which they do not interfere with overhead cranes, forklifts or other material-handling equipment, they can be mounted at low levels and between fluorescent-light fixtures. In certain situations they provide a total sound absorption per pound of acoustic material that is 10 times that of other acoustic treatments.[11] The suspension of space absorbers at low heights prevents some sound from radiating to remote parts of the enclosure, and the surface area of even small diameter spheres or cylinders is large, resulting in a high absorption per unit space absorber. Also, their *effective* area is much greater than the area of the ceiling covered since sound is incident from many directions. They have the further advantage of easy relocation and reuse. Figure 9 shows a typical installation.

NOISE LEVELS IN INDUSTRIAL PLANT ENVIRONS

In urban areas, most noise generated inside a plant will not have a marked effect on the noise levels of the surrounding community.[12] Exceptions will occur; but for manufacturing plants—as opposed to outdoor

Figure 9   Suspended absorbers are used to reduce machine shop noise. (Courtesy of United States Gypsum Co.)

process plants—community noise levels will be conditioned more by ancillary activities accompanying an industrial area. In-plant generated noise at the industrial plant boundary will usually be masked by traffic and construction noise. In many locations, noise from trains or aircraft will also be dominant. Some noise may be generated by building ventilating systems, associated power substations, and testing yards; these are likely sources of community annoyance at night.

In suburban communities, manufacturing plant noise is largely controlled through zoning ordinances. Some communities make substantial efforts to attract industrial plants within its boundaries whereas others prefer to remain bedroom communities and supply workers to industries outside their town line. In either case, the environmental noise impact is largely determined by increased passenger and heavy vehicle flow through a town, rather than by a new plant. In addition, most suburban towns are tending to develop industrial parks, so that it is possible to place specific limits on noise at the industrial park boundary, and locate the noisiest industrial activity well within the industrial park area rather than at its periphery.

Figure 10 shows results of a survey of noise levels in suburban communities not penetrated by major highways and the range of values is consistent with other measurements taken over a five-year period.[13] The numbers on each curve indicate the percentile of observations that gave octave-band values below the spectrum represented by the curve. The 95% curve roughly approximates typical suburban zoning codes; the 5% curve approximates night conditions in quiet suburban areas. The type of industrial plant that is likely to locate in suburban residential areas can usually match the upper curve at its boundaries by taking such measures as (1) purchasing a sufficiently extensive plot (2) designing structures that are adequate to contain in-plant noise, and (3) taking special precautions to shield exhaust fans and power transformers. Noise from

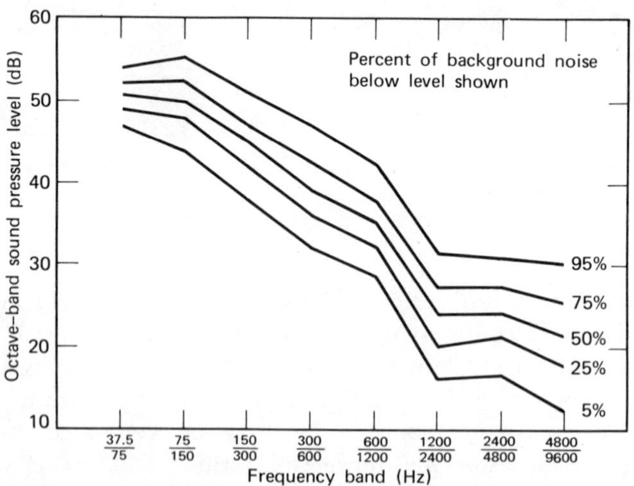

Figure 10 Percentile values of sound pressure levels found in suburban communities not penetrated by major highways. [From P. B. Ostergaard and R. Donley, *J. Acoust. Soc. Am.*, **36**, 409 (1964).]

pick-up and deliveries must be restricted on the basis of traffic noise control for the area.

INDUSTRIAL PRODUCTS

Of the many industrial products that are objectionable noise sources, only a few will be cited. They include power transducers (e.g., transformers, fans, pumps and motors) whose operation will necessarily be accompanied by the radiation of sound. The generation of pure tone components can be a special source of annoyance when such products must be utilized on a continuous duty basis.

Transformer Noise

Iron-core transformers connected to alternating current power lines produce a humming noise. The hum consists of a harmonic series of pure tone components based on a fundamental tone of twice the supply frequency. Thus for conventional 60 cycle/sec power, the tones will be 120, 240, 360 Hz, and so on. Figure 11 shows the relative levels of transformer core harmonics, together with an "A" weighted contribution, reported for a number of power transformers.[14] (The 7th and 8th harmonic are

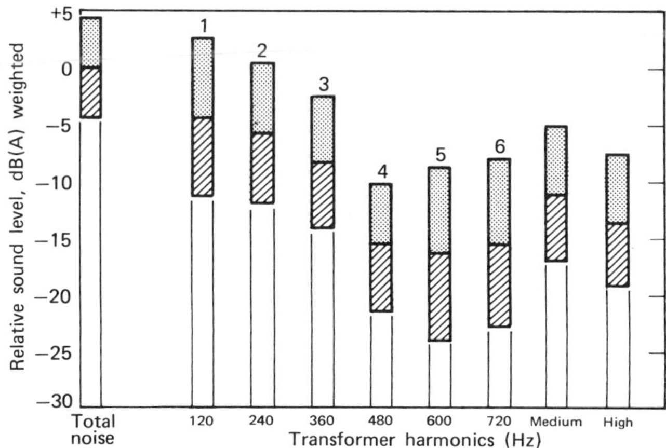

Figure 11   Relative levels of transformer noise harmonics. Limits include 95% of transformers tested. (From M. W. Schulz, Jr., and R. J. Ringlee, IEEE, *Power Apparatus and Systems,* June 1960.)

called a medium band, and the 9th, 10th, and 11th are called the high band). The spread in the data represents 95% of the transformers tested, and it is clear that the 1st, 2nd, and 3rd harmonics are the dominant frequencies, even though many higher harmonics are present. The primary cause of this harmonic noise is the magnetostrictive effect in the iron cores. This magnetostrictive phenomenon is a small change in the lengths of core materials upon magnetization, so that transformer noise is produced by alternate extensions and contractions of the core that are in phase with the exciting current.

In general, the smaller the transformer, the higher the frequency of the loudest harmonic; conversely, larger transformers tend to have louder low frequency components. This trend is largely because sound waves are most efficiently transmitted when the transformer dimensions are comparable to the wave length. The overall sound pressure, in dB(A) increases with power transformer rating. In the immediate vicinity of a 300 kVA transformer the noise level may measure 55 dB(A), rising to 80 dB(A) for a transformer of 100,000 kVA. A general rough rule for the *attenuation* of transformer noise with distance is given by*

$$dB_{att} = 4.4 + 20 \log \frac{D}{\sqrt{WH}} \tag{3}$$

where D is the distance from transformer to the measurement point in feet, W is transformer width, and H is transformer height (in feet). Equation 3 however, does not take into account the high degree of sound field asymmetry that accompanies many large substation transformers. Figure 12 is a profile of measurements made in the vicinity of such a transformer, which suggests that neighboring residents will not necessarily be exposed to equally loud sounds on the basis of equal distances.

Transformer noise will also include noise from cooling fans, but at large distances, only the pure tone harmonics will be discerned.

Fan Noise

Fans are among one of the most widely used industrial products, being essential for cooling, ventilating, and air circulating purposes. They are divided into two general classifications, regardless of size and power rat-

---

* The equation is valid only for $D > \sqrt{WH}$, and the form of the equation results from measuring D from the transformer *surface* rather than from the transformer's "acoustic center."

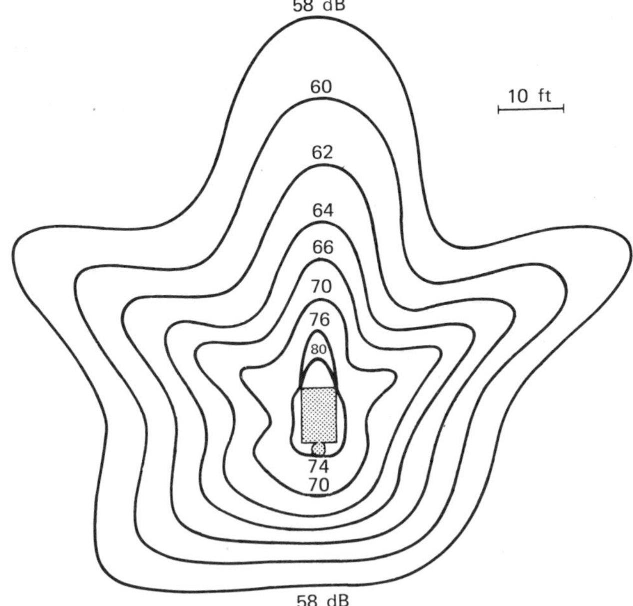

Figure 12   Contours of equal sound pressure level around a power distribution transformer. (From A. P. G. Peterson and E. E. Gross, Jr., *Handbook of Noise Measurement*, General Radio Co., 1972.)

ing: (1) *axial-flow* or propeller fans and (2) *centrifugal* fans. Figures 13*a* and 13*b* are sketches of these two types.

Axial-flow fans operate against little or no static pressure, require simpler housings, often operate at high speeds (>1000 rpm), and they have a wide variety of blade shapes. Noise is generated by the blades, and because of their high rotational velocity they are likely to produce more noise at mid and high frequency than the lower speed centrifugal type. The fundamental frequency of the axial fan is the "blade-passing frequency," which depends on the number of propeller blades and the rotational velocity, that is,

$$f = \frac{Nn}{60} \text{ Hz} \tag{4}$$

where N is the number of blades of the fan passing a given point per revolution, and n is the rotational speed in rpm. The overall fan noise will consist of this fundamental frequency plus its harmonics, bearing noise, motor noise, and vibrational noise generated from fan unbalance.

In general, a doubling of the fan power will result in approximately a 3 dB increase in noise level.

In some cases, 1/3 octave analysis is not narrow enough to define many of the blade harmonics so that a constant bandwidth filter is used. Figure 13c is frequency spectrum of an axial fan, obtained with a 100 Hz filter.[15]

In centrifugal fans, an axle is surrounded by a large number of blades that are contained in a circular-shaped and concentric rotor. Upon rota-

Figure 13a   Schematic diagram of axial-flow fan.

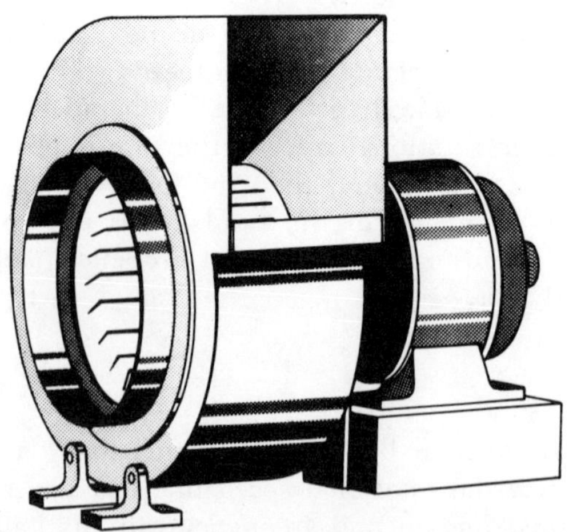

Figure 13b   Schematic diagram of centrifugal-type fan.

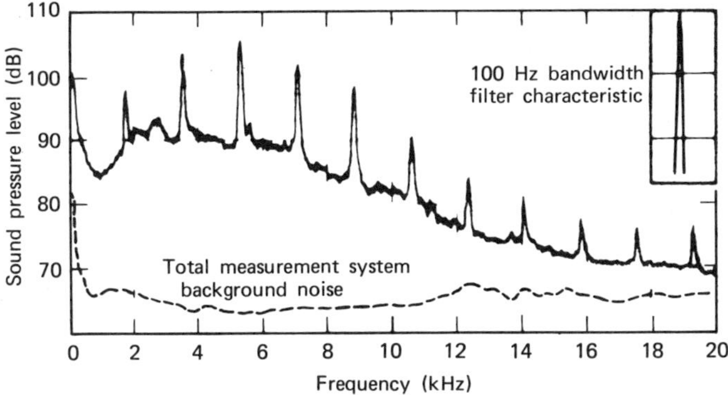

Figure 13c   Frequency spectrum of an axial fan obtained with a 100 Hz filter. [From D. B. Hanson, *J. Acoust. Soc. Am.*, **54**, 1571 (1973).]

tion, centrifugal force causes a flow and compression of the mass of air enclosed within the rotor. Noise levels and spectra are dependent on rotor speed, number of blades, blade configuration, and overall fan power rating. Two major parameters that have been related to total radiated sound power for a given centrifugal fan configuration are (1) the pressure differential generated by the fan and (2) the fan speed in rpm. Sound power is increased by the third power of the pressure, and by the fifth power of the rpm.[16] Centrifugal fans can move substantial amounts of air at low speeds and much of the radiated sound power is below 200 to 500 Hz. Hence they are employed in home heating and ventilating systems, and in many industrial situations that require high volume air flow.

For fan-diffuser combinations, the frequencies that will be generated will depend not only on the blade-passing frequency, but the number of stationary diffuser vanes, that is,[17]

$$f = \frac{(N_r \times N_s)}{K} \cdot \frac{n}{60} \tag{5}$$

where f = the frequency in Hz
n = is the fan or impeller speed in rpm
N_r = the number of rotating (impeller) blades
N_s = the number of stationary (diffuser) vanes
K = the highest common factor of N_r and N_s

Thus, as the impeller rotates, certain combinations of rotating and stationary vanes coincide during each revolution. The number of such coincidences is equal to the product of the number of rotating vanes and stationary vanes divided by the highest common factor. This consideration suggests that it is advantageous to use unequal numbers of rotating and stationary vanes, and that noise can be controlled, in part, by a judicious choice of impeller-stationary vane combinations.

Pumps and Motors

Noise from pumps is analogous to transformer and fan noise in that it contains fundamental and harmonic frequencies. Figure 14a shows a typical spectrum of a 150 HP, 3600 rpm induction motor, and Fig. 14b is a combined spectrum of both electric motor and pump noise. Pump and motor combinations range from fractions of a horsepower to greater than 5000 HP, depending upon the application. Large circulating pumps, such as are used in nuclear power plants, generate very high noise levels, but they are located in controlled areas remote from most operating personnel. High capacity pumps produce cavitation and fluidborne noise and mechanical vibrations of large amplitude. Noise from such pumps can only be suppressed by enclosures, isolation, and vibration control. Special vibration sensors are employed to provide an on-line monitoring of bearings and impellers for reasons of safety rather than noise.

Small pump-motor assemblies can sometimes be "canned" to reduce noise when designed for use in hospital or other quiet-area applications.

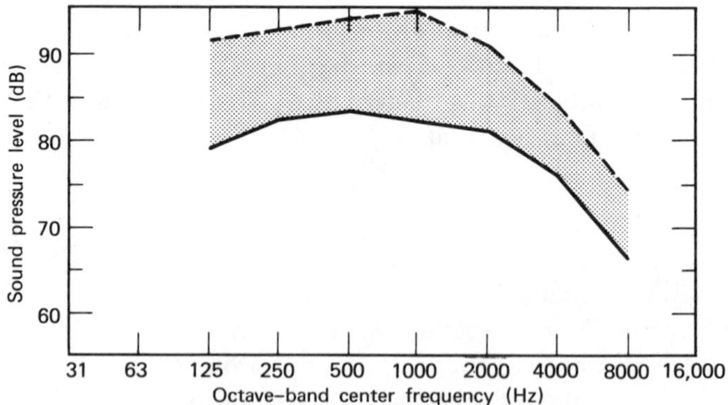

Figure 14a   Envelope of typical radiated noise spectrum of a totally enclosed, fan-cooled induction motor, 150 HP and 3600 rpm, at a distance of 3 ft. (From American Petroleum Institute, Medical Research Report, EA7301, *Guidelines on Noise,* 1973.)

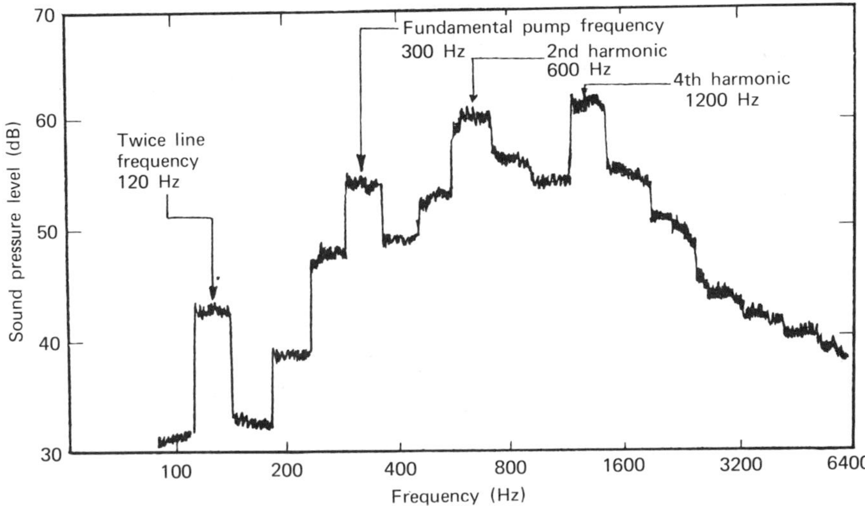

Figure 14*b*   Typical spectrum of pump noise showing electrical line frequency, and the fundamental and higher harmonics. (Courtesy of M. J. Crocker.)

Conversely, often motor noise is increased when attempts are made to reduce costs by producing thinner and less massive motor casings. These thin motor shrouds greatly amplify the noise of an otherwise acceptable motor. Differences of 10 dB(A) have been observed from motors of the same horsepower, based solely on engineering and production methods.

The origin of motor noise in electric motors and generators (so-called magnetic noise) is due to periodic forces that exist between the stator and the rotor. These forces are proportional to the square of the magnetic flux density, and it is the tangential components of these forces that provide torque to turn the motor. The radial components, however, produce noise at the harmonics of the line frequency.

Other sources of motor noise include vibrations in the rotor that modulate the air flow between components, noise from brushes and bearings, and noise whose origin can be attributed to eccentricities of machined parts and mechanical unbalance.

Power Equipment Specifications

The noise problem associated with electrical power equipment has been of sufficient concern that a general noise level rating scheme has been defined by the National Electrical Manufacturers Association (NEMA). Figure 15 shows this NEMA rating* in terms of sound pressure levels for octave bands. Thus a *"D"* NEMA rating has a sound pressure level upper limit of 70 dB at 63 Hz and a limit of 35 dB at 8000 Hz, and the curves reflect the fact that hearing is less sensitive to low than to high frequencies. Hence, in both the manufacture and purchase of electrical equipment, industrywide ratings can be applied.

Gears and Bearings

Gears and bearings provide the essential link between a driving motor or engine and the application of useful torque to an axle or power-consuming system. In many applications, gear and bearing noise will be masked by other sounds, and often gear housings surround gear assemblies. The massive housing for an automobile differential gear, for example, provides rear axle support, encloses the space around the gears to confine the lubricant, and provides an effective noise shield. Under normal driving conditions, this gear noise is masked by exhaust and engine noise. Gear noise is sometimes evident, however, because (1) the tooth-meshing frequency is often within the audible range, (2) these pure tone frequencies plus higher harmonics can be amplified through other parts of the drive system, and (3) gear noise almost always increases with increasing wear. The magnitude of the generated noise is dependent upon the load transferred by the gear system, gear design, flexing of gear shafts, lubricant, and so on. The tooth-meshing frequency for a simple gear arrangement will have the same form as equation 4, where N is the number of teeth on the gear and n is the rotational speed in rpm. Complex gear systems will generate a large number of discrete frequencies.

Bearings are classified into two general categories: (1) plain bearings and (2) rolling-contact bearings. Plain bearings include so-called *journal* and *thrust* bearings in which a shaft rotates within a lubricated circular bearing surface, and *guide* bearings, such as the inner surface of a piston engine that provides a linear constraint. Ball and roller bearings are of the rolling-contact type; the balls or rollers (cylinders) are the rolling element and provide precise alignment. There are formulas that relate discrete frequencies to bearing eccentricities and defects, and with sophis-

---

* The dB(A) equivalents are: NEMA $B = 42$ dB(A), NEMA $C = 47$ dB(A), NEMA $D = 52$ dB(A). The gas turbine industry now favors a dB(A) rating at 400 ft or at a property line, and the dB(A) scale may eventually replace the NEMA nomenclature.

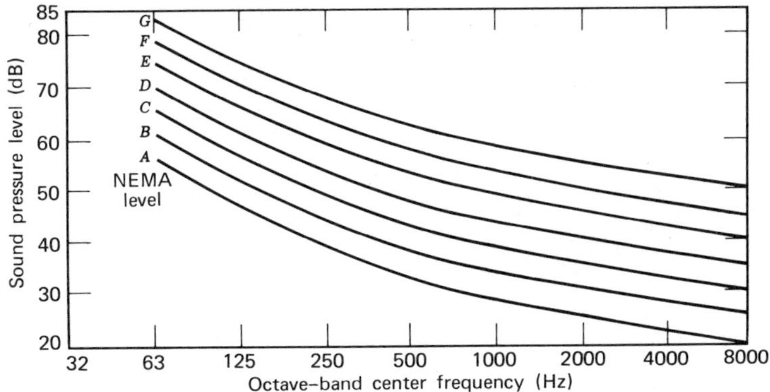

Figure 15   Power equipment noise ratings defined by the National Electrical Manufacturers Association (NEMA).

ticated equipment it is now possible to obtain a mechanical "signature" of a bearing. Figure 16 is such a "signature" of a ball bearing displaying many discrete frequencies that can be related to various surface defects.[18] This correlation of both amplitude and frequency of an acoustic spectrum with bearing condition provides a new diagnostic tool for fault location, wear, and quality control.

The Development of "Quiet" Components

An interesting new development relates to the development of alloy metals that can be used in products ranging from industrial machinery to household appliances. These alloy metals exhibit normal metallic properties such as reasonably high tensile strength, hardness, and rigidity (nearly equal to mild steel), and the ability to absorb shock. Nevertheless, they have the property of damping vibrations induced in them very quickly, thus attenuating substantially the radiation of sound that is emitted from components such as gears and similar components. An illustration of the effectiveness of this metallurgical development has been reported in preliminary testing of "treated" and "untreated" gears.[19] Figure 17a shows the sound decay pattern emitted subsequent to impact from a gear utilizing a high-damping alloy; Fig. 17b shows the relative decay rate of vibrations produced in a poor damping alloy. The use of polymeric materials as substitutes for metals where strength and wear criteria permit is another alternative.

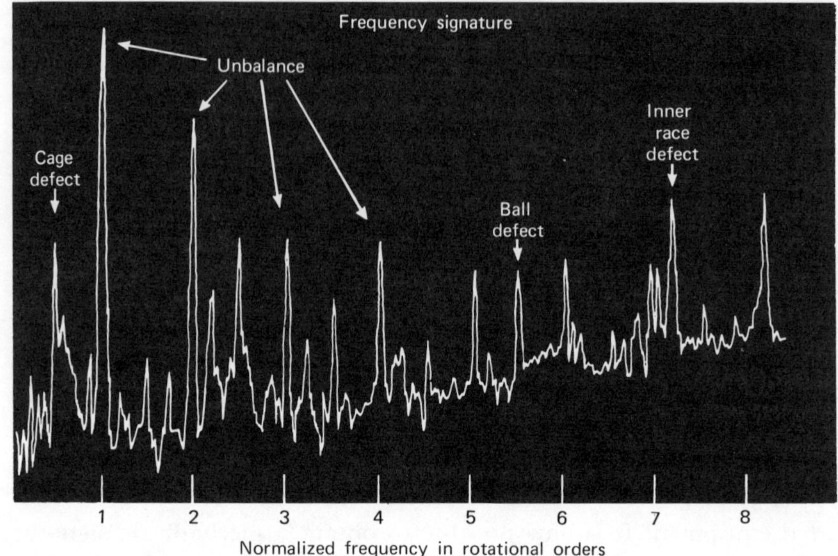

Figure 16   Mechanical "signature" of a ball bearing showing the discrete frequencies that can be related to surface defects. (From A. S. Babkin and J. L. Anderson, *Sound Vib.*, April 1973.)

In a corollary development, the use of steel-viscoelastic composites to replace ordinary sheet steel is being considered for potential applications to automobiles, railway cars, building partitions, and ventilating ducts. For example, a cold-rolled steel sheet of 0.048 in. thickness has been compared in the laboratory to a laminated sheet consisting of two 0.024 inch (24 gauge) sheets between which is bonded a viscoelastic layer of approximately the same thickness. This type of matrix affords both extensional and shear damping so as to dissipate a large amount of vibrational energy into heat. Whereas the single sheet of steel has a reverberation time (60 dB vibration decay period) of over 5 sec, the three-layer laminate has a reverberation time of approximately 0.1 sec.[19]

CONSUMER PRODUCTS

Current efforts are underway to reduce the noise from the broad class of consumer products that includes everything from lawn mowers to electric clocks. Recently, there was some reticence about reducing the noise from such items as vacuum cleaners because manufacturers found that users equated lower sound levels with less powerful products. This concept is

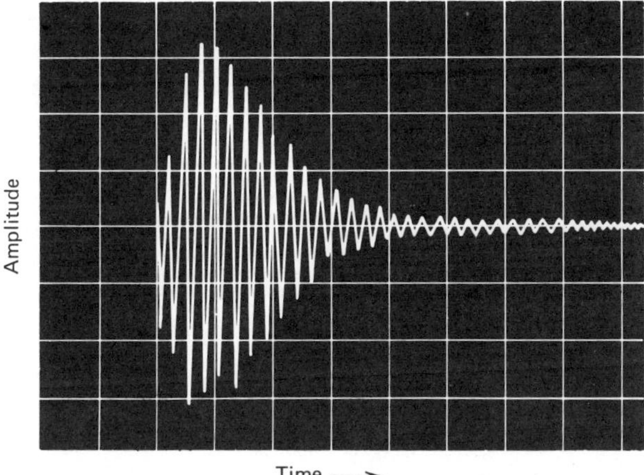

Figure 17*a* Decay sound pattern in a gear utilizing a high damping alloy. (Courtesy of General Radio Co.)

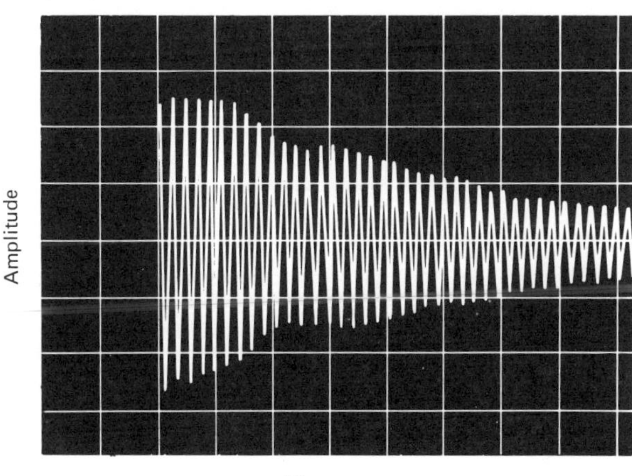

Figure 17*b* Decay sound pattern in a poor damping alloy. (Courtesy of General Radio Co.)

fast disappearing, and there is now a definite trend toward quieter air conditioners, garbage disposers, blenders, and automatic washers, for example. Nevertheless, consumers are not willing to pay a large premium for a quiet product, especially if the "duty cycle" or on-time is short. Because in some instances a high noise level will last for less than a minute, the noise is tolerable.

The noise spectrum of a vacuum cleaner will usually contain many harmonics,[20] as indicated in Fig. 18. This type of spectral analysis is prerequisite to identifying the source of noise and undertaking appropriate design changes. Almost all motor driven devices contain pure tone harmonics of varying amplitudes, some of which change with the age and condition of the product. Most appliances can be improved by one or several of the following: (1) reducing the amplitude of the structural resonance of sheet metal by damping, (2) using resilient gaskets, (3) surrounding the noise source by a sound attenuating housing, (4) using improved mufflers, and (5) improving motor design and manufacture. Substantial noise reduction was achieved in an electric clock by modifications to the clock case, modifications of the rotor and reduction gears, and the addition of some friction to the gear train to reduce backlash or rebound between impacting gear teeth.[21]

A list of home appliances from various American manufacturers is provided in Table 5, together with a range of dB(A) levels measured at a distance of 3 ft. Some refrigerators are very quiet; those with "frost-free" features and devices for making ice cubes may have a longer duty cycle and thus generate more noise than earlier models. The long duty cycles and high noise levels of many window-type air conditioners have made central air conditioning systems an attractive alternative.

NOISE CONTROL IN PROCESS PLANTS

Noise control in process plants poses some special problems. This is especially true in the large outdoor process plants of oil refineries where high

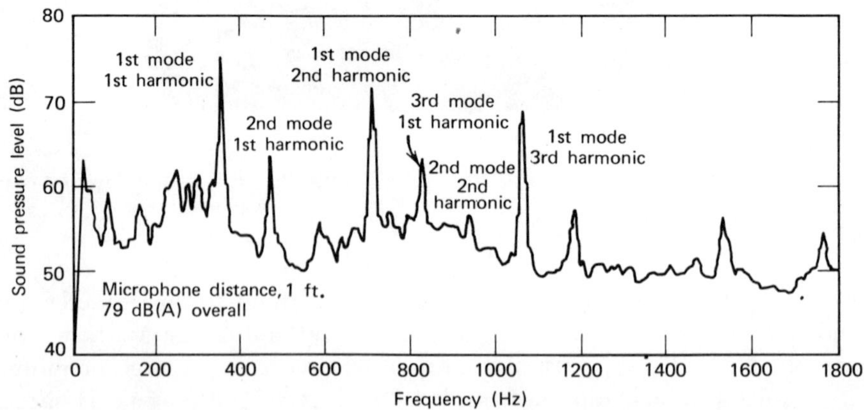

Figure 18   Noise spectrum of a vacuum cleaner. (Courtesy of R. S. Musa, Westinghouse Research Laboratories.)

Table 5   Range of Noise Levels for Appliances (at three feet)[22]

Appliance	Sound Level, dB(A)
Freezer	39–45
Refrigerator	34–52
Toothbrush, electric	48–55
Humidifier	41–51
Fan	37–69
Dehumidifier	52–62
Clothers dryer	51–66
Air conditioner	50–67
Shaver, electric	47–69
Hair dryer	59–65
Clothes washer	47–71
Dishwasher	54–72
Can opener, electric	54–76
Food mixer	53–79
Sewing machine	70–74
Vacuum cleaner	63–85
Food blender	62–88
Food disposer	67–74
Home shop tools	63–97
Lawn mower, electric	81–89

noise levels are generated by furnaces, heat exchangers, control valves, compressors, piping systems, and flares. Not only will the noise level vary at the site, but sound levels remote from a refinery will change because of variations in atmospheric conditions. Hence it has been suggested that "in the long run, distance may be the cheapest form of community noise control."[23]

Furnace or burner noise will contain[24]

1. Injector noise
2. Mixing tube (or system) resonances
3. Combustion (or flame) noise

On the basis of classical jet theory and experimental measurements the acoustic power P of gas injector noise from cylindrical orifices has the form:[25]

$$P = \frac{K\rho^2 d^2 V^8}{\rho_0 c^5} \tag{6}$$

where K is a sound power coefficient, ρ is the density in the jet flow, ρ_0 is the density of the atmosphere, d is the diameter of the injector nozzle, c is the velocity of sound in air, and V is the exit velocity of the gas. The exit velocity of the gas jet is seen to be an exceedingly important parameter. Resonances may occur in the mixing tubes or chamber, depending on burner design.

Combustion noise, per se, is not wholly understood. It is a function of the reactivity of the air and gas mixture, and a substantial source of combustion noise is attributed to turbulence in the air and gas mixture approaching the flame front. Thus any scheme that tends to suppress this turbulence (i.e., time-random pressure fluctuations) will tend to reduce the noise level.[26] Other benefits of a more stable combustion include reduced vibration and fatigue in mechanical components of the system, and a higher combustion efficiency.

For large process plant furnaces, combustion roar is suppressed from the firebox by acoustically treated air-intake plenums, such as shown in Fig. 19. The accompanying octave-band spectrum also shows the substantial low frequency noise reduction that is achieved by the use of such a plenum.[27]

Noise from heat exchangers is produced from high-powered fans, although this noise is usually less than furnace noise. Control valves generate most of the noise associated with piping systems, the noise level being related to the pressure reduction across the valve. Compressors and

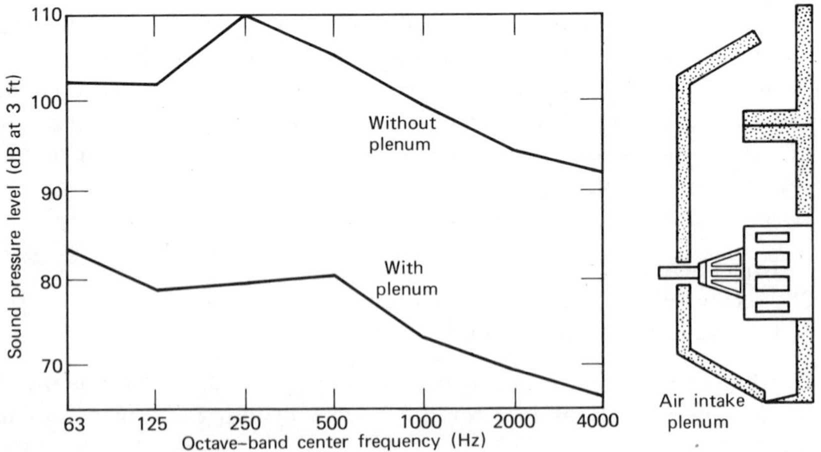

Figure 19 Reduction of furnace noise achieved by use of lined air intake baffles. (From J. G. Seebold, *Sound Vib.*, June 1973, p. 16.)

turbo-machinery are another important noise source; in oil refinery installations, these components have power ratings from 1000 to 30,000 HP operating at 4000 to 12,000 rpm.[28] In the near field of these units the noise level can exceed 100 dB(A).

Flares, which are used to burn excess gases, can be sources of community annoyance.[29] Flare stacks can reach heights from 100 to 250 ft so that noise is effectively broadcast. Steam is sometimes injected into the exhaust as it helps control the weight ratio of carbon to hydrogen in the combustion zone, and attenuates smoke. However, additional noise results. Some decrease in flare noise has been effected by the use of multiport steam nozzles.

POWER PLANT NOISE CONTROL

Coal, oil-fired, nuclear, and hydroelectric power plants supply the bulk of the nation's electrical power. An additional source of electrical power is the modern gas turbine plant, which borrows its technology from the jet engine industry. Because of its function, sheer size, and power, electrical generating apparatus is inherently noisy. The range of noise levels within a plant and in the immediate vicinity of typical power plant components, is given in Table 6.

In most new large power plants, the shielding afforded by the plant structure and its isolation from the community are adequate from a gen-

Table 6 Typical Power Plant Equipment and Noise Levels[30]

Equipment	Noise Level Range dB(A)
Coal pulverizers	86–100
Flue dust exhausters	85–103
Boiler feed pumps	85–100
Forced draft fans	85–110
Induced draft fans	77–97
Condenser rooms below turbine generators	83–100
Pressure-reducing stations	82–109
Turbine generators	83–100
Auxiliary exciters	87–93
DC generators	93–103
Demineralizers	85–101

eral environmental standpoint. In fact, extensive environmental impact studies, including those on noise, are mandatory prior to construction. Inside the power plant, however, operating and maintenance personnel must be afforded special protection. The apparatus is too large to be acoustically shielded, so that service personnel must be provided with individual ear protectors, or they are required to work in acoustically designed enclosures. Furthermore, the interiors of power plants, with highly reflective concrete walls and floors, have long reverberation times, and high noise levels exist throughout the main building areas.

The spectral characteristics of several important facilities have recently been reported;[31] taped noise recordings were made in the immediate zone of the various types of equipment, and subsequent analysis was done in the laboratory. Figure 20 is a 1/3 octave spectrum of noise from two types of coal pulverizers. One is a so-called "ball mill" type with contrarotating races, the other is a "bowl mill" coal crusher. Both have approximately the same production capacity, and both exhibit broad band noise with pure tone components. High frequency noise extends to 16 kHz. Figure 21 is a 1/3 octave spectrum of noise from an 800 MW turbine generator. A substantial amount of low frequency noise is apparent. The "A"-weighting overlay on the graph suggests that a simple A-weighted noise meter reading would provide no clue to its low frequency content.

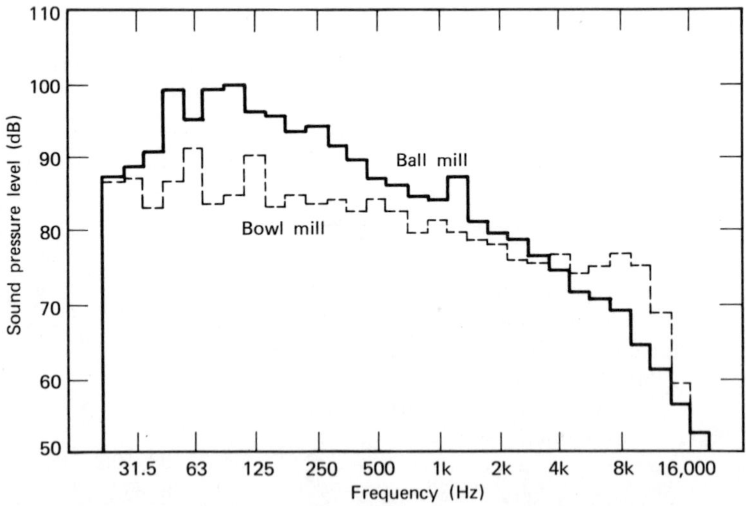

Figure 20   One-third octave spectra of noise from two types of coal mills. (Courtesy of Detroit Edison Co., Ref. 30.)

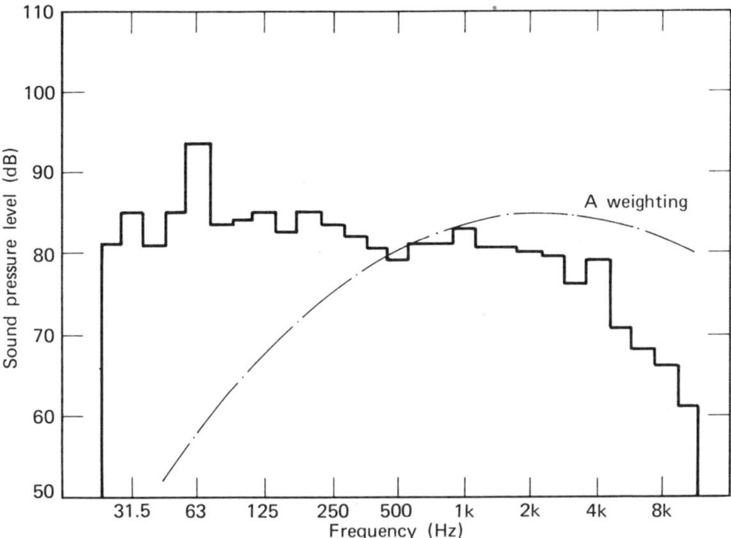

Figure 21   One-third octave spectrum from an 800 MW turbine generator. (Courtesy of Detroit Edison Co., Ref. 30.)

Figure 22 shows a 1/3 octave spectrum of both a turbine-driven and motor-driven boiler feed pump. In the motor-driven pump the pure tone component at 120 Hz is readily apparent. In some systems, spectra will also reflect pressure pulsations due to resonances in associated piping, so that both the pump and piping constitute the dual radiators of noise.

The gas turbine electric generating plant poses some special problems. Advantages of the gas turbine plant include relatively prompt installation and the flexibility of such facilities to provide power for peak loads. Constraints include high operating costs and noise. The latter factor, however, has been successfully resolved by the use of massive silencing equipment that is tailored so that a power plant can meet federal, state, and community standards.

Sources of noise in a complete turbine power plant include (1) inlet noise, (2) compressor noise, (3) exhaust noise, (4) piping noise, (5) broad band windage noise, (6) harmonic components from the electrical generator, (7) accessory noise, and (8) casing noise. Because the turbine inlet and exhaust are part of the gas path, total enclosure is impossible so that recourse is taken to specially designed silencers of the type shown in Figures 23a and 23b.[32] Figure 23a is a parallel baffle consisting of fibrous materials that can withstand operating temperatures and gas velocities.

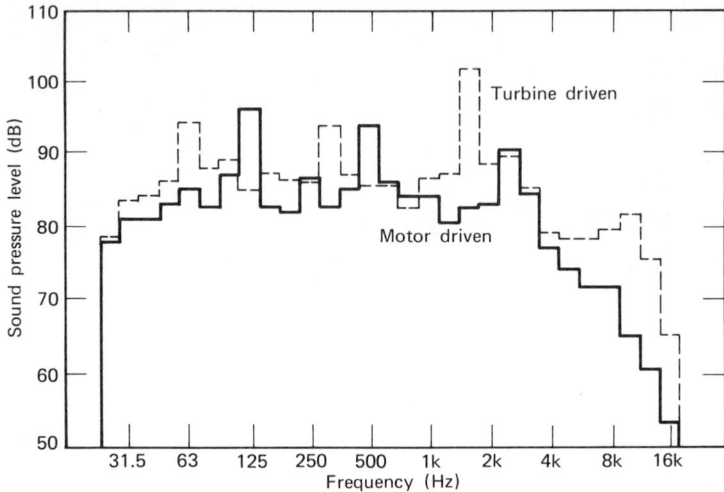

Figure 22   One-third octave spectrum of a turbine-driven and motor-driven feed pump. (Courtesy of Detroit Edison Co., Ref. 30.)

Figure 23*b* shows a "switchback" silencer so named because it uses a folded-path principle that permits exhaust gas to flow freely, but the noise encounters a series of discontinuities that attenuate sound propagation. With such type of silencers, the exceedingly intense noise from a gas turbine can be reduced to a D NEMA level (Fig. 15) at a distance of 400 ft. With additional baffles, walls, and surrounding building structure, a NEMA rating between B and C, ~45 dB(A) is obtained at the same distance. Without silencing such a turbine installation would be about 80 dB(A).

It is appropriate to mention another contributor to noise in the vicinity of some nuclear-powered plants: cooling tower noise. In an attempt to minimize the "thermal pollution" of lakes and rivers, some regulatory agencies have required that waste heat be transferred to the atmosphere. As a result, very large cooling towers have been erected in England as well as in the United States. Even in natural-draft evaporation towers a substantial amount of broad band noise is produced by the large mass of water falling back into the cooling pond. At the rim of cooling towers of 250 MW capacity,[33] noise levels of 80 dB(A) and greater have been measured; at 30 meters the sound level has been approximately 70 dB(A), and at 120 meters the sound level is roughly 60 dB(A). While noise levels around some power stations is dominated by such cooling tower noise,

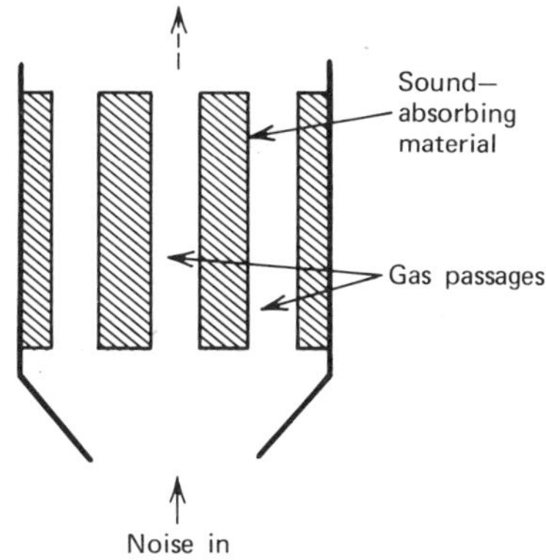

Figure 23*a* Schematic diagram of parallel baffle silencer for a large gas turbine. (From R. B. Tatge, *Sound Vib.*, June 1973, p. 23.)

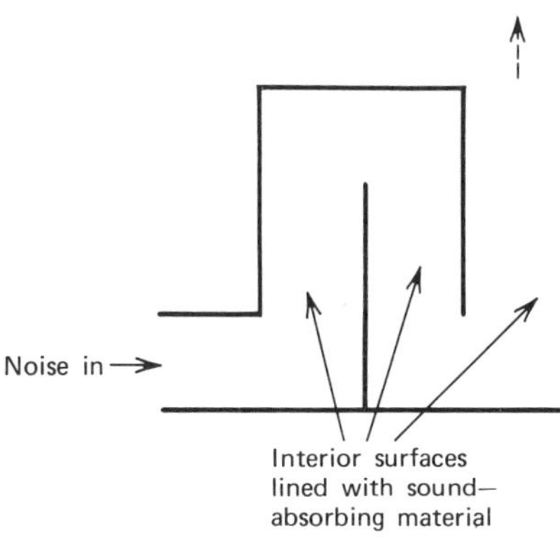

Figure 23*b* Schematic diagram of "switchback" type silencer for gas turbine exhaust. (From R. B. Tatge, *Sound Vib.*, June 1973, p. 23.)

communities are usually sufficiently remote from the site so as not to be annoyed. Also, problems of fogging from such towers are likely to be of greater environmental concern than noise.

NOISE CONTROL IN OFFICES

Noise control in offices is important in facilitating speech communication, and in enhancing freedom from fatigue and annoyance. Rarely do noise levels in offices attain those of industrial shops where consideration must be given to possible hearing damage. Accordingly, criteria have been established that permit a numerical rating to be assigned to offices and other comparable spaces, based on their acceptability for speech and general freedom from broad band noise. It is presumed that high level impact or pure tone noise will not be present, and that the overall noise level is generated from a number of noise sources (e.g., conversations, telephones, office equipment, air conditioning, and general office activities).

In otherwise quiet offices, noise from air-conditioning systems is objectionable; low noise levels from air conditioners require low air velocity flow from ceiling or wall diffusers. Careful acoustic treatment of interconnecting corridors is also as important as that of the separate offices when office doors remain open. Considerations of speech communication in "open office" arrangements are discussed in Chapter 14.

Noise Criterion (NC)

One rating scheme that has been widely used by architects and acousticians is termed an NC or *noise criterion set* of values. These values are published in graphical form as a set of curves, or in the tabular form of Table 7. An office has a noise criterion (NC) of 35 if none of the octave band levels for NC-35 is exceeded, and noise at one or more bands exceeds an NC-30 limit. Such a rating provides an alternative acoustical figure of merit of an office environment to that of a dB(A) number. Furthermore, octave-band measurements suggest the types of absorbing materials that may be required for noise reduction.

Table 7   Octave-Band Levels Associated with Noise Criteria (NC) Curves[34]

NC Value	63 Hz	125 Hz	250 Hz	500 Hz	1000 Hz	2000 Hz	4000 Hz	8000 Hz
NC-15	47	36	29	22	17	14	12	11
NC-20	51	40	33	26	22	19	17	16
NC-25	54	44	37	31	27	24	22	21
NC-30	57	48	41	35	31	29	28	27
NC-35	60	52	45	40	36	34	33	32
NC-40	64	56	50	45	41	39	38	37
NC-45	67	60	54	49	46	44	43	42
NC-50	71	64	58	54	51	49	48	47
NC-55	74	67	62	58	56	54	53	52
NC-60	77	71	67	63	61	59	58	57
NC-65	80	75	71	68	66	64	63	62

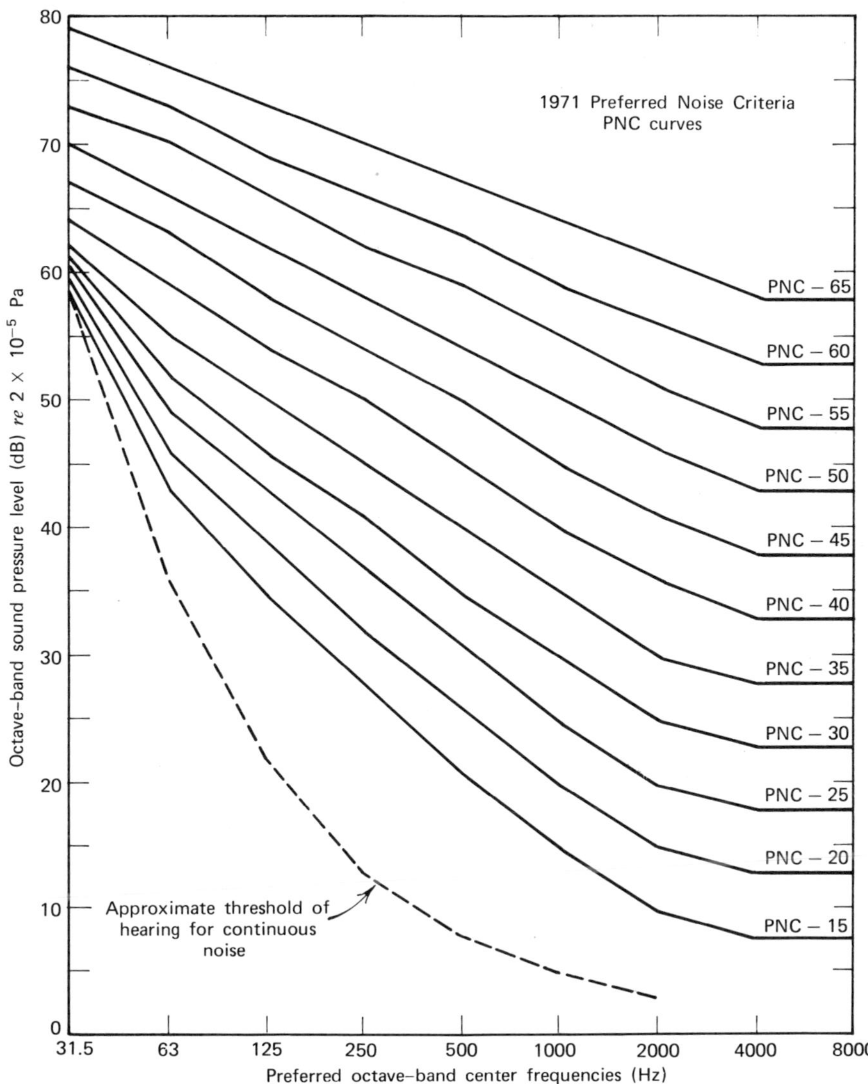

Figure 24   Preferred noise criterion curves proposed for rating offices and other interior building spaces. (From L. L. Beranek, *Noise and Vibration Control,* Copyright 1971 McGraw-Hill, used with permission of McGraw-Hill Book Co.)

Preferred Noise Criterion (PNC)

Recently, a new set of values has been proposed (although not formally accepted) termed preferred noise criterion (PNC) curves.[35] These curves are shown in Fig. 24. The new PNC numbers are about 1 dB lower than the NC values at the four octave bands 125, 250, 500, and 1000 Hz for the same rating numbers. In the 63 Hz band, and in the three highest bands, they are 4 or 5 dB lower. These modifications have been made in an attempt to improve on the subjective response to the NC contours that (when produced by accurately calibrated noise generators) were judged to be both "hissy" and "rumbly." A PNC-30 rating for an office environment is judged to be very satisfactory in many modern buildings.[36] [A very rough dB(A) equivalent to a PNC rating is to add 6 to the PNC number.]

REFERENCES

1. W. J. Jend, M.D., medical director, Michigan Bell Telephone Co., private communication.

2. *Conservation of Hearing in Industry,* American Academy of Opthalmology and Otolaryngology, 3819 Maple Avenue, Dallas, Texas, 75219.

3. E. K. Bender, *Sound & Vib.,* 33 (May 1973).

4. W. A. Meisen, *Sound & Vib.,* 24 (September 1973).

5. E. M. Diehl, *Machinery Acoustics,* Wiley, New York, 1973, p. 19.

6. L. F. Yerges, *Sound & Vib.,* 18 (August 1973).

7. *Acoustical Materials,* The Acoustical and Insulating Materials Assn., Park Ridge, Ill., 1972, p. 12.

8. *Ibid.*

9. V. O. Knudsen and C. M. Harris, *Acoustical Designing in Architecture,* Wiley, New York, 1950, p. 123.

10. *Ibid.,* p. 86.

11. *Ibid.,* p. 131.

12. U. S. Environmental Protection Agency, *Noise from Industrial Plants,* Washington, D. C., 1971, p. 207.

13. P. B. Ostergaard and R. Donley, *J. Acoust. Soc. Am.,* **36,** 409 (1964).

14. M. W. Schulz Jr. and R. J. Ringlee, *Power Apparatus and Systems,* 1 (June 1960).

15. D. B. Hanson, *J. Acoust. Soc. Am.,* **54,** 1571 (1973).

16. R. J. Wells and R. D. Madison, *Fan Noise,* in *Handbook of Noise Control,* edited by C. M. Harris, McGraw-Hill, New York, **25,** 12 (1957).

17. Diehl, *op. cit.*, p. 126.

18. A. S. Babkin and J. L. Anderson, *Sound & Vib.*, 35 (April 1973).

19. R. C. Miles, *Sound & Vib.*, 27 (July 1973).

20. M. C. C. Tsao and R. S. Musa, *Sound & Vib.*, 31 (March 1973).

21. W. M. Viebrack and M. J. Crocker, *Sound & Vib.*, 22 (March 1973).

22. U. S. Environmental Protection Agency, *Noise from Construction Equipment and Operations, Building Equipment and Home Appliances,* Washington, D C. 1971, p. 31.

23. J. G. Seebold, *Proceedings of the Purdue Noise Control Conference,* edited by M. J. Crocker, 173 (July 1971).

24. J. P. Roberts, *Appl. Acoust.*, 4, 103 (1971).

25. H. E. von Gierke, in *Handbook of Noise Control,* edited by C. M. Harris, McGraw-Hill, 33, 32 (1957).

26. J. S. Arnold, *J. Acoust. Soc. Am.*, 52, 5 (1972).

27. J. G. Seebold, *Sound & Vib.*, 16 (June 1973).

28. *Ibid.*

29. *Ibid.*

30. R. A. Popeck, *Sound & Vib.*, 28 (June 1974).

31. *Ibid.*

32. R. B. Tatge, *Sound & Vib.*, 23 (June 1973).

33. R. M. Ellis, *J. Sound & Vib.*, 14, 171 (1971).

34. L. L. Beranek, *Noise and Vibration Control,* McGraw-Hill, New York, 1971, p. 566.

35. *Ibid.*, p. 567.

36. L. L. Beranek, W. E. Blazier, and J. J. Figiver, *J. Acoust. Soc. Am.*, 50, 1223 (1971).

12

LEGAL ASPECTS OF NOISE ABATEMENT

NATIONAL ENVIRONMENTAL POLICY ACT

The National Environmental Policy Act* of 1969 (NEPA) has been termed "the most important and far-reaching environmental and conservation measure ever enacted by Congress." The act defined general policies and goals, and it paved the way for the implementation of subsequent measures by various federal agencies and state legislatures. Specifically outlined in the act is a requirement that all agencies of the federal government must include, in every recommendation for legislation or action "affecting the quality of the human environment," statements relating to the following:

1. The environmental impact of the proposed action.
2. Any adverse environmental effects that cannot be avoided should the proposed action be implemented.
3. Alternatives to the proposed action.
4. The relationship between local short term uses of man's environment and the maintenance and enhancement of long term productivity.
5. Any reversible and irretrievable commitments of resources that would be involved in the proposed action should it be implemented.

* 42 USC SS4321-4347.

304

The federal government is further obliged to use "all practicable means" to implement courses of action that include "appropriate consideration" to general environmental values, in addition to economic and technological considerations.

Thus, the Act was a broad declaration of a new national policy; it led to the establishment of a Council on Environmental Quality, and subsequently to the creation of the EPA.

ENVIRONMENTAL PROTECTION AGENCY

Under Title IV of Public Law 91-604, signed on December 31, 1970 by the President, the Environmental Protection Agency (EPA) was directed to conduct "a full and complete investigation and study of noise and its effect on public health and welfare." Furthermore, under specific provisions of the Noise Control Act of 1972, "all Federal agencies are required to consult with the EPA administrator in prescribing standards or regulations respecting noise." . . . "and all Federal Agencies shall, to the fullest extent consistent with their authority under Federal laws administered by them, carry out the programs within their control so as to . . . promote an environment for all Americans free from noise that jeopardizes their health or welfare." Today, the EPA, through its Office of Noise Abatement and Control, is the major agency for developing noise standards and advising on all aspects of federal noise control legislation. Nevertheless, many other agencies have provided technical and legal input into noise abatement programs over an extended period of time. Outlined below is only a partial list of federal agencies and their noise control concerns.[1]

Agency	Program
Department of Agriculture	Research on noise reduction from vegetative screens, etc. Clark-McNary Act of 1942, McSweeny-McNary Forest Act of 1928, and Agricultural Experimental (Smith-Lever) Act of 1955.
Department of Commerce	Noise standards; materials research and subjective studies are undertaken by the National Bureau of Standards.
Department of Defense U.S. Air Force	Noise abatement in propulsion systems, and bioacoustics.

(Continued)

Agency	Program
U.S. Army	Hearing conservation and occupational noise control. Biomedical effects of noise, auditory perception and psychoacoustics, and siting of military activities.
U.S. Navy	Noise reduction of operating and testing aircraft.
Department of Health, Education and Welfare	Occupational health, industrial safety, diagnosis of hearing loss, and training.
Department of Housing and Urban Development	Housing site selection, structural characteristics of buildings, and noise emission ratings for appliances and equipment.
Department of Interior	Bureau of Mines training program and study of sonic booms.
Department of Labor	Walsh-Healey Contracts Act and Occupational Safety and Health Act (OSHA).
Department of Transportation	Transportation noise, including aircraft (PL. 89-670; Section 4; PL. 91-605).
National Aeronautics and Space Administration	Aircraft noise research, audiometric testing programs; The Aircraft and Noise Reduction Laboratory (Hampton, Va.).

Specific legislative control measures, administered by various agencies are outlined in subsequent sections.

FEDERAL AVIATION ADMINISTRATION

The FAA* possesses broad regulatory powers with respect to aircraft and airport noise. It includes initial certification of new aircraft, operational procedures and flight patterns, control over ground servicing, and the retrofitting of new engines on existing aircraft; it also determines the manner in which specific noise tests shall be conducted and evaluated.

This authority was granted, in part, by Public Law 90-411 that added Section 1431 (1969) to the FAA Act of 1958:[2]

§ 1431. Control and abatement of aircraft noise and sonic boom—consultations; standards; rules and regulations.

(a) In order to afford present and future relief and protection to the public from unnecessary aircraft noise and sonic boom, the Administrator of

* An agency of the Department of Transportation.

the Federal Aviation Administration, after consultation with the Secretary of Transportation, shall prescribe and amend standards for the measurement of aircraft noise and sonic boom and shall prescribe and amend such rules and regulations as he may find necessary to provide for the control and abatement of aircraft noise and sonic boom, including the application of such standards, rules, and regulations in the issuance, amendment, modification, suspension, or revocation of any certificate authorized by this subchapter.

Considerations determinative of standards, rules, and regulations.

(b) In prescribing and amending standards, rules, and regulations under this section, the Administrator shall—

(1) consider relevant available data relating to aircraft noise and sonic boom, including the results of research, development, testing, and evaluation activities conducted pursuant to this chapter and chapter 23 of this title;

(2) consult with such Federal, State, and interstate agencies as he deems appropriate;

(3) consider whether any proposed standard, rule, or regulation is consistent with the highest degree of safety in air commerce or air transportation in the public interest;

(4) consider whether any proposed standard, rule or regulation is economically reasonable, technologically practicable, and appropriate for the particular type of aircraft, aircraft engine, appliance, or certificate to which it will apply; and

(5) consider the extent to which such standard, rule, or regulation will contribute to carrying out the proposes of this section.

Amendment, modifications, suspension, or revocation of certificate; notice and appeal rights

(c) In any action to amend, modify, suspend, or revoke a certificate in which violation of aircraft noise or sonic boom standards, rules, or regulations is at issue, the certificate holder shall have the same notice and appeal rights as are contained in section 1429 of this title, and in any appeal to the National Transportation Safety Board, the Board may amend, modify or reverse the order of the Administrator if it finds that control or abatement of aircraft noise or sonic boom and the public interest do not require the affirmation of such order, or that such order is not consistent with safety in air commerce or air transportation.

Under the above broad powers, and subsequent amendments, the FAA now prescribes noise limits for all new commercial aircraft. However, because some of today's operational aircraft were designed and operating many years prior to 1969, the only means for achieving a lower operating noise level is by retrofitting them with new engines. The cost of retrofitting the existing United States jet fleet is estimated at between one and two billion dollars,[3] so there is some question of such a program being "economically reasonable."

Noise certification measurements have been specified in some detail. A series of measurements must be made for aircraft according to the diagram of Fig. 16 (Chapter 10) and upper noise limits for takeoff are indicated in Fig. 17 (Chapter 10). In these measurements a 0.25 nautical mile distance (sideline) is specified for aircraft of less than four engines, while a 0.35 nautical mile distance is specified for four or more engines. Other factors such as glide path of the plane are controlled in the aircraft certification tests.

The regulations are written in terms of "effective perceived noise level," EPNL or EPNdB rather than dB (A). This noise level is to be normalized to sea level pressure, 77°F, 70% relative humidity, and 0 mph wind velocity.

The FAA has imposed a ban on supersonic flights of civilian aircraft over the United States or its territorial waters (effective April 27, 1973). This action, however, will have little effect on foreign airlines operating SST planes on Trans-Atlantic and Pacific routes as they will be programmed to fly subsonic on their approach. A written permit must be obtained from the FAA if a sonic boom is to result from a flight operation and then only for research and development.

Further proposed FAA legislation would require the modification or phaseout of many existing aircraft. This proposal would apply to all jets weighing 75,000 pounds or more, including those owned by corporations and other general aviation operators, and would become effective July 1, 1978. Among the aircraft affected would be the Boeing 707, 727, 737, and the McDonnell Douglas DC-8 and DC-9. These aircraft would have to comply with the noise standards specified in Part 36 of the Federal Aviation Regulations that went into effect in 1969 and that have been applied to the certification of new aircraft types such as the DC-10, L1011, Boeing 747, F-28, Cessna Citation, and the Dessault Falcon-10.

AIRPORT AND AIRWAYS DEVELOPMENT ACT

The Airport and Airways Development Act of 1970[4] supplements the authority of the FAA and Public Law 90-411. This Development Act grants further authority for noise control at both the federal and state levels. Specifically, the Secretary of Transportation is required to undertake a *National Airport System Plan*

    —to determine the extent, type, nature, location, and timing of airport development needed in a specific area to establish a viable and balanced system of public airports. It includes identification of the specific aeronautical role of each airport within the system, development of estimates of system-wide costs, and the conduct of such studies, surveys, and other planning actions as may be necessary to determine the short, intermediate, and long range aeronautical demands required to be met by a particular system of airports.

The Secretary also is directed to consult with the Council on Environmental Quality and other Department Secretaries with respect to the preservation of environmental quality. Consideration must also be given to the local communities affected by site location and airport layout. In the selection of a new airport the local communities in the environs have influence on the approval of the Secretary for a grant application.

An additional salient aspect of this Act is a provision that requires approval from the governor of the state in which a new airport is to be located. Thus any statutes or regulations relating to noise control within a state would be factored into land use and new airport designs.

FEDERAL-AID HIGHWAY ACT

Through the Federal-Aid Highway Act of 1970 (Public Law 91-605), the Secretary of Transportation, was authorized to "develop and promulgate standards for highway noise levels compatible with different land uses and after July 1, 1972, shall not approve plans and specifications for any proposed project on any Federal-aid system for which location approval has not yet been secured unless he determines that such plans and specifications include adequate measures to implement the appropriate noise level standards." Subsequent to the passage of this legislation the Federal Highway Administration adopted Noise Standards and Procedures.[5] These standards require (1) an assessment of noise effects in all federal highway projects, (2) the implementation of appropriate noise abatement measures during road construction, and (3) cooperation with local officials regarding future land use.

WALSH-HEALEY PUBLIC CONTRACTS ACT

The Bureau of Labor Standards, U.S. Department of Labor, in May 1969, added several regulations to its existing Walsh-Healey Public Contracts Act.[6] This act requires that contracts entered into by any agency of the United States for the manufacture or furnishing of materials, supplies, articles, and equipment in any amount exceeding $10,000 must contain, among other provisions, a stipulation that "no part of such contract will be performed nor will any of the materials, supplies, articles, or equipment to be manufactured or furnished under said contract be manufactured or fabricated in any plants, factories, buildings, or surroundings or under working conditions which are unsanitary or hazardous or dangerous to the health and safety of employees engaged in the performance of said contract."

A specific section 50-204.10 relates to the protection of employees against noise that is deemed hazardous.

> Occupational Noise Exposure; (a) Protection against the effects of noise exposure shall be provided when the sound levels exceed those shown in Table 1 of this section when measured on the A scale of a standard sound level meter at slow response. When noise levels are determined by octave band analysis, the equivalent A-weighted sound level may be determined from Figure 1, (b) When employees are subjected to sound exceeding those

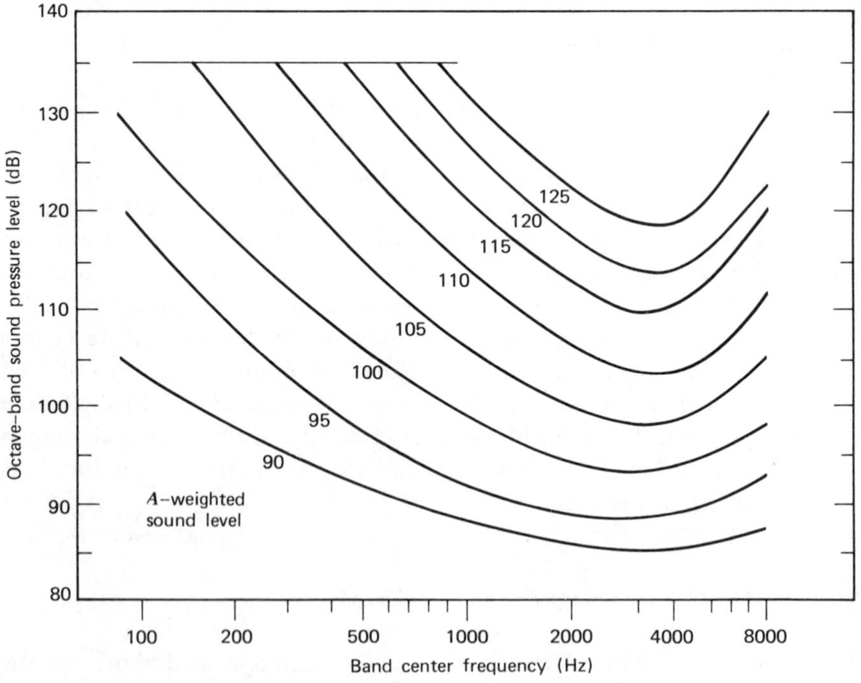

Figure 1   Graphical representation of the Walsh-Healey Act noise limits, when noise measurements are made in octave bands having center frequencies of 100, 200, 500, 1000, 2000, 4000, and 8000 Hz.

listed in Table 1 of this section, feasible administrative or engineering controls shall be utilized. If such controls fail to reduce sound levels within the levels of the table, personal protective equipment shall be provided and used to reduce sound levels within the levels of the table, (c) If the variations in noise level involve maxima at intervals of 1 second or less, it is to be considered continuous, (d) In all cases where the sound levels exceed the values shown herein, a continuing, effective hearing conservation program shall be administered.

Exposure to impulsive or impact noise should not exceed 140 dB peak sound pressure level. When the daily noise exposure is composed of two or more periods of noise exposure of different levels, their combined effect should be considered, rather than the individual effect of each. If the sum of the following fractions: $C_1/T_1 + C_2/T_2 \ldots C_n/T_n$ exceeds unity, then the mixed exposure should be considered to exceed the limit value. C_n indicates the total time of exposure at a specified noise level, and T_n indicates the total time of exposure permitted at that level.

Table 1   Permissible Noise Exposures

Duration per Day (hours)	Sound Level dB(A), (slow response)
8	90
6	92
4	95
3	97
2	100
$1\frac{1}{2}$	102
1	105
$\frac{1}{2}$	110
$\frac{1}{4}$ or less	115

It should be noted that the Walsh-Healey Act specifies that feasible *engineering* or *administrative* controls must be applied to protect the hearing of employees. The meaning of engineering controls is that the noise level at the work area of each employee must be controlled to achieve a noise exposure rating of unity or less. Possible procedures include quieter tools, machinery, fabricating processes, enclosures, or use of sound-absorbent materials in the work zone. The meaning of administrative controls is that hazardous noise is controlled by limiting the exposure time of workers by rotation of personnel or other administrative procedures. If these procedures are not feasible, hearing protection devices are mandatory.

Even though the noise level in the regulation is complied with, a corporation will still be held responsible if an appreciable number of workers develop a substantial hearing loss. Thus the clear intent of the regulation is to promote a hearing conservation program for employees regardless of the specific noise exposure levels. If the contractor fails to comply with the Act, the regional administrator may recommend to all government agencies that the contractor be prohibited from bidding on future government contracts for a period of three consecutive years.

OCCUPATIONAL SAFETY AND HEALTH ACT

The Occupational Safety and Health Act of 1970 (OSHA) broadened the authority of the Department of Labor "by authorizing the Secretary of Labor to set mandatory occupational safety and health standards applicable to businesses affecting interstate commerce," and by "creating an Occupational Safety and Health Review Commission for carrying out adjudicatory functions under the Act." By this Act the Walsh-Healey standards were made applicable to all businesses affecting interstate commerce, rather than being limited solely to contractor-government situations. Penalties to insure compliance with noise and other standards are also specified. Civil fines up to $10,000 can be imposed for willful and repeated violations, and criminal penalties can accompany flagrant disregard for the statute.[7]

The Act also contains a provision whereby an individual State may assume primary regulatory responsibility for occupational health and safety.

Section 18

(b) Any State which, at any time, desires to assume responsibility for development and enforcement therein of occupational safety and health standards relating to any occupational safety or health issue with respect to which a Federal standard has been promulgated under section 6 shall submit a State plan for the development of such standards and their enforcement.

(c) The Secretary shall approve the plan submitted by a State under subsection (b), or any modification thereof, if such plan in his judgment—

(1) designates a State agency or agencies as the agency or agencies responsible for administering the plan throughout the State,

(2) provides for the development and enforcement of safety and health standards relating to one or more safety or health issues, which standards (and the enforcement of which standards) are or will be at least as effective in providing safe and healthful employment and places of employment as the standards promulgated under section 6 which relate to the same issues, and which standards, when applicable to products which are distributed or used in interstate commerce, are required by compelling local conditions and do not unduly burden interstate commerce,

(3) provides for a right of entry and inspection of all work-places subject to the Act which is at least as effective as that provided in section 8, and includes a prohibition on advance notice of inspections,

(4) contains satisfactory assurances that such agency or agencies have or will have the legal authority and qualified personnel necessary for the enforcement of such standards,

(5) gives satisfactory assurances that such State will devote adequate funds to the administration and enforcement of such standards,

(6) contains satisfactory assurances that such State will, to the extent permitted by its law, establish and maintain an effective and comprehensive occupational safety and health program applicable to all employees of public agencies of the State and its political subdivisions, which program is as effective as the standards contained in an approved plan.

(7) requires employers in the State to make reports to the Secretary in the same manner and to the same extent as if the plan were not in effect, and

(8) provides that the State agency will make such reports to the Secretary (of Labor) in such form and containing such information, as the Secretary shall from time to time require.

THE CLEAN AIR ACT

The Clean Air Act was amended in 1970 "to provide a more effective program to improve the quality of the Nation's air." The specific bill was Public Law 91-604 dated December 31, 1970. While the Act focused primarily on the control and removal of combustion byproducts of fuel, it is noteworthy that the term "hazardous air pollutant" was employed in a broader context than heretofore. In a section titled "Development of Low-Emission Vehicles," (e.g., automobiles and trucks), a Certification Board was charged with considering the following criteria:

1. The safety of the vehicle.
2. Its performance characteristics.

3. Its reliability potential.
4. Its serviceability.
5. Its fuel availability.
6. *Its noise level.*
7. Its maintenance costs.

A specific section of the Act also deals with the general problem of environmental noise:

TITLE IV—NOISE POLLUTION

Sec. 401.   This title may be cited as the Noise Pollution and Abatement Act of 1970.

Sec. 402.   (a) The Administrator shall establish within the Environmental Protection Agency an Office of Noise Abatement and Control, and shall carry out through such Office a full and complete investigation and study of noise and its effect on the public health and welfare in order to (1) identify and classify causes and sources of noise, and (2) determine—

    (i) effects at various levels;

    (ii) projected growth of noise levels in urban areas through the year 2000;

    (iii) the psychological and physiological effect on humans;

    (iv) effects of sporadic extreme noise (such as jet noise near airports) as compared with constant noise;

    (v) effect on wildlife and property (including values);

    (vi) effect of sonic booms on property (including values); and

    (vii) such other matters as may be of interest in the public welfare.

(b) In conducting such investigation, the Administrator shall hold public hearings, conduct research, experiments, demonstrations, and studies. The Administrator shall report the results of such investigation and study, together with his recommendations for legislation or other action, to the President and the Congress not later than one year after the date of enactment of this title.

(c) In any case where any Federal department or agency is carrying out or sponsoring any activity resulting in noise which the Administrator determines amounts to a public nuisance or is otherwise objectionable, such department or agency shall consult with the Administrator to determine possible means of abating such noise.

NOISE CONTROL ACT OF 1972

The Noise Control Act of 1972 (Public Law 92-574) was passed by Congress on October 18, and signed by the President on October 28, 1972. This legislation designated the Environmental Protection Agency with the primary responsibility for controlling environmental noise. The complete text is contained in the *Congressional Record*—Senate S18638, October 18, 1972. Some of the salient features of this legislation are outlined below.

1. It establishes a United States policy to eliminate from our environment noise sources that would jeopardize health and welfare, and implements the policy by establishing noise emission standards for products and by providing the public with appropriate information on these products.

2. It directs federal agencies to administer noise abatement programs commensurate with their authorities, and requires the Environmental Protection Agency to coordinate all federal programs relating to noise research and control. EPA is also required to report periodically on the status and progress of such activities.

3. It requires EPA to publish identifiable effects on the public health and welfare resulting from different quantities and qualities of noise.

4. EPA is granted authority to prescribe and amend standards relating to noise emissions from new products distributed in interstate commerce, including construction equipment, transportation equipment (including recreational vehicles), motors, and electrical or electronic products. The agency is directed to take into account both technological and economic factors in arriving at standards. Furthermore, states and other local governments are prohibited from establishing standards different from those of the EPA, although they may independently regulate the use or operation of certain products.

5. Authority of the FAA to deal with aircraft noise is maintained, but EPA may submit to FAA for consideration new proposals deemed necessary to protect the public health and welfare.

6. Manufacturer's labeling of products (with respect to noise or noise reduction) must conform to the EPA code. Manufacturers or importers of nonconforming or mislabeled products are subject to fines and imprisonment. The U.S. District Courts have jurisdiction over violations of the act.

7. The EPA is also authorized to conduct research on physiological and psychological effects of sound on humans, animals and wildlife, and

the effect of noise on property values. The agency must also assume limited responsibility for assisting state and local governments on environmental monitoring and the training of personnel.

8. In collaboration with the Department of Transportation, EPA is required to promulgate regulations for carriers engaged in interstate commerce (railroad and motor vehicles).

Other portions of the act relate to (1) the federal procurement of "low-noise emission products," (2) a provision for citizen suits for violations of the act, (3) provisions for judicial review, and (4) administrative matters. The concept of "health and welfare" is given special emphasis in the overall intent of this legislation.

STATE STATUTES

Since the individual states have very broad legislative powers, it is to be expected that noise control measures would be included in the context of general legislation, or through special environmental control laws. The California Environmental Quality Act of 1970 was patterned after the National Environmental Policy Act of 1969 (NEPA). Connecticut, Indiana, Massachusetts, New Mexico, Puerto Rico, Washington, and Wisconsin also have "purpose" and "policy" provisions reflecting NEPA; hence, state statutes are sometimes called "little NEPAs."[8] Furthermore, a comprehensive State Noise Control Act has recently been drafted to give the states the maximum authority to regulate noise pollution consistent with the preemptive provisions of the Federal Noise Control Act of 1972. This proposed legislation is patterned closely after federal objectives on general environmental noise, product noise and labeling requirements, setting of ambient standards, and noise insulation standards for new construction. The U.S. Department of Commerce is also currently drafting a Model State Building Code. Table 2 indicates the scope of existing legislation* that varies widely both with respect to the degree of control and potential penalties for nonconformance.

Several states have prescribed, in detail, "zoning" measures that are usually in the province of municipal law. New Jersey, for example, has adopted noise control regulations relative to industrial and commercial operations. When measured at any residential property line, continuous airborne sound shall not exceed 65 dB(A) from 7 a.m. to 10 p.m.; from 10 p.m. to 7 a.m the sound level (starting January 1, 1976) shall not exceed 50 dB(A) Table 3 gives the alternative limits of this regulation in terms of octave bands.

* Effective 1973; many new proposed measures are pending.

Table 2 State Noise Statutes

	Aircraft	Automobile-Truck	Boats, Motorcycles, Snowmobiles	Commercial	Construction	Occupational	Disturbance of the Peace	General
Alabama		x						
Alaska							x	
Arizona		x					x	
Arkansas		x						
California	x	x	M		x	x	x	x
Colorado		x	M/S	x	x	x		x
Connecticut		x		x			x	
Delaware		x		x		x		
Florida		x				x		x
Georgia		x						
Hawaii			M			x		x
Idaho		x				x		
Illinois		x	B				x	x
Indiana		x		x		x		
Iowa		x					x	
Kansas		x	B	x		x	x	
Kentucky		x				x	x	
Louisiana		x				x	x	
Maine		x						
Maryland		x						
Massachusetts		x						
Michigan		x	M			x		x
Minnesota	x	x	M					x
Mississippi		x		x		x	x	
Missouri		x				x	x	
Montana		x	S					
Nebraska		x	B				x	
Nevada			M	x				
New Hampshire		x				x		
New Jersey		x		x		x		
New Mexico		x				x		
New York		x				x		
North Carolina		x						
North Dakota		x						
Ohio		x					x	
Oklahoma		x					x	
Oregon		x				x		

(*Continued*)

Table 2   State Noise Statutes (continued)

	Aircraft	Automobile-Truck	Boats, Motorcycles, Snowmobiles	Commercial	Construction	Occupational	Disturbance of the Peace	General
Pennsylvania		x				x	x	x
Rhode Island		x					x	
South Carolina		x						
South Dakota		x						
Tennessee		x				x	x	
Texas		x		x			x	
Utah		x				x	x	
Vermont							x	
Virginia		x				x		
Washington		x				x	x	
West Virginia		x				x		
Wisconsin		x	B/S			x	x	
Wyoming		x					x	

The manner in which the various states administer legislation relating to noise differs widely. Typical administrative groups include the State Department of Health, the Department of Transportation, the State Environmental Protection Agency, and the Department of Environmental Conservation. In California, the Department of Aeronautics was granted authority to set noise standards for the state's airports.[9]

Motor vehicle regulations in most of the nation's states are enforced by a State Highway Patrol, the Commissioner of Motor Vehicles, or similar authority.

REGULATION OF VEHICULAR NOISE

Regulations concerning noise from motor vehicles have been established by many states and a reasonable number of municipalities. Eventually, legislation can be expected for all states and most major cities.

California was the first among the states to pass a comprehensive vehicle noise ordinance, and the city of Chicago amended its municipal code to specify noise limits for the 1970-1980 decade. Because these two codes are typical of noise abatement ordinances, selected sections from these two statutes are reproduced or paraphrased below.

Table 3   New Jersey Industrial Noise Limits at Residential Property Lines[a]

7:00 a.m. to 10:00 p.m.	
Octave-Band Center Frequency (Hz)	Octave-Band Sound Pressure Level (dB)
31.5	96
63	82
125	74
250	67
500	63
1000	60
2000	57
4000	55
8000	53

10:00 p.m. to 7:00 a.m.	
31.5	86
63	71
125	61
250	53
500	48
1000	45
2000	42
4000	40
8000	38

[a](Effective 1976)

State of California (Vehicle Code 27160)

No person shall sell or offer for sale a new motor vehicle that produces a maximum noise exceeding the following noise limit at a distance of 50 ft from the centerline of travel under test procedures established by the department.

(1) Any motorcycle manufactured before January 1, 1970........92 dB(A)
(2) Any motorcycle, other than a motor-driven cycle, manufactured on or after January 1, 1970, and before January 1, 1973......88 dB(A)
(3) Any motorcycle, other than a motor-driven cycle, manufactured on or after January 1, 1973...............................86 dB(A)

(4) Any motor vehicle with a gross vehicle weight rating of 6000 pounds or more manufactured on or after January 1, 1968, and before January 1, 1973.................................88 dB(A)

(5) Any motor vehicle with a gross vehicle rating of 6000 pounds or more manufactured on or after January 1, 1973...86 dB(A)

(6) Any other motor vehicle manufactured on or after January 1, 1968, and before January 1, 1973............................86 dB(A)

(7) Any other motor vehicle manufactured after January 1, 1973..84 dB(A)

A subsequent amendment relating noise level and vehicle speed was enacted in August 1970 (Section 23130):

No person shall operate either a motor vehicle or combination of vehicles of a type subject to registration at any time or under any condition of grade, load, acceleration or deceleration in such a manner as to exceed the following noise limit for the category of motor vehicle based on a distance of 50 ft from the center of the lane of travel within the speed limits specified in this section:

	Speed Limit of 35 mph or less	Speed Limit of more than 35 mph
(1) Any motor vehicle with a manufacturer's gross vehicle weight rating of 6000 pounds or more, and any combination of vehicles towed by such motor vehicle:		
(a) Before January 1, 1973	88 dB(A)	90 dB(A)
(b) On and after January 1, 1973	86 dB(A)	90 dB(A)
(2) Any motorcycle other than a motor-driven cycle	82 dB(A)	86 dB(A)
(3) Any other motor vehicle and any combination of vehicles towed by such motor vehicles	76 dB(A)	82 dB(A)

City of Chicago Municipal Code (Section 17-4.7)

(a) It shall be unlawful for any person to operate any motor vehicle of a weight in excess of four tons (8000 lbs) for a consecutive period longer than two minutes while such vehicle is standing on private property and located within 150 ft of property zoned and used for residential purposes except where such vehicle is standing within a completely enclosed structure.

This section shall not apply to buses operated for the transportation of passengers while standing in established bus turnarounds, bus terminals, bus parking lots, and bus-storage yards.

(b) No person shall sell, or offer for sale a new motor vehicle that produces a maximum noise exceeding the following noise limit at a distance of 50 ft from the center line of travel under test procedures established by Section 17–4.24 of this chapter:

Type of Vehicle	Date of Manufacture	Noise Limit
(1) Motorcycle	before 1 Jan. 1970	92 dB(A)
Same	after 1 Jan. 1970	88 dB(A)
Same	after 1 Jan. 1973	86 dB(A)
Same	after 1 Jan. 1975	84 dB(A)
Same	after 1 Jan. 1980	75 dB(A)
(2) Any motor vehicle with a gross vehicle weight of 8000 pounds or more	after 1 Jan. 1968	88 dB(A)
Same	after 1 Jan. 1973	86 dB(A)
Same	after 1 Jan. 1975	84 dB(A)
Same	after 1 Jan. 1980	75 dB(A)
(3) Passenger cars, motor-driven cycle and any other motor vehicle	before 1 Jan. 1973	86 dB(A)
Same	after 1 Jan. 1973	84 dB(A)
Same	after 1 Jan. 1975	80 dB(A)
Same	after 1 Jan. 1980	75 dB(A)

The manufacturer, distributor, importer, or designated agent shall certify in writing to the Commissioner that his vehicles sold within the City comply with the provisions of this section.

(c) No person shall operate within the speed limits specified in this section either a motor vehicle or combination of vehicles of a type subject to registration at any time or under any condition of grade, load, acceleration, or deceleration in such manner as to exceed the following noise limit for the category of motor vehicle, based on a distance of not less than 50 ft from the center line of travel.

	Noise Limit in Relation To Posted Speed Limit	
	35 mph or less	Over 35 mph
Type of Vehicle		
(1) Any motor vehicle with a manufacturer's GVW rating of 8,000 lbs, or more, and combination of vehicles towed by such motor vehicle		
before 1 Jan. 1973	88 dB(A)	90 dB(A)
after 1 Jan. 1973	86 dB(A)	90 dB(A)
(2) Any motorcycle other than a motor-driven cycle		
before 1 Jan. 1973	82 dB(A)	86 dB(A)
after 1 Jan. 1973	78 dB(A)	82 dB(A)
(3) Any other motor vehicle and any combination of motor vehicles towed by such motor vehicle		
after 1 Jan. 1970	76 dB(A)	82 dB(A)
after 1 Jan. 1973	70 dB(A)	79 dB(A)

This section applies to the total noise from a vehicle or combination of vehicles and shall not be construed as limited or precluding the enforcement of any other provisions of this code relating to motor vehicle mufflers for noise control.

(d) No person shall modify or change the exhaust muffler, intake muffler, or any other noise abatement device of a motor vehicle in a manner such that the noise emitted by the motor vehicle is increased above that emitted by the vehicle as originally manufactured.

REGULATION OF PUBLIC UTILITIES

Until the present decade, environmental statutes relating to corporations engaged in the generation of electrical power were primarily concerned with gaseous effluents or, in the case of nuclear power plants, with the potential discharge of nuclear radiations. Today, the siting and operation of all new generating plants (nuclear, fossil fuel, or hydroelectric) are also being appraised with respect to the impact of environmental noise. In most instances the utilities must furnish evidence that the generating station, transmission lines, and distribution substations will not have an adverse effect upon the acoustic environment of the areas that they serve. For example, the Nevada Utility Environmental Protection Act requires that such facilities will have "the minimum adverse environmental impact" considering the state of technology, economics, and alternatives.

Representative of the recent legislation that has been enacted is an amendment to the Public Service Law, State of New York, dated March 7, 1972. This amendment requires that a utility must file an environmental impact statement, prior to constructing and operating any new facility. The utility must do the following:

(a) Describe the ambient noise levels at the proposed site and in areas adjacent to the proposed facility (average and peak noise levels, pure tone levels, etc. for daytime and nights).

(b) Identify noise sensitive facilities' activities near the site (e.g., hospitals, churches, outdoor amphitheaters, residential dwellings, schools and colleges).

(c) Identify standards, criteria, rules, regulations, ordinances, or statutes of the federal, state or its political subdivisions that may limit environmental noise levels caused by the construction or operation of the proposed facility.

(d) Describe environmental noise levels that will be produced during construction and operation of the proposed facility.

It is further required that all monitoring programs be planned in consultation with both the New York State Public Service Commission and the Department of Environmental Conservation.

LOCAL REGULATION OF MACHINE NOISE

A few large cities have passed legislation to limit the noise of construction equipment and miscellaneous machine noise. Table 4 lists the limits on equipment for Chicago, Boston, and New York City. As a result of such legislation, substantial noise reductions are envisioned for equipment designed for either commercial or residential use.

Table 4   Machine Noise [dB(A) Levels at 50 ft Unless Noted Otherwise]

| | City | | |
Description	Chicago	Boston	New York City
Sale or lease of new construction and industrial equipment	After 1/1/72 94 After 1/1/73 88 After 1/1/75 86 After 1/1/80 80	After 1/1/72 94 After 1/1/73 88 After 1/1/80 80	
Air compressors (operate or permit to be operated) at one meter			12/31/72 90 6/30/74 80
Air compressors (sell, offer for sale, permit to be operated) at one meter			12/31/72 85 6/30/74 75 12/31/75 70
Agricultural tractors and equipment	After 1/1/72 88 After 1/1/75 86 After 1/1/80 80	After 1/1/72 88 After 1/1/75 86 After 1/1/80 80	
Powered commercial equipment intended for infrequent use in residential area (chain saws, pavement breakers, log chippers, etc.)	After 1/1/72 88 After 1/1/73 84 After 1/1/80 80	After 1/1/72 88 After 1/1/73 84 After 1/1/80 80	
Powered equipment intended for repetitive use in residential area (lawn mower, garden tools, etc.)	After 1/1/72 74 After 1/1/75 70 After 1/1/78 65	After 1/1/72 74 After 1/1/75 70 After 1/1/78 65	

NOISE NUISANCE ORDINANCES

Quite distinct from the problem of traffic noise has been that of placing limits on miscellaneous "excessive and unnecessary" noise within cities and towns. Noises arising from aircraft, traffic, and construction operations can at least be partially justified because such activities relate directly to essential business, health, and welfare. Many other sources of noise, however, accompany activities that cannot be regarded generally as essential. Thus, nuisance noise can be generally defined as noise that infringes on the desire for quiet of others and for which there is no clear justification on the basis of being an essential activity. Legislation to control such noise is difficult both to formulate and to enforce. Too strict a code is met with strong reactions by people who claim their constitutional liberties are being abrogated. A code that is either very vague or very permissive hardly merits enactment.

Nuisance noise ordinances that have been included in municipal or town law have usually applied criteria exemplified by the general noise ordinance of the City of Beverly Hills, California:[10]

The standards which shall be considered in determining whether a violation of this section exists shall include, but shall not be limited to, the following:

a) the volume of the noise.
b) the intensity of the noise.
c) whether the nature of the noise is usual or unusual.
d) whether the origin of the noise is natural or unnatural.
e) the volume and intensity of the background noise, if any.
f) the proximity of the noise to residential sleeping facilities.
g) the nature and zoning of the area within which the noise emanates.
h) the time of the day or night the noise occurs.
i) the duration of the noise.
j) whether the noise is recurrent, intermittent or constant.
k) whether the noise is produced by a commercial or non-commercial activity.

Here are a few examples of specific sounds that are deemed objectionable in various communities.[11]

a. *The use of any horn.*   The following is declared to be loud, unnecessary and in violation:

The sounding of any horn or signal . . . on any automobile, motorcycle, bus . . . or other vehicle while stationary except as a danger signal when an approaching vehicle is apparently out of control . . . or the sounding of any such device for an unnecessary and unreasonable period of time.

—New York, New York

b. *Hawkers, peddlers.*

It is illegal for hawkers or peddlers to disturb the peace and quiet by shouting or crying their wares.

—Dayton, Ohio

c. *The keeping of domestic pets.*

No person within the City shall keep . . . any animal or fowl the type, behavior or quantity of which shall cause discomfort or annoyance to a reasonable person of normal sensitiveness.

—Beverly Hills, California

d. *Radios, phonographs.* The following is declared to be loud, unnecessary and in violation:

The playing of any radio or T.V. . . . or device for producing or reproducing of sound in such manner as to disturb the peace, quiet, and comfort of the neighboring inhabitants.

—Seattle, Washington

e. *Sound trucks.* Typical codes restrict the use of sound vehicles to broadcast only matters of public concern, for example, voting and special local events. Limitations are also imposed upon the time of day and the maximum distance of audibility (~300 to 500 ft)[12]

A "MODEL" CITY ORDINANCE

The specific noise problems of a community will vary to some degree, and as evidenced in the preceding section, local codes will differ. Nevertheless, some 30 cities have or are adopting a general ordinance patterned after the document drafted by the *National Institute of Municipal Law Officers* (NIMLO).* This document does not concern itself with traffic noise, which is often treated under separate regulatory acts, but it does provide general guidelines that are designed to discourage unnecessary noise in both commercial and residential areas of a city.

NIMLO MODEL ORDINANCE PROHIBITING UNNECESSARY NOISES

Be It Ordained by the City Council of the City of . . . . :
Section 8-301. It is found and declared that:
(a) The making and creation of loud, unnecessary or unusual noises within the limits of the City of . . . . is a condition which has existed for some time and the extent and volume of such noises is increasing;
(b) The making, creation or maintenance of such loud, unnecessary, unnatural or unusual noises which are prolonged, unusual and unnatural

* 839 17th Street NW, Washington, D. C. 20006.

in their time, place and use affect and are a detriment to public health, comfort, convenience, safety, welfare and prosperity of the residents of the City of ....; and

(c) The necessity in the public interest for the provisions and prohibitions hereinafter contained and enacted, is declared as a matter of legislative determination and public policy, and it is further declared that the provisions and prohibitions hereinafter contained and enacted are in pursuance of and for the purpose of securing and promoting the public health, comfort, convenience, safety, welfare and prosperity and the peace and quiet of the City of :.............. and its inhabitants.

Section 8-302.   It shall be unlawful for any person to make, continue, or cause to be made or continued any loud, unnecessary or unusual noise which either annoys, disturbs, injures or endangers the comfort, repose, health, peace or safety of others, within the limits of the city.

Section 8-303.   The following acts, among others, are declared to be loud, disturbing and unnecessary noises in violation of this ordinance, but said enumeration shall not be deemed to be exclusive, namely:

(1) *Horns, Signaling Devices, etc.*   The sounding of any horn or signaling device on any automobile, motorcycle, street car or other vehicle on any street or public place of the city, except as a danger warning; the creation by means of any such signaling device of any unreasonably loud or harsh sound; and the sounding of any such device for an unnecessary and unreasonable period of time. The use of any signaling device except one operated by hand or electricity; the use of any horn, whistle or other device operated by engine exhaust; and the use of any such signaling device when traffic is for any reason held up.

(2) *Radios, Phonographs, etc.*   The using, operating, or permitting to be played, any radio receiving set, musical instrument, phonograph, or other machine or device for the producing or reproducing of sound in such manner as to disturb the peace, quiet and comfort of the neighboring inhabitants or at any time with louder volume than is necessary for convenient hearing for the person or persons who are in the room, vehicle or chamber in which such machine or device is operated and who are voluntary listeners thereto. The operation of any such set, instrument, phonograph, machine or device between the hours of eleven o'clock P.M. and seven o'clock A.M. in such a manner as to be plainly audible at a distance of fifty (50) feet from the building, structure or vehicle in which it is located shall be prima facie evidence of a violation of this section.

(3) *Loud Speakers, Amplifiers for Advertising.*   The using, operating or permitting to be played, used, or operated of any radio receiving set, musical instrument, phonograph, loudspeaker, sound amplifier, or other machine or device for the producing or reproducing of sound which is cast upon the public streets for the purpose of commercial advertising or attracting the attention of the public to any building or structure.

(4) *Yelling, Shouting, etc.*   Yelling, shouting, hooting, whistling, or singing on the public streets, particularly between the hours of 11 P.M. and
7 A.M. or at any time or place so as to annoy or disturb the quiet, comfort,
or repose of persons in any office, or in any dwelling, hotel and other type
of residence, or of any persons in the vicinity.

(4) *Animals, Birds, etc.*   The keeping of any animal or bird which by
causing frequent or long continued noise shall disturb the comfort or repose of any persons in the vicinity.

(6) *Steam Whistles.*   The blowing of any locomotive steam whistle attached to any stationary boiler except to give notice of the time to begin
or stop work or as a warning of fire or danger, or upon request of proper
city authorities.

(7) *Exhausts.*   The discharge into the open air of the exhaust of any
steam engine, stationary internal combustion engine, motor boat, or motor
vehicle except through a muffler or other device which will effectively prevent loud or explosive noises therefrom.

(8) *Defect in Vehicle or Load.*   The use of any automobile, motorcycle,
or vehicle so out of repair, so loaded or in such manner as to create loud
and unnecessary grating, grinding, rattling or other noise.

(9) *Loading, Unloading, Opening Boxes.*   The creation of a loud and
excessive noise in connection with loading or unloading any vehicle or the
opening and destruction of bales, boxes, crates, and containers.

(10) *Construction or Repairing of Buildings.*   The erection (including
excavating), demolition, alteration or repair of any building other than
between the hours of 7 A.M. and 6 P.M. on week days, except in case of
urgent necessity in the interest of public health and safety, and then only
with a permit from the Building Inspector, which permit may be granted
for a period not to exceed three (3) days or less while the emergency continues and which permit may be renewed for periods of three days or less
while the emergency continues. If the Building Inspector should determine
that the public health and safety will not be impaired by the erection,
demolition, alteration or repair of any building or the excavation of streets
and highways within the hours of 6 P.M. and 7 A.M., and if he shall further determine that loss or inconvenience would result to any party in interest, he may grant permission for such work to be done within the hours
of 6 P.M. and 7 A.M., upon application being made at the time the permit
for the work is awarded or during the progress of the work.

(11) *Schools, Courts, Churches, Hospitals.*   The creation of any excessive
noise on any street adjacent to any school, institution of learning, church
or court while the same are in use, or adjacent to any hospital, which unreasonably interferes with the workings of such institution, or which disturbs or unduly annoys patients in the hospital, provided conspicuous signs
are displayed in such streets indicating that the same is a school, hospital
or court street.

(12) *Hawkers, Peddlers.* The shouting and crying of peddlers, hawkers and vendors which disturbs the peace and quiet of the neighborhood.

(13) *Drums.* The use of any drum or other instrument or device for the purpose of attracting attention by creation of noise to any performance, show or sale.

(14) *Metal Rails, Pillars and Columns, Transportation Thereof.* The transportation of rails, pillars or columns of iron, steel or other material, over and along streets and other public places upon carts, drays, cars, trucks, or in any other manner so loaded as to cause loud noises or as to disturb the peace and quiet of such streets or other public places.

(15) *Street Railway Cars, Operation Thereof.* The causing, permitting or continuing any excessive, unnecessary and avoidable noise in the operation of a street railway car.

(16) *Pile Drivers, Hammers, etc.* The operation between the hours of 10 P.M. and 7 A.M. of any pile driver, steam shovel, pneumatic hammer, derrick, steam or electric hoist or other appliance, the use of which is attended by loud or unusual noise.

(7) *Blowers.* The operation of any noise-creating blower or power fan or any internal combustion engine, the operation of which causes noise due to the explosion of operating gases or fluids, unless the noise from such blower or fan is muffled and such engine is equipped with a muffler device sufficient to deaden such noise.

Section 8-304. Penalties. Any person who violates any provision of this ordinance shall be deemed guilty of a misdemeanor and upon conviction thereof shall be fined not exceeding $......, or by imprisonment for not more than .... days, or by both said fine and said imprisonment.

Section 8-305. Separability. It is the intention of the City Council that each separate provision of this ordinance shall be deemed independent of all other provisions herein, and it is further the intention of the City Council that if any provision of this ordinance be declared to be invalid, all other provisions thereof shall remain valid and enforceable.

Adopted this .... day of ...., 19.......

INTERNATIONAL EFFORTS TOWARD NOISE CONTROL

Mention should be made of noise abatement efforts outside the United States especially since many of the pioneering environmental studies originated in Europe and Great Britain. The urban population density is high in Europe, and even farmers tend to reside in towns rather than on the land they cultivate. The massive reconstruction program following World War II stimulated concern with construction noise, and airport and traffic noise has often become a more serious problem than in

North America. Regard for the environmental quality of life is also reflected in ordinances that severely limit noise at night, for example; a West German statute even provides a precise definition: "the night is understood here to be 8 hours long, it begins at 10 p.m. and ends at 6 a.m."

Many countries now have specific national or local laws regulating noise from motor vehicles. A European Conference of Ministers and the Economic Commission for Europe has been active in attempting to obtain international agreement for the maximum emission levels for all vehicles. London spent many years debating the site of its third, and Tokyo its second airport. Building codes have explicit specifications on sound proofing. In East European countries all apartments must be separated longitudinally by double walls, and several countries recommend floating floors for control of impact noise and the use of lead-based foundations to attenuate ground-transmitted vibrations. Noise control measures often are applicable to concert halls, libraries, museums, and public administrative buildings as well as residential units. In Bonn, as in Chicago, church bells have even been given the label of a noise disturbance, indicating a trend toward reducing all loud sounds that impinge upon the environment of others.

An assessment of noise concern of many nations has been prepared by the U.S. Environmental Protection Agency and the summary of the following paragraphs is derived from this document.[13]

Switzerland

The Swiss regulations for traffic noise are sometimes viewed as models for international codes. All types of vehicles are subject to a standard noise test. Measurement is made of noise from a *stationary* vehicle at full throttle. Microphones are placed at a distance of 7 meters on each side of the vehicle and at a height of 1.2 meter above ground. The maximum permissible noise emission levels are cited in Table 5. If a vehicle on the road seems to be unduly noisy, it may be stopped and tested for noise level. If the legal level is exceeded, the operator must take steps to make his vehicle conform to standards, and failure to do so can result in deprivation of his license.

Swiss airports also have well-developed noise monitoring systems, and land use planning is keyed to noise contours in airport environs. Limits on construction noise has also been legislated in some of the cantons.

Table 5 Maximum Permissible Motor Vehicle Noise Emissions in Switzerland

Type of Vehicles	Maximum Noise Level dB(A)[a]
Motorscooters	70
Light motorcycles, up to 50 cc	73
Heavy motor cycles, above 50 cc	82
Cars with diesel engines ($\sim$50 HP)	82
Other personal cars	78
Heavy trucks, tractors, other	85

[a] Measured at a distance of 7 meters.

Great Britain

The first Act of Parliament that specifically focused on the noise abatement problem was a November 28, 1960 Noise Abatement Act, which widened the scope of Part III of the Public Health Act of 1936. Subsequent action has included the Public Health (Recurring Nuisances) Act of 1969; the formation of the Royal Committee on Environmental Pollution in February 1970 accompanied by the formation of a Noise Advisory Council; the Department of the Environment was formed on October 15, 1970. A Road Traffic Act of 1960 also empowered the Ministry of Transport to make regulations relative to "equipment and the use of motor vehicles." Proposed limits for new vehicles after October 1973 are comparable to the Swiss vehicular codes.

The problem of aircraft noise has not been solved. Heathrow Airport near London has problems representative of most of the international airports. Operations are directly over a major city, traffic has increased beyond all original design expectations, and jet engines have superceded propeller type craft for which the airports were originally designed. From 1960 to 1970 flight operations increased from approximately 135,000 to 270,000 per annum, and the percentage of night departures increased. Because there have been no practical alternatives to substantially lower noise levels in the immediate vicinity of the airport, legislation was passed to assist nearby residents via grants for sound insulation. Householders whose dwellings were completed prior to January 1, 1966 became entitled to a maximum of 150 pounds ($375) to install sound insulation. The sound treatment was required to meet certain specifications, and it included a provision for an alternative means of ventilation. Noise control in building construction has been effected through the Ministry of Public Buildings and Works. One of the institutes that it sponsors is the Building Research Station in Garston, England. This government research institute has provided specific recommendations and information on (1) building codes, (2) sound transmission and insulation within buildings, and (3) detailed design schemes to minimize sound levels with buildings subject to excessive aircraft and roadway noise.

Federal Republic of Germany (West Germany)

Of all the continental European countries, West Germany has had the most extensive program relating to noise abatement. As a highly industrialized republic, it has enacted the following legislation:

> Law on Protection Against Construction Noise (1965)
> The Air Pollution Control Act (1960)
> The Federal Zoning Code (1965)
> The Street Traffic Code (1960)
> Law for Protection Against Aviation Noise (1971)
> "Technische Anleitung Laerm" (1968)

The Technische Anleitung Laerm is a general administrative directive that defines zoning categories, denotes permissible levels of noise in such zones, and specifies codes for new construction. It is very specific with respect to noise that intrudes upon adjacent areas. An *immission* is defined as "an effect of a noise issuing from premises upon neighbors or third parties." Immission standards are established as follows:

Zone	*Noise Level dB(A)*
(a) purely industrial areas	70
(b) primarily commercial	65 daytime 50 night
(c) areas with commercial and residential buildings	60 daytime 45 night
(d) primarily residential	55 daytime 40 night
(e) exclusively residential	50 daytime 35 night

(f) rest homes, hospitals, and	45 daytime
health resorts	35 night
(g) residences attached to an	40 daytime
establishment	30 night

In addition to federal statutes, the individual West German states and city-states have environmental protection laws: for example, North Rhine-Westphalia, Bavaria, Lower Saxony, the Palatinate, and Berlin. The Ruhr industry region has very strict standards.

France

A new Ministry of the Environment was created in 1971 with some authority that was formerly reserved to local authority. A Technical Commission for the Study of Noise (Commission technique d'étude du bruit) in the Ministry of Health also furnishes expertise for the several other agencies concerned with noise. Factory noise is regulated by a statute passed in 1969 and in the same year a decree was implemented with respect to building noise—requiring compliance with soundproofing standards established by the Ministry of Logistics and Housing.

The Ministries of Public Works, Transportation, and Public Health instituted a set of highway regulations "Code de la Route" requiring adequate mufflers and the observance of noise levels. Offenders are subject to fines and detention (for repeated offenses). Automobile horn blowing is prohibited in cities, except in emergencies. Noise from ships and boats has been restricted since 1966; a limit of 75 dB(A) is specified at a distance of 25 meters.

Italy

National legislation pertaining to noise is primarily limited to vehicular noise. Testing of vehicles is similar to the Swiss regulation, but noise limits are higher. Nevertheless, the number of fines issued for noncompliance has been substantial in recent years. Police are empowered to inspect vehicles on the road and all public transport vehicles and vehicles for hire are subject to annual inspections for both safety and noise standards.

Japan

Until 1968, noise-related regulations remained within the jurisdiction of regional or municipal governments. In 1968, the National Government enacted a "Noise Abatement Law" concerned with industrial and construction noise, and in 1971 a new environmental protection agency was created reporting directly to the Prime Minister.

Currently there are nine individual laws on noise abatement administered by one or more of the following agencies:

1. Japanese Environmental Protection Agency
2. Ministry of Transportation
    (a) Aviation Bureau
    (b) Highway Transportation Division
3. Ministry of International Trade and Industry
4. Food Agency
5. Ministry of Health and Welfare
6. Ministry of Construction
7. Agency for Defense Equipment
8. Ministry of Postal Service

The various laws apply to zoning, construction and industrial noise emission, vehicular noise, broadcasting, and aviation. The "Noise Abatement Law on Public or Private Airports and Vicinities (1966) established maximum exposures for persons in schools and hospitals due to aircraft flyovers; the law was also made applicable to U.S. Air Force bases in Japan. The country is also in the early stages of a large scale program that includes retrofitting commercial aircraft, reducing the noise of their railway system, sound-proofing houses, and purchasing land to reduce noise exposure.

The general noise problem has been especially severe in Tokyo. Because of the density of traffic and available manpower, approximately 75% of all public construction work in Tokyo (streets, water lines, sewers, etc.) has been undertaken at night in recent years. Much of industrial construction has also been scheduled outside of the usual daytime hours.

Scandinavian Countries (Denmark, Norway, Sweden)

Noise regulation or recommendations in the Scandinavian countries are concerned primarily with problems of town planning and outdoor noise from traffic and aviation.

In Denmark, tolerance levels have been set for motor vehicles registered after 1969, and workers are protected from noise hazards under the Protection of Workers Act. The first Building Act (1960) and subsequent national building regulations have specified requirements for the total transmission of noise between dwellings, for noise emitted by external technical installations, and for other noise sources. Sound level measurements are undertaken by the National Building Research Institute, the National Institute of Industrial Hygiene, and the Institute of Sound Technology (Academy of Technical Sciences).

In Norway, the Road Traffic Act enables authorities to impose restrictions on motor vehicle noise, and maximum noise limits have been established for motorcycles. A permanent Commission on Air Noise serves as an advisory body to the Government on aircraft-related noise.

Sweden passed an environmental protection act in 1969 (Miljoskyddslagen: Svensk Fortattningssammling 1969:387). The law incorporates provisions for noise control from factories, shipyards, traffic on highways, streets, railroads, and airports. At regional levels, city councils have special boards that work with local housing committees and public health boards. As early as 1967 the Swedish government also declared that supersonic civil air traffic would not be permitted over Swedish territory if the noise from such aircraft would be adverse to health.

The U.S.S.R.

Most of the Soviet law dealing with noise is administered by various ministries such as the Ministry of Health, the Ministry of Railroad Transport, and by local city or regional authorities. In some respects, the Soviet Union has adopted stricter personnel standards than those recommended by the International Standards Organization, which have been adopted by other nations. Specific laws have been promulgated by the Ministry of Health that deal with (1) railroad workers, (2) sailors on maritime, river, and lake vessels, (3) flight crews on passenger aircraft, (4) noise emission from machinery, and (5) the determination of noise-induced deafness from occupational hazards.

The most important recent legislation is the 1969 Sanitary Norm in Industrial Noise (SNiP No. 785-69) that was developed by the Academy of Medical Sciences under the Ministry of Health and approved by the Council of Ministers. The norms give primary emphasis to noise limits for workers in industry, but they also cover industrial noise emissions to adjacent neighborhoods.

REFERENCES

1. U. S. Environmental Protection Agency, *Summary of Noise Programs in the Federal Government,* Washington, D. C., 1971, p. 11.

2. United States Congress, SS1431 (Supp. IV, 1969).

3. U. S. Environmental Protection Agency, *Laws and Regulatory Schemes for Noise Abatement,* Washington, D. C., 1971, pp. 1-19.

4. 49 U.S.C.A. Section 1701 et seq. (Supp. 1971).

5. H. M. Rupert and J. E. Wesler, *Sound & Vib.,* 34 (October 1973).

6. Walsh-Healey Public Contracts Act, *Federal Register,* May 20, 1969.

7. Public Law 91-596, Section 2193, December 29, 1970.

8. N. C. Yost, "State Environmental Policy Acts," *Env. Law Rep.,* **3,** 50090-50098.

9. California Public Utility Code, Section 21669, 1970.

10. W. S. Gately and E. E. Frye, *Regulation of Noise in Urban Areas,* University of Missouri, Rolla, **5,** 3 (1971).

11. *Ibid.,* **5,** 4,5.

12. *Ibid.,* **5,** 12.

13. U.S. Environmental Protection Agency, *An Assessment of Noise Concern in Other Nations,* Washington, D. C., 1971, Vol. I.

13

COMMUNITY PLANNING

FACTORS AFFECTING COMMUNITY RESPONSE TO NOISE

Community planning with respect to noise abatement and control must begin with noise exposure measurements and take into account the subjective factors relating to loudness and annoyance. Some of these factors have been cited previously and a further discussion of peoples' psychological responses to noise is given in Chapter 17. Results of several community studies, however, point up general trends and conclusions that are of interest to the professional planner or environmental engineer whose responsibility is to predict or evaluate community response to noise. The factors cited below are not necessarily independent and the list does not include other considerations of community dynamics that may be relevant.

1. *Overall Noise Level.* The physical amplitude of the noise and noise duration time is directly related to community response. Figure 1 suggests that this response ranges from mild annoyance at low noise levels to legal action for sustained high levels. The actual shape of the noise spectrum is also important; the higher frequencies are more annoying than lower frequencies of the same amplitude.

2. *Pure Tones and Impulsive Noise.* Both laboratory and community studies show that noise containing strong pure-tone components is very annoying (fan, transformer noise, etc.). Also, noise having a rapid rise time and of a highly intrusive character (pneumatic hammer, construc-

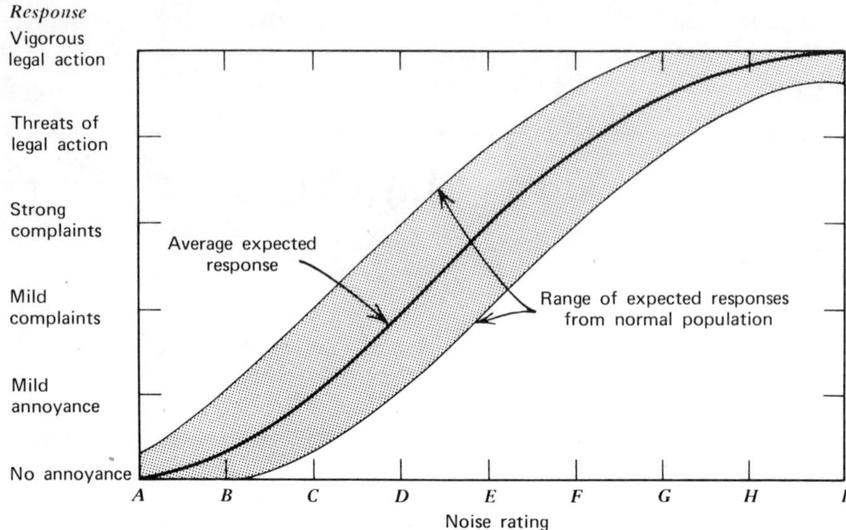

Figure 1   Relation between noise exposure and community response. The noise exposure increases from *A* to *I*. (From J. D. Miller, Ref. 2.)

tion noise, barking dogs, etc.) is subjectively more annoying than steady-state noise.

3. *Time of Day or Year.*   Quite apart from the character of the noise, response to nighttime noise is consistently more unfavorable than to daytime disturbance. Seasonal differences can also be very important. Traffic and general outdoor noise, in New England, for example, causes less annoyance in winter when people live substantially indoors. However, "indoor" populations tend to be less tolerant of general noise exposure than those who live in "outdoor" climates where windows are open most of the year.[1] This difference has been noted especially in Europe where populations in the warmer climates of Italy and Southern France seem to be 5 to 10 dB(A) more tolerant of noise than people in England and Scandinavia.

4. *Relevance to Community Life.*   In some instances noises are less annoying if they are clearly identified with the special nature or activity of a community. Noises from a harbor, a shipyard, or a military air field are considered to be indigenous to community goals and are therefore more acceptable than otherwise. Highly intrusive noises from ambulance sirens will even be tolerated in a community that provides essential hospital facilities for the area.

5. *Attitudinal-Psychological Factors.* A population exposed to noise is somewhat more tolerant if it believes that the persons responsible for it are concerned with the public welfare and that all reasonable measures are being employed to minimize annoyance. Highly annoyed groups are more likely to believe that those responsible for the noise are relatively unconcerned with the public welfare; such groups also tend to believe that the source of noise is not of great importance to the economic and social success of the community.[2]

6. *Socioeconomic Status.* Community response to noise has been shown to be conditioned by the socioeconomic and cultural status of the community group.[3] Neighborhoods where loud noises have existed for years are more tolerant to continued noise than in communities where the onset of increased noise levels has occurred over the short term.

It also appears that a rental population differs appreciably from the home owner population. A partial explanation is that an apartment dweller is more mobile and can "leave the noise"; the non home owner also does not have to be concerned about noise decreasing the value of his property.[4]

Overall, people's response to noise depends strongly on their values, beliefs, and attitudes, and these factors are often more important than the characteristics of the noise.[5]

It should also be noted that only a small percentage of the population of any community of those who are highly annoyed will actually register a formal complaint to local authorities. A rule-of-thumb relationship that is useful in predicting the number of highly annoyed households per thousand, (H), compared to the actual number of registered complaints (letters, telephone calls, etc.) per thousand (C) in an area has been reported to be[6]

$$H = 5C \tag{1}$$

Thus, if there are 100 persons who complain in a community of 1000 households, there will be about 500 highly annoyed households, 400 of whom do not complain. The relationship above, of course, cannot expect to hold after a community has gotten used to or "accepted" a new noise source.

LAND USE AND ZONING

Perhaps the most important and also the most difficult task relating to environmental noise control is compatible land use. Land use and zoning involve long-term commitments, the programs are costly, and such planning is at variance with traditional rights of private landowners to utilize properties without restriction. But now it is generally agreed that appropriate land use furnishes the only base for accommodating the diverse interests of a small community or major city, while retaining an environmental quality that is reasonably acceptable to all. The basic problem is to provide distinct areas compatible for the functions of industry, commerce, residential living and recreation, and to interconnect these zones (and other cities) with a transportation system that moves people and products at reasonable cost and convenience.

Table 1 indicates the general concept of community land use in which distinct geographical areas are reserved for specific functions. The range of acceptable outdoor noise levels in dB(A) for each area are those taken from a Swiss study.[7] These values represent basic noise level ranges; the lower and upper numbers corresponding to night and day values, respectively. Frequent noise peaks must be expected to occur that are 10 dB(A) greater, and infrequent intrusive noise levels can be expected to be 15 dB(A) higher than these base levels. Recommended *indoor* noise levels from a British study[8] are shown in Table 2.

Table 1   Tolerated Outdoor Noise Levels in dB(A)[7]

Zone	Basic Noise Level		Frequent Peaks		Infrequent Peaks	
	Night	Day	Night	Day	Night	Day
Hospital	35	45	45	50	55	55
Quiet residential	45	55	55	65	65	70
Mixed	45	60	55	70	65	75
Commercial	50	60	60	70	65	75
Industrial	55	65	60	75	70	80
Main road	65	70	70	80	80	90

Table 2   Recommended Indoor Noise Levels[8]

Situation	Noise Level, dB(A)	
	Night	Day
Country areas	30	40
Suburban areas (away from main traffic routes)	35	45
Busy urban areas	35	50

HIGHWAY PLANNING

Several studies have pointed up the importance of adequate planning for a highway right-of-way. When a highway must pass through a populated community, one of the most effective means for achieving sound reduction is via the "depressed-highway," that is, a location of the roadbed substantially below the general elevation of the landscape.

At an intersection this depressed highway may take the form of an underpass or a tunnel (Fig. 2). In such an instance not only will vehicle noise be attenuated because of the lowered grade, but the substantial noise arising from traffic stopping and starting at a controlled intersection will be eliminated. Quantitative measurements have been made at a location along the Baltimore (Maryland) Beltway at a point where the highway is depressed· approximately 25 ft. Figure 3a shows the land contour and the measured traffic noise at various distances from the highway. Figure 3b shows a corresponding plot of the sound level attenuation

Figure 2   Photo of a depressed roadway through a town. (Courtesy New York State Department of Transportation.)

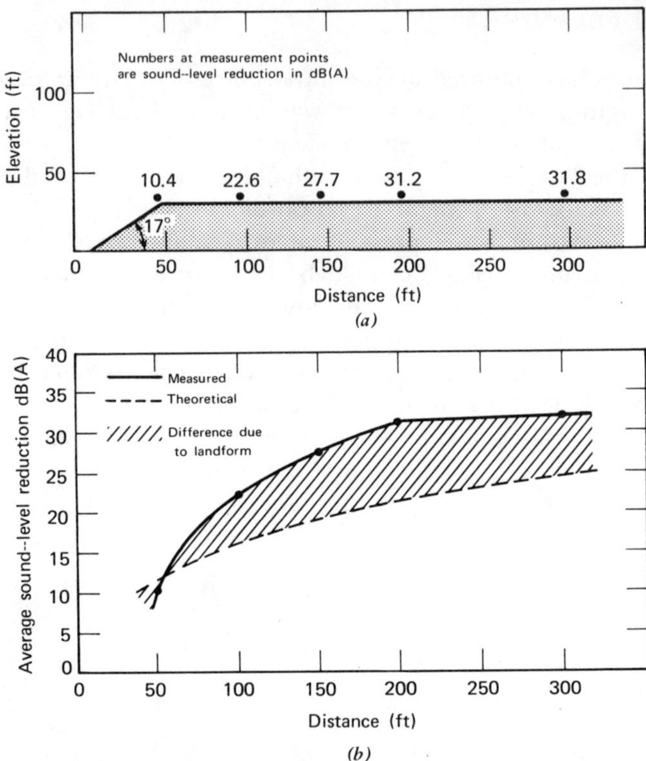

Figure 3   Average sound level reduction in dB(A) as a function of distance, from a deeply depressed highway. (Ref. 9.)

in dB(A) and the theoretical curve (dotted line) representing the calculated noise attenuation due to distance alone.[9]

It will be noted that the measured sound level reduction at 50 ft is below the theoretical value and is attributed to near-reflections. Beyond 50 ft, however, the bank serves as an effective barrier to line-of-sight propagation.

One of the most extensive surveys relative to highway planning has been sponsored by the Highway Research Board in which the following expressways and freeways were investigated.

Northeast

Connecticut —Connecticut Turnpike
New York   —Long Island Expressway
               Northern State Parkway

Pennsylvania—Schuylkill Expressway
              Pennsylvania Turnpike
New Jersey —New Jersey Turnpike
              Interstate 295
Maryland   —Baltimore Beltway

Midwest

Ohio   —Detroit-Toledo Expressway

West

California   —San Diego Freeway (Los Angeles)
              Ventura Freeway (Los Angeles)
              San Bernardino Freeway (Los Angeles)

Acoustic, economic, and esthetic considerations were evaluated on a nationwide basis resulting in the following general conclusions.[10]

1. Sound from trucks is the most objectionable highway disturbance to persons living in homes, apartments, and farms next to limited-access highways, regardless of geographic location. A sound-disturbance threshold level exists, above which most people living next to expressways will perceive the sound as disturbing.

2. Lack of proper maintenance of highway right-of-way was the second most objectionable highway annoyance.

3. Presence of a limited-access highway does not devalue adjacent properties.

4. Presence or absence of landscaping on the right-of-way of a limited-access highway does not affect the value of adjacent properties. However, people living next to such highways indicated that they would accept the presence of the highway more readily if it were concealed from view by landscaping.

5. Attitudes relating to disturbance factors of people living next to a highway, even in the same geographic location, vary greatly. From the interviews, it was learned that people living in older, relatively less expensive homes next to limited-access highways tended to accept associated disturbances more readily than people in more expensive homes; however, Los Angeles homeowners interviewed accepted the highway regardless of property value.

6. A supplementary study indicated that highway depression and elevation affect sound levels emanating from highways markedly, while tree plantings affect sound levels only slightly. Highway depression is potentially the greatest single reducer of sound levels.

7. Sound disturbance at high-rise apartment buildings located adjacent to limited-access highways is highly objectionable and caused apartment owners to demand economic adjustments for units facing the highway.

Other studies have attempted to provide guidelines for "acceptable" noise levels from automotive traffic in residential areas. Table 3 is one such recommendation. In this listing the levels are the ambient *outdoor* suggested values.[11]

Table 3   Design Goals for Auto and Truck Noise, dB(A) in Outdoor Environments[11]

Area	Auto Levels	Truck Levels
Very quiet suburb	43	51
Suburb	49	56
Residential urban	55	61
Some industry	61	66
Heavy industry	67	71

LAND USE SURROUNDING AIRPORTS

As has been discussed in Chapter 10, it is not feasible to change radically the noise environment near airports unless severe restrictions are imposed on the size and type of aircraft. Private airfields designed solely for small planes can operate in rural or urban areas without serious environmental impact. Most airports, however, will continue to be noisy, and one of the few available options is to reduce the sensitivity of the noise-affected areas. Some of the considerations or courses of action listed below may apply.[12]

1.   Open Land, Commercial, and Industrial Use

Although little can be achieved over the short term, long range planning can be effective in modifying land use from a noise-sensitive to noise-insensitive pattern. Approach and takeoff corridors should be acquired so as not to impose unduly steep glide paths upon aircraft.* In some in-

---

* For New York City's proposed fourth major jetport, 65 miles north of Manhattan, a 10,000-acre woodland "buffer zone" will surround the airfield.

stances an approach pattern may be over water that provides an additional degree of safety. In areas exposed to less than 90 PNdB the noise reduction achieved by ordinary building construction may be sufficient to provide an acceptable residential interior noise environment when all windows are closed. Thus, whether residential units should continue to be located near such an airport will depend on the climate and general living styles. For noise levels greater than 90 PNdB, it may be feasible to locate industrial and commercial activities that inherently generate high interior noise environments. In airport areas of the highest noise levels, it may be possible to eventually locate relatively unmanned sewage disposal plants, utility substations, and other facilities. Parking areas are also a possibility. If new construction is contemplated, it is also possible to build some commercial structures having acceptable interior noise levels, provided that careful acoustic planning is undertaken and future exterior noise levels can be predicted.

Some specific open-land, industrial, and commercial uses for airport environs are listed in Table 4.

Table 4   Open-Land, Industrial, and Commercial Uses for Airport Environs

Open Land	Industrial and Commercial
Reservoirs	Air freight terminals
Cemeteries	Trucking terminals
Utility substations	Aviation schools
Water-treatment plants	Warehouses
Sewage disposal plants	Airport motels and hotels
Truck farming	Bus terminals
Landscape nurseries	Wholesale distribution centers
Golf courses	Industrial parks (if buildings
Botanical gardens	are suitably sound proofed)
Recreational areas	Food processing plants
	Parking lots

2.   "Eminent Domain"

The right of "eminent domain" relates to the power of the government to acquire private land or property for public use. Airport planners, when empowered, can arrange for the acquisition of needed airport land at a "fair market price" if an owner refuses to sell or if an exorbitant price is asked. Furthermore, if owners of property in noise-affected areas try to restrict airport operations to alleviate noise, authorities can employ eminent domain to purchase these properties. The airport may then resell such properties with a noise easement, thereby obtaining relief from claimants at minimum cost.[13]

This type of legal action may be used, for instance, to acquire clear zones beyond runways or remove buildings or towers that would endanger aircraft operations. The Housing Act of 1954 further broadened the use of eminent domain to include ameliorating adverse environmental conditions via rehabilitation rather than the total demolition of existing structures.

3.   Taxing Powers

By offering preferential tax treatment, tax laws may be used (a) to attract noise insensitive activities that are compatible with an airport environs and (b) to encourage the sound proofing of either existing or new structures.

RECREATIONAL AREAS

The problem of noise generated in recreational areas is an acute one. Many persons have moved from the center of the city to the suburbs in search of more living space and a quiet environment, but often the latter has not been realized. In some instances it has been the recreational area, constructed without adequate acoustic consideration, that has been responsible for loud intrusive sounds. At times, property values have been adversely affected. Outdoor movie theaters have been placed close to residential zones, generating increased traffic noise in the environs, if not increased direct sound. More frequent sources of noise in suburban acreage include:

1. Swimming pools
2. Ice skating rinks
3. The local school sports stadium
4. Neighborhood playgrounds
5. Area amusement parks, marinas, etc.

Realistically, these accoutrements to suburban living are desirable, but often less than adequate attention is given to the immediate neighborhood adjacent to them.

Yet in the initial design of such facilities, some consideration should be given to the "air rights" of people at the boundaries. A local sports stadium can actually be constructed as an amphitheater, so that the stadium itself acts as a partial shield; the orientation should be situated so as to protect the most noise sensitive area. A neighborhood swimming pool or sports ground should be located with a park area of substantial width surrounding it. Also, in some instances, a belt of dense trees or properly designed barriers will be beneficial. If space is limited, then a pool can be located asymmetrically within the recreational plot, with an acoustic barrier interposed between the pool area and the nearest boundary, as shown in Fig. 4.

In many suburban parks, the most intrusive sounds come from loud speaker systems. Such sound amplification systems are often unnecessary, or are over utilized. Restraint on the use of these systems is a prerequisite

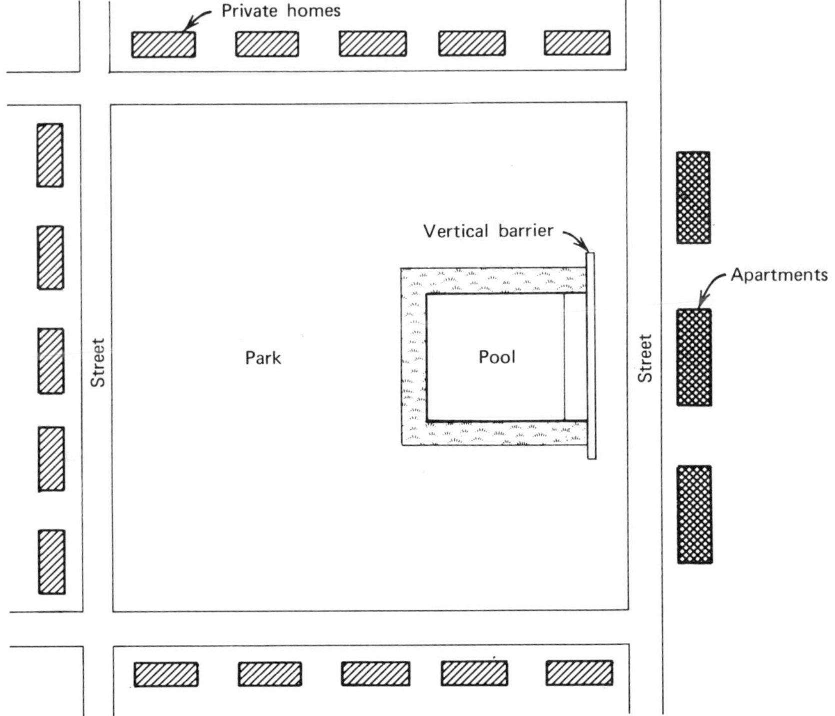

Figure 4   Location of swimming pool near roadway on a small recreational plot. A barrier can be constructed between the pool and roadway; the pool is placed as far as possible from residential dwellings.

for providing both desirable recreational activities and giving due consideration to residents in the environs.

Ultimately, some consideration might also be given to developing public parks that would provide people with an acoustic option. Admittedly, the major public demand is for recreational areas that include activities that are inherently noisy. But at least some areas of municipal, state, and national parks could be allocated for those who might choose to experience only the sights and sounds of nature. These areas, isolated from all traffic, and other noise generating equipment (including portable radios and TV's) would provide an acoustic oasis. In such areas, it would then be possible to capture not only the photographic images of birds and other wildlife, but to record the infinite variety of their sounds.

UTILITIES AND MUNICIPAL SERVICES

Sound radiated from power transformers occasionally places a lower limit to community noise. Although an increasing number of power lines are being buried and some modern substations in residential areas are being totally enclosed, many transformer substations will continue to be "exposed" with only fencing surrounding them. These transformers normally are not a source of complaint during daytime hours, but at night when the ambient sound level is low they are sometimes objectionable. As noted in Chapter 11, the frequencies generated from the transformer core will be 120 Hz, 240 Hz, 360 Hz, and higher harmonics.[14]

These power transformers must usually be sited close to their electrical load centers in both residential and commercial areas, in order to minimize losses in power distribution lines, although restrictions are imposed, by some cities, on the continuous radiation of transformer noise. Usually the ordinances will specify the maximum permissible sound pressure levels (in octave bands) at residential zone boundaries. In any event, a judicious choice of location is necessary to minimize citizen complaints, and in certain special situations a partial or total enclosure may be the only satisfactory solution. Figure 5 is a photo of the older type "exposed" power substation, which is satisfactory if located at a reasonable distance from residential dwellings. Figure 6 shows an ideal transformer substation enclosure that is located in a residential community designed to blend in with the architectural pattern of the community, and it provides acoustic shielding.[15]

Audible noise from high voltage transmission lines is a phenomenon that becomes important during adverse weather conditions. For clear

Figure 5   A standard "exposed" power substation.

conditions, transmission line noise will be limited to harmonics of the
fundamental 60 Hz frequency for which the most noticeable tone is 120
Hz. Figure 7 shows a typical octave-band spectrum of audible transmission line noise under good weather conditions.[16] Under conditions of fog,
rain, and wet snow, a corona discharge can occur from conductors operating at ultra high voltage (UHV). This electrical discharge generates
acoustic pressure waves, much of whose energy falls within the audio frequency band. The noise is a broad band hiss and for conditions of substantial corona discharge, the noise can be very annoying.

The problem is especially important since utilities are experimenting
with continually higher voltages. Much of the cross-country transmission
network currently operates at the 345 kV level. However, some lines are
now operating at 525 and 765 kV, and test lines have been developed for
operation as high as even 1100 kV. The reasons for proceeding to these
higher voltages are both engineering and environmental. The higher the
transmission line voltage, the lower will be the power losses. Furthermore, if very high transmission voltages can be attained in practice, fewer

Figure 6 A modern enclosed electric power substation in a suburban environment. (Courtesy of Rochester Gas & Electric Corp.)

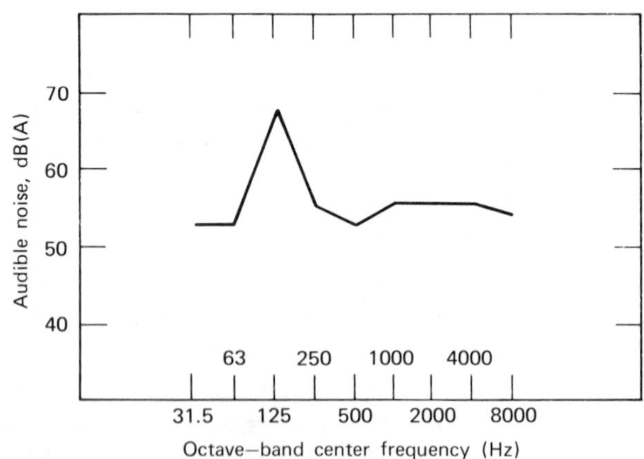

Figure 7 Octave-band spectrum of noise from a 765 kV transmission line. (From D. E. Perry, Ref. 16.)

additional transmission towers will need to be erected, coast-to-coast, to supply the nation's power requirements. Some noise will accompany these higher voltages and adequate "rights-of-way" for these lines must be provided to circumvent large scale citizen complaints.

However, electric power is a basic necessity for homes, hospitals, agriculture, communication, industry, and commerce. It is even a prerequisite for achieving other environmental goals, that is, clean air and pure water. Thus, in applying a "public health and welfare" criterion, the courts must balance the rigid enforcement of a noise code with the need of society for the service that generates the noise. At some future date high voltage lines may be buried underground, but the technology of buried cables has not developed to the stage that such cables can be installed at a reasonable cost. The burial of low voltage cables (distribution within urban and suburban areas) is, of course, proceeding in many parts of the country.

Another important development related to municipal concerns for both air and noise pollution, is the potential use of electric power for fleet and service vehicles. The San Francisco Municipal Railway presently operates over 300 electric trolley buses, and the Department of Transportation is now experimenting with electric cars. These vehicles will permit substantial reductions of noise in urban areas if problems of batteries and costs can be solved. Initial wide scale adoption of electric cars will probably occur with service vehicles (post office delivery trucks, municipal fleet cars, etc.) where the daily mileage is consistently low. Over 70,000 electric powered vehicles are now in service in England alone,[17] where they are used for milk and mail delivery and as ambulances.

ARCHITECTURAL PLANNING

It is beyond the scope of this book to engage in any detailed discussion relative to the acoustic design of buildings, and this information is available elsewhere.[18-22] It is appropriate, however, to include several general guidelines that are used by architects to achieve acoustically acceptable environments. Basic considerations will include one or more of the following.

a. The attenuation of noise from exterior sources.
b. The limitation of noise to property line boundaries.
c. The sound insulation from noise within different parts of a building, or building complex.

In practice, the architect must then attempt to optimize (1) site utilization and (2) building design.

Site Utilization

Site utilization is of primary importance for both the attenuation of noise from exterior sources and the restriction of noise to property line boundaries. Figure 8 is an example of a common situation in an urban area where tall buildings immediately adjacent to the street result in "noise corridors." Traffic noise is contained between the buildings, and in the event that the area is in an aircraft takeoff or landing pattern, a futher buildup of noise occurs through direct and reflected sound. Noise

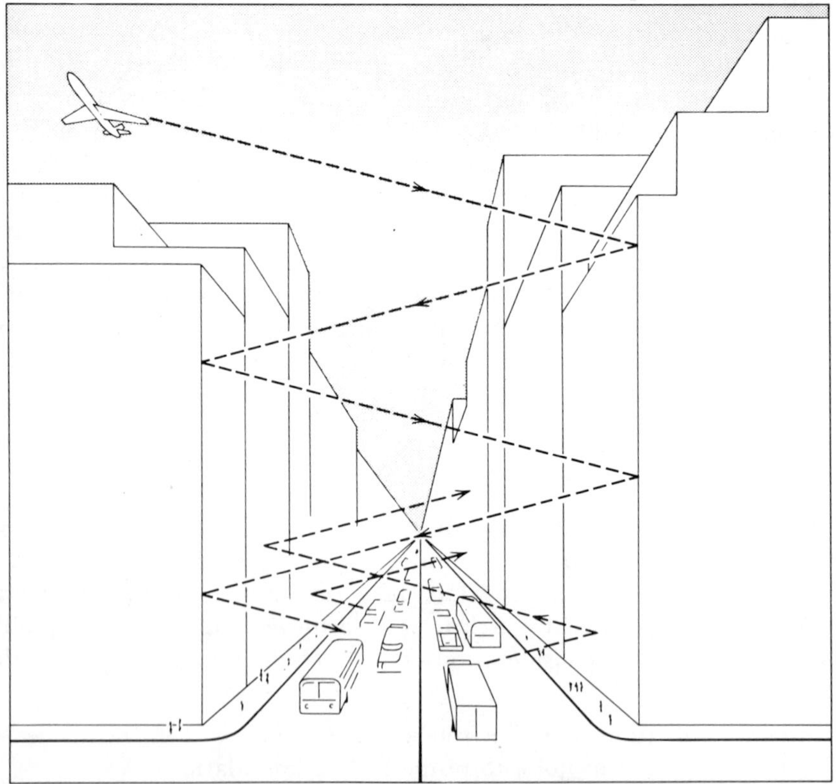

Figure 8   "Noise corridors" produced by multiple reflections between high-rise, closely spaced buildings.

at the street level and for occupants of the adjacent buildings is substantially greater than in the "town square" arrangement where a central park is surrounded by low-profile buildings.

In other midtown situations, architects and acousticians are recommending wider sidewalks, sound-absorbent street surfacing, sound-absorbent exterior paneling for the lower portion of buildings, and a departure from traditional parallel walls. High-rise structures of the "bellbottom" design are now appearing in many cities; the lower stories are tapered sharply away from the street (Fig. 9). The net result is a much improved site utilization both visually and acoustically.

The growth of "industrial parks" in suburban communities has been significant, and this trend has been helpful in industrial noise control.

Figure 9   A "bell-bottom" type structure that prevents buildup of traffic noise at street level. (Courtesy of Lincoln First Bank of Rochester, N.Y.)

Compatible commercial and industrial activities can be merged within a zone that is within close proximity to, but effectively isolated from, residential areas where employees live. Noise levels at the boundaries can be maintained through zoning ordinances and with intelligent planning; the least noisy activities of an industrial park can be located on its perimeter.

The layout of school sites to minimize disturbances generated within its own locale is shown in the plot plan of Fig. 10. The athletic field should be located in such a manner so that loud intrusive noises will reach neither school classroom nor the properties of neighborhood resi-

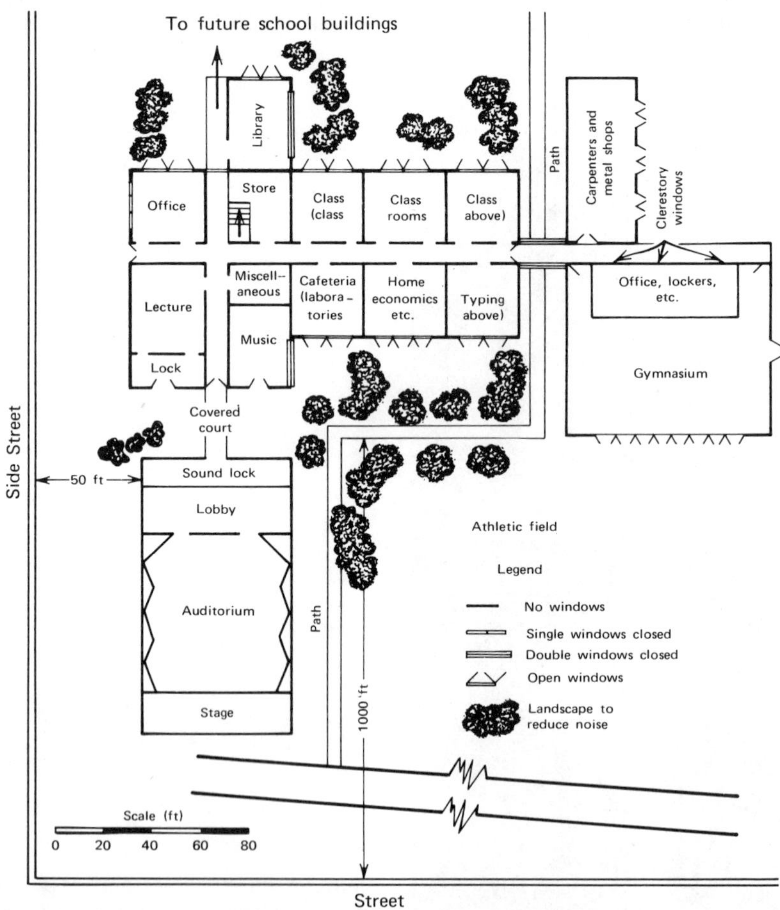

Figure 10   Plot plan of school buildings, playgrounds, and plantings to minimize intrusive noise. (From V. O. Knudsen and C. M. Harris, *Acoustical Designing in Architecture*, Wiley, 1950.)

dents. Classroom windows should also not open toward other classrooms or onto noisy streets. Churches and hospitals should be placed well back from roads and highways, and both the site and the structures should be planned for maximum isolation from noise. The same considerations apply to libraries, museums, cultural centers, etc.

Building Design

Because building areas are often restricted, the architect must frequently achieve desired interior sound levels primarily through building design. In residential dwellings, living and sleeping spaces can be oriented away from traffic noise, and a portion of the house can be utilized as an acoustic baffle (Fig. 11). In commercial buildings the upper floors and "quiet" exposures can be allocated to functions requiring the minimum sound levels. But the interior sound levels will be largely determined by the quality of construction of walls, doors, foyers, roofs, windows, and freedom from acoustic "leaks." The attenuation of outdoor noise afforded by the exterior shell of an average house is in the range of 25 dB. Average values of airborne sound insulation of exterior walls are given in Table 5.

In most buildings the windows and doors will represent the weakest acoustic barriers, as they are lightweight portions of an otherwise relatively high density barrier. A single door may reduce the sound insulation of a $4\frac{1}{2}$ in. wall by about 10 dB, and a single glazed window of average dimensions will usually determine the total noise reduction that can be realized. Average sound insulation values for some door and window constructions are listed in Table 6.

It should be noted that so-called thermal double glazing (windows with two panes of glass separated by a very small air space) provide no greater

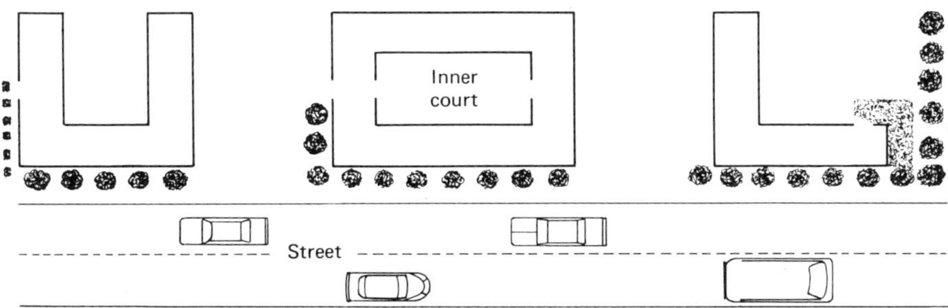

Figure 11   Building design to minimize the annoyance of traffic noise.

Table 5 Average Values of Airborne Sound Insulation of Walls[23]

Approximate Insulation (dB)	Weight (lb/ft²)	Construction
50	90	9-in. solid brick plastered. 7-in. dense concrete plastered.
	64	Two leaves 4-in. cinder block with 2-in. cavity, wire ties, plastered.
	38	Two leaves, 2-in. cinder block with 6-in. cavity, wire ties, plastered.
45	50	$4\frac{1}{2}$-in. brick plastered. 4-in. dense concrete plastered. 8-in. hollow dense concrete block plastered.
	38	Two leaves, 2-in. cinder block with at least 1-in. cavity, wire ties, plastered.
40	30	3-in. cinder block plastered both sides.
	25	2-in. dense concrete.
	20	Two leaves, 2-in. wood-wool slab with 2-in. cavity, wire ties, plastered both sides.
35	50	$4\frac{1}{2}$-in. brick unplastered.
	22	2-in. cinder block plastered both sides.
	20	$2\frac{1}{2}$-in. hollow clay block plastered both sides. 2-in. solid gypsum plaster, reinforced.
	16	3-coat plaster on wood or metal laths on both sides of 4-in. timber frame. 1-in. wood-wool slab with $\frac{1}{2}$-in. plaster both sides.
30	6	$\frac{3}{8}$-in. plasterboard with skim coat of plaster, both sides of 4-in. timber frame.
	3	$\frac{1}{4}$-in. asbestos wall board both sides of timber frame, absorbent quilt between.
25	$2\frac{1}{2}$	$\frac{3}{8}$-in. plasterboard. $\frac{3}{4}$-in. chipboard. $\frac{1}{4}$-in. asbestos cement sheet. $\frac{7}{8}$-in. tongued and grooved boarding.
20	Less than 1	$\frac{1}{4}$-in. plywood. $\frac{1}{8}$-in. hardboard. $\frac{1}{16}$-in. aluminum sheet.
	$1\frac{1}{2}$	$\frac{1}{2}$-in. fiber insulation board both sides of timber frame.

Table 6   Average Airborne Insulation of Windows and Doors[24]

Approximate Insulation (dB)	Construction
45	Two 2-in. solid wood doors with edges adequately sealed and a sound lock.[a]
40	Double windows of 32-oz glass, 8-in. airspace, tightly sealed, absorbent in reveals.
35	Double windows, 26- or 32-oz glass, 4-in. airspace tightly sealed, absorbent in reveals.
	Double windows, 8-in. airspace, absorbent in reveals; opening lights closed but not sealed.
	Two framed flush-paneled doors with edges adequately sealed and a sound lock.
30	Single $\frac{1}{4}$-in. plate glass window sealed. Single 2-in. solid wood door with edges sealed.
25	Single 26- or 32-oz glass window sealed. Single 2-in. solid wood door, no sealing precautions taken.
20	Single 26- or 32-oz glass openable window closed with no sealing precautions taken.
	Framed, flush-paneled door with edges adequately sealed.
15	Framed flush-paneled door with cracks at the edges.

[a] A sound lock should provide a minimum distance of 8 in. between the doors and incorporate as much sound absorption as practicable.

airborne sound insulation than a single pane of glass of equivalent mass per unit area.

Sound isolation in building interiors is achieved by either massive walls and flooring or by discontinuous construction. In commercial buildings with concrete floors, it is common to separate rooms by nonbearing concrete block walls. In residences, sound transmission between rooms is improved by having the discontinuous wall structure indicated in Fig. 12. In this case 2 × 4 in. studs are offset and make contact with one wall surface, and an insulating blanket is included. The attenuation of low frequency sound is especially aided by this technique. Freedom from "acoustic leaks" is as important as basic construction; a 1 in.^{2} hole through a 40 dB wall will transmit as much acoustical energy as 100 ft^2 of the wall.[25]

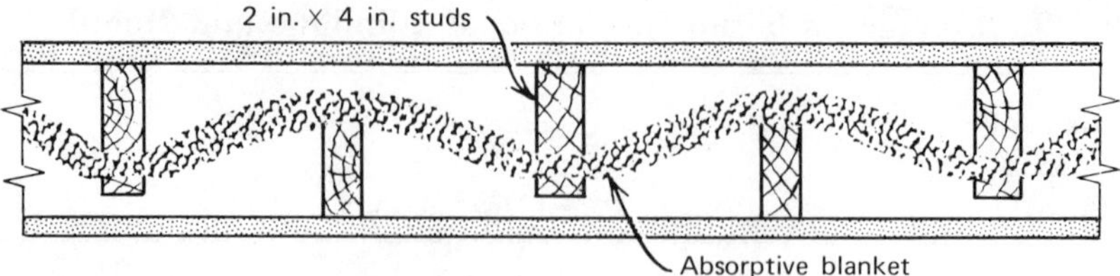

Figure 12   Use of staggered 2 × 4 in. studs to improve the sound attenuation properties of interior walls.

Some recent studies have been reported that provide special design guidelines for houses that must be placed near an airport.[26] Acoustic insulation calculations for windows, transmission losses for ceiling roof spaces, and a rating for walls has been made specifically. By using attic spaces as shields and by giving careful consideration to windows, doors, foyers, air vents, chimneys, and so on, it is suggested that indoor levels may be achieved that are substantially lower than can be attained by standard construction methods.

BUILDING CODES

Although the architect has the technical responsibility for designing housing that includes noise insulation, the general upgrading of both commercial and residential construction will probably require the adoption of new building codes. In Europe many cities have building codes that specify noise transmission limits for airborne and, sometimes, impact noise. High population densities and narrow streets, for example, are factors. In England, new constructon codes followed the Great Fire of London in 1666. Massive brick walls were specified to isolate apartments from fire hazard, and the net result was that urban construction became both fireproof and relatively soundproof.

In the United States major cities can be expected to adopt building codes that will tend to isolate a residence from outdoor traffic noise, and that will also tend to contain the sounds generated by a "noisy neighbor," whether the neighbor is an industrial or commercial enterprise or a residential dweller. Another important factor relates to the nation's energy problem and the need for conserving our limited reserves of oil and natural gas. Existing building codes are primarily concerned with structural safety and fire prevention, but do not treat problems of heat loss and thermal insulation. In view of current energy concerns, we may expect future building codes to have specifications on heat losses from walls and windows, and hopefully noise transmission losses can also be included. A noise-isolated structure will usually have good thermal insulation properties, although not necessarily vice-versa. In any event, there exist straightforward engineering techniques that can be applied to new building construction, that can result in vast savings of fossil fuel as well as a substantial noise reduction. High density, double-glazed windows (with a substantial air gap between panes) can furnish both thermal and sound insulation, and massive or composite walls can also be designed that will provide 50 dB sound insulation. This type of construction is not cheap, but in northern climates and over a period of several years, it can be amortized on the basis of fuel savings alone.

ENVIRONMENTAL NOISE—A NEW RATING SCHEME (L_{eq})

Whereas dB(A) numbers provide a simple and convenient scale for measuring noise, a new rating scheme has recently been proposed for quantifying noise exposure under varying conditions and over long periods of time. Termed an *equivalent sound level* L_{eq},[27] it is designed to give the equivalent steady noise level that in a given period, would contain the same noise energy as the time-varying noise. Such a scale is thus useful in integrating the fluctuating noise from traffic, aircraft, playgrounds, and other community sources. The tradeoff of level and duration to noise exposure seems to apply also to hearing loss, and it even applies to annoyance in some situations.

The three versions of equivalent sound level listed below have been identified, corresponding to the time periods during which the noise is averaged.

1. $L_{eq(8)}$. This is the equivalent A-weighted sound level computed over an 8-hour workday.

2. L_{dn}. The 15-hr day time integration from 7 a.m. to 10 p.m. is identified as L_d; the 9-hr integration is identified as L_n. L_{dn} is defined as the A-weighted average sound level in decibels during a 24-hr period with a 10 dB weighting applied to nighttime sound levels.

3. $L_{eq(24)}$ is used to demote the A-weighted average sound level for a 24-hr period, with no day or night consideration.

Figure 13 shows examples of outdoor community noise levels in terms of the L_{dn} scale.

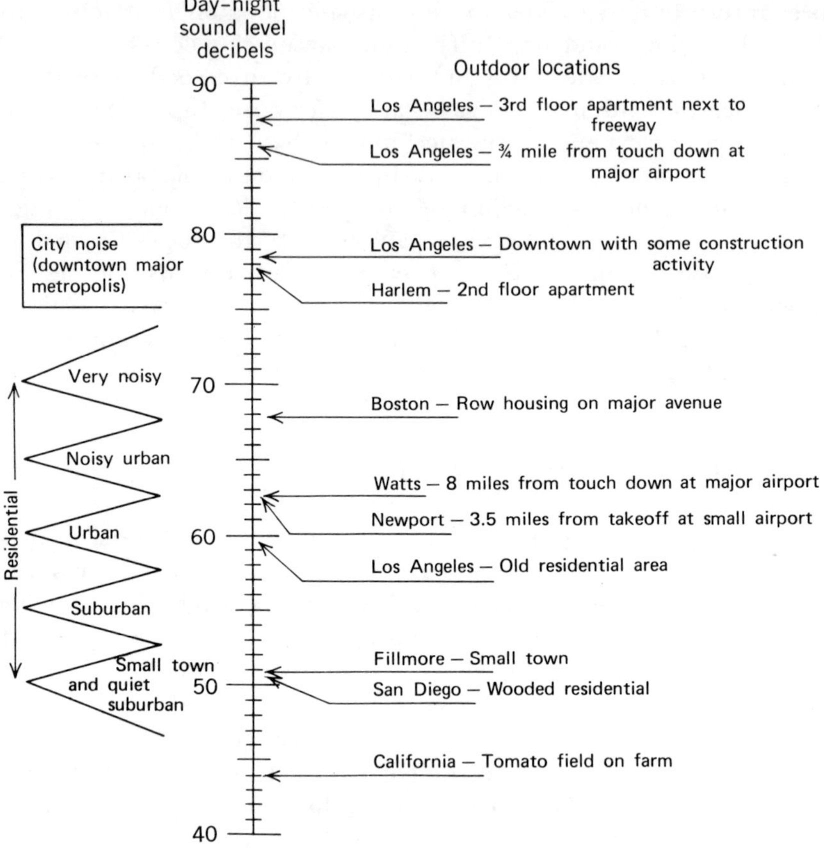

Figure 13   Examples of outdoor day-night sound level in decibels (re 20 μPa) measured at various locations. (Ref. 27)

CONSERVATIVE LEVELS OF COMMUNITY NOISE

Recently, the U. S. Environmental Protection Agency has sought to provide information on levels of noise "requisite to protect health and welfare *with an adequate margin of safety."* A statistical determination has been made of the lowest noise level at which some deleterious effect might occur, and then an additional "margin of safety" has been introduced. These levels are subject to considerable qualification, and "they are not to be construed as standards as they do not take into account cost or feasibility."[28] Table 7 lists these recommendations, however, in terms of the new noise level nomenclature that is outlined in the preceding paragraph.

Table 7   Summary of Noise Levels Identified as Requisite to Protect Public Health and Welfare With an Adequate Margin of Safety

Effect	Level	Area
Hearing loss	$L_{eq(24)}. \leq 70$ dB	All areas.
Outdoor activity interference and annoyance	$L_{dn} \leq 55$ dB	Outdoors in residential areas and farms and other outdoor areas where people spend widely varying amounts of time and other places in which quiet is a basis for use.
	$L_{eq(24)}. \leq 55$ dB	Outdoor areas where people spend limited amounts of time, such as school yards, and playgrounds.
Indoor activity interference and annoyance	$L_{dn} \leq 45$ dB	Indoor residential areas.
	$L_{eq(24)}. \leq 45$ dB	Other indoor areas with human activities such as schools, etc.

REFERENCES

1. L. L. Beranek, *Noise and Vibration Control,* McGraw-Hill, New York, 1971, p. 573.

2. J. D. Miller, *Effects of Noise on People,* U. S. Environmental Protection Agency, Washington, D. C., 1971, p. 102.

3. T. J. Schultz, *Community Noise Ratings,* Applied Science Publications, London, 1972, p. 64.

4. C. R. Bragdon, *Noise Pollution,* University of Pennsylvania Press, Philadelphia, 1971, p. 169.

5. Miller, *op. cit.,* p. 107. Also, J. D. Miller, *J. Acoust. Soc. Am.,* **56**, 729 (1974).

6. R. J. Wells, private communication.

7. U. S. Department of Commerce, *The Noise Around Us,* Washington, D. C., 1970, p. 63.

8. *Ibid.*

9. *Highway Report 75,* National Cooperative Highway Research Program, National Academy of Science, Washington, D. C., p. 30.

10. *Ibid.,* p. 2.

11. *Highway Report 117,* National Cooperative Highway Research Program, National Academy of Science, Washington, D. C., 1971, p. 29.

12. NASA Report, CR-410, *A Study of the Optimum Use of Land Exposed to Aircraft Landing and Take-off Noise,* Washington, D. C., 1966, p. 35.

13. *Ibid.,* p. 43.

14. M. W. Schulz, Jr. and R. J. Ringlee, *Power Apparatus and Systems,* 1 (June 1960).

15. Rochester Gas & Electric Company, private communication.

16. D. E. Perry, *Power Apparatus and Systems,* IEEE, 1972, p. 854.

17. *Electric Vehicle News,* 25 (November 1972).

18. A. B. Lawrence, *Architectural Acoustics,* Elsevier, London, 1970.

19. L. F. Yerges, *Sound, Noise and Vibration Control,* Van Nostrand Reinhold, New York, 1969.

20. P. H. Parkin and H. R. Humphreys, *Acoustics Noise and Buildings,* Praeger, New York, 1958.

21. B. F. Day, R. D. Ford, and P. Lord, *Building Acoustics,* Elsevier, London, 1969.

22. H. J. Purkis, *Building Physics: Acoustics,* Pergamon Press, London, 1966.

23. *Ibid.,* p. 68.

24. *Ibid.,* p. 65.

25. Yerges, *op. cit.,* p. 43.

26. R. J. Donato, *P. Acoust. Soc. Am.,* **53,** 1025 (1973).

27. "Information on Levels of Environmental Noise Requisite to Protect Public Health and Welfare with an Adequate Margin of Safety," U.S. Environmental Protection Agency, Washington, D. C., 1974).

28. *Ibid.*

III

SPEECH, MUSIC, AND PHYSIO-PSYCHOLOGICAL EFFECTS OF SOUND

14

THE SOUNDS OF SPEECH

The discrete elements of speech are characterized by relative sound pressure levels, duration, and spectral content. They may or may not have a fundamental frequency, depending upon how the vocal cords modulate the outgoing air stream. In male speech, frequencies can extend as low as 50 Hz; high frequency components of speech may extend to 8000 Hz or more. However, no single acoustic parameter ever completely identifies a speech sound, and speech recognition is conditioned by the expectations of the listener, the context, the rules of grammar, and other subjective elements.

THE VOCAL APPARATUS

The precise manner in which speech sounds are produced depends upon the complex interplay of the many components that comprise the vocal tract. Figure 1 is a simplified schematic diagram of the system. This system includes the lungs, the trachea, vocal cords, a larynx tube and pharynx cavity, mouth, and nasal cavities. There are also the *articulators* whose specific position or action modulates sound, that is, the jaw, tongue, lips, and velum.

The lungs act as the reservoir and, in combination with their respiratory muscles, they represent the vocal power supply. The production of

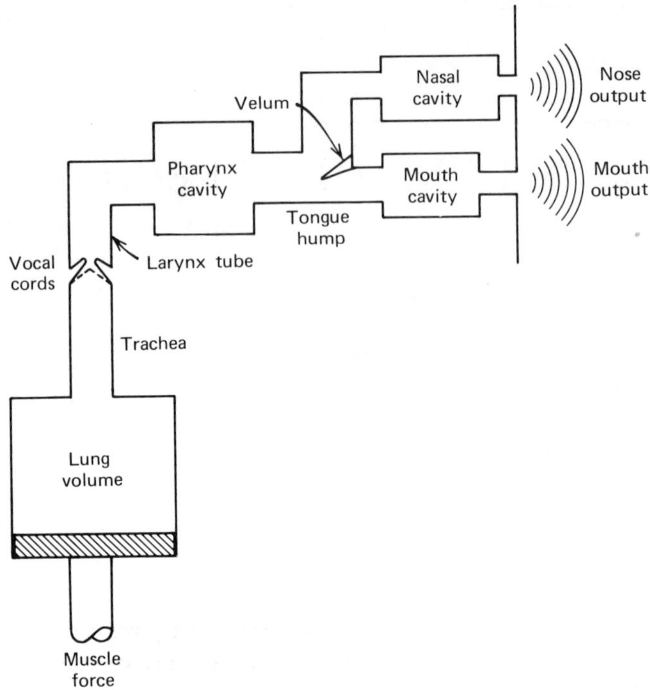

Figure 1   Schematic diagram of the power supply and resonant cavities of the human voice.

vowel sounds at the softest possible level requires a lung pressure of about 4 cm of H_2O; for very loud speech sounds, pressures of five times or greater than this value are required.[1] In any event, the differential pressure above atmospheric pressure is very small. Air is drawn into the lungs by increasing the volume of the chest cavity and lowering the diaphragm. Air is expelled upon contraction of the respiratory muscles and action of the diaphragm. During speech, a steady lung pressure is maintained by a gradual and controlled contraction. When a vowel sound is uttered, 40 to 200 cm^3 of air are forced from the lungs through the larynx each second.[2]

The vocal tract proper may be considered as an acoustic tube about 17 cm in length. It extends from the vocal cords to the lips, and its cross section is highly nonuniform. This cross section of the vocal tract at the mouth is variable from complete closure to about 20 cm^2, depending upon the movement of the articulators. The nasal cavity provides a parallel path for sound transmission, extending from the velum to the nostrils. An acoustic coupling between the vocal and nasal tracts is controlled by the action of the velum.

The source of *voiced* sounds is the vibratory motion of the vocal cords, which has been studied in great detail with the aid of high speed motion pictures and stroboscopic illumination. Basically, the vocal cords function like the vibrating lips of a musician who is blowing a trumpet. Lip vibrations allow quasiperiodic pulses of air to generate resonances in the instrument. In an analogous fashion, the vocal cords permit pulses of air to pass on to the several cavities and constrictions of the vocal tract. The production of *unvoiced* or *breath* sounds occur without the vocal cords. A steady forcible exhaling of breath can be modulated by the articulators (as in whispering). An analysis of this sound shows that it is broad band and that the upper frequencies are dominant.

Virtually all of the sections of the vocal tract play some role in the production of speech sounds, that is, (1) the pharynx cavity, (2) the narrow region where the tongue is humped, (3) the constriction of the velum, (4) the nasal cavity, (5) the large oral cavity, and (6) the radiating apertures of the lips and nostrils.

Collectively, we can consider the overall vocal apparatus as comprising (1) a power supply (the lungs), (2) a vibrator (vocal cords), and (3) a system of valves and resonators (the vocal tract).

ACOUSTIC POWER OF SPEECH

The acoustic power of speech varies over many orders of magnitude. The power associated with whispering is only about 10^{-3} μW (10^{-9} W). A person speaking at a normal conversational level might generate 10 to 20 μW; a loud voice may have a power output of 100 μW (10^{-4} W). Upon shouting, a speaker may develop a power output of 10^{-3} W or greater. Long term average power for male talkers is about 34 μW, and for women, 18 μW. Short term levels can be much higher; 1% of all 1/8 sec intervals in a long term average indicate an excess of 230 μW for men and 150 μW for women.[3] However, the acoustic power of speech is exceedingly small. If the entire population of the world were whispering simultaneously, the total acoustic power would only be about 10 W; at normal conversational levels the entire human race might develop a total acoustic power of less than 100 kW.

Sizable fluctuations occur, of course, in the level of speech power over short intervals for different speech sounds. The duration of a syllable may be 0.2 sec in normal conversational speech, and the power level will vary markedly from syllable to syllable. For example, the acoustic power of the vowel *o* as in *low* is about 50 μW, whereas the weak consonant *v* has an average power of only 0.03 μW.[4]

There are 42 such basic linguistic elements, called *phonemes* in the English language, comprising vowels, diphthongs, and consonants. The relative acoustic power associated with a few of these *phoneme* utterances are listed in Table 1.

Table 1   Relative   Acoustic Power   of   Speech   Sounds   in English   Speech[5]

Sound	Relative Power (dB)[a]
ah (ma)	28
ō	27
ē	23
R	23
L	20
SH	19
M	17
S	12
K	11
B, D	8
F	7
TH	0

[a] The weakest sound, TH, is arbitrarily represented as 0 dB.

Measurements have also been made of the distribution of speech power with frequency, for speech averaged over long time intervals.[6] Figure 2 shows the average spectrum for male and female; the ordinate represents the *pressure spectrum level*, defined as the sound pressure level in bands of 1 Hz width, measured at a distance of one meter from the mouth of the speaker. At this distance the integrated overall sound pressure level of all frequencies from 10 Hz to 10 kHz is about 65 dB, based upon a total acoustic output from the speaker of 20 μW. While the curve of Fig. 2 shows something about the power distribution vs. frequency averaged over long intervals, it tells little about the specific spectrum of discrete speech sounds. For vowels, the *mean* fundamental frequency, F_0, of the

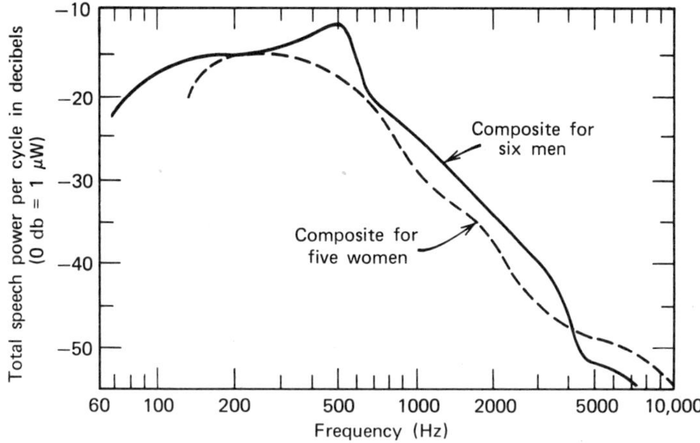

Figure 2   Relative speech power as a function of frequency, for men and women. (From H. K. Dunn and S. D. White, *J. Acoust. Soc. Am.*, **11**, 278 (1940) and V. O. Knudsen and C. M. Harris, *Acoustical Designing in Architecture*, Wiley 1950).

male voice (a lowest dominant tone) is about 124 Hz; for females this fundamental frequency is about 244 Hz.[7] However, the human voice contains several strong frequency bands, and it is these bands that characterize vowel sounds.

SPEECH FORMANTS

Those portions of the voiced spectrum in which the frequency components have relatively large amplitudes are called *formants*. Formants are of particular importance in connection with the utterance of vowels and diphthongs, for example. In such sounds the characteristic spectra will depend on the shape or configuration of the vocal tract, and there will be *frequency regions* that can be regarded as broad frequency resonances for a particular configuration. Figure 3 shows a speaker talking into a microphone that is connected to a spectral analyzer, and these resonances or formants are clearly evident.

Every vowel sound has three or more formants. These formants are labeled F1, F2, and F3, and their maxima will usually be contained within frequency bands of 250 to 700 Hz, 700 to 2000 Hz, and 2000 to 3000 Hz, respectively, in adult voices. The center frequencies of formant

Figure 3 The formants or resonances of speech are characteristic of the shape and configuration of the vocal tract. (Courtesy of Computer Signal Processors, Inc.)

bands of 10 vowel sounds, are listed in Table 2. Figure 4 shows the formant structure of the vowel [æ], as in "had" for a male and child's voice.[8] The exact formant structure will, of course, depend upon not only the speech sound but upon the individual speaker, and the phonetic environment in which it is produced. Formants thus give speech a characteristic quality in much the same sense that broad frequency bands are associated with the spectral output of various musical instruments (Chapter 15). In fact, we could say that there is a very rough analogy between speech formants and musical chords. A chord is recognizable independent of its pitch position, as the several frequencies comprising the chord have well-defined frequency ratios. We might also say that, for a given individual, the various vowel sounds will have spectral band frequencies whose ratios are substantially invariant to the voice pitch.

Using X-ray cine-radiography and voice spectrograms,[9] it has been found that female speakers have larger average frequency differences be-

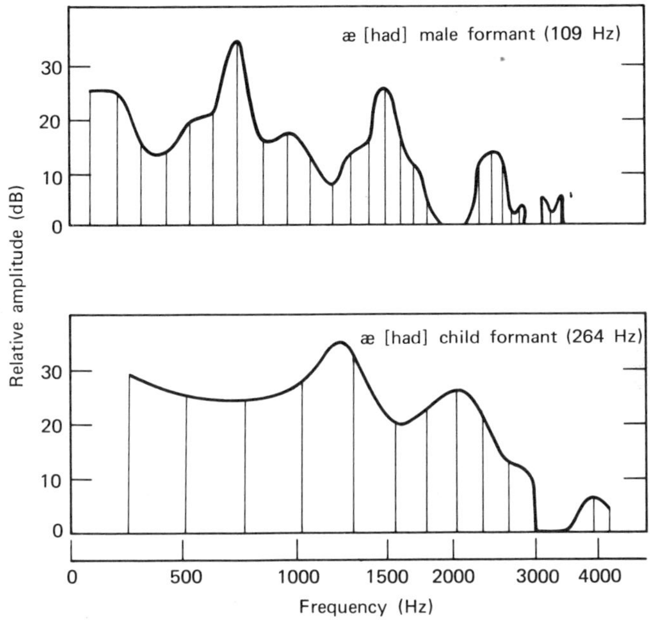

Figure 4 Formant pattern of the vowel [æ] as in the word "had," for a male and a child's voice; the fundamental frequencies are 109 Hz and 264 Hz, respectively. [From R. K. Potter and J. C. Steinberg, *J. Acoust. Soc. Am.*, **22**, 812 (1950).]

Table 2 Center Frequencies of Formant Bands[7]

	ee [i]	I [I]	e [e]	æ [æ]	ah [a]	aw [ɔ]	u [ʊ]	u [u]	v [ʌ]	er [ɝ]
First Formant Frequency (Hz)										
Male:	270	390	530	660	730	570	440	300	640	490
Female:	310	430	610	860	850	590	470	370	760	500
Second Formant Frequency (Hz)										
Male:	2290	1990	1840	1720	1090	840	1020	870	1190	1350
Female:	2790	2480	2330	2050	1220	920	1160	950	1400	1640
Third Formant Frequency (Hz)										
Male:	3010	2550	2480	2410	2440	2410	2240	2240	2390	1690
Female:	3310	3070	2990	2850	2810	2710	2610	2670	2780	1960

tween formants than do males; the length of the vocal tract is correlated with the frequency separation, and the somewhat shorter cavity length in females results in the greater separation. There is also some difference in the *spectral width* of a single formant in children and adults. For example, the first formant, Fl, of [i] was found to extend from 254 to 610 Hz and 250 to 430 Hz for children and adults, respectively, indicating that the specific nature of formants is reflected in the dimension of the vocal tract.[10]

SPEECH INTELLIGIBILITY AND BANDWIDTH

We have seen in Fig. 2 that much of the acoustic power of speech is concentrated in the frequency range of 100 to 1000 Hz. Despite this fact, the higher frequencies are of great importance, for it is the higher frequency bands of speech that make the consonants of speech intelligible.

Experimental tests to determine an intelligibility-frequency relationship require (1) special filters so that certain frequencies may be suppressed and (2) some type of speech recognition or articulation tests. Syllable or phoneme tests are usually classified as articulation tests; word, phrase, and sentence tests are classified as intelligibility tests. The marked dependence of syllable recognition on bandwidth is shown in Fig. 5. The data were obtained by having a speaker read meaningless syllables (vowel-consonant combinations) to a group of observers having normal hearing.[11] When the speaker read the list of syllables through an electrical transmission system having a variable upper frequency cutoff, it was noted that if all frequencies above 250 Hz were removed, the syllables were unintelligible. With an upper cutoff frequency of 2000 Hz, about half of the syllables were identified. Only when the frequency band of the electrical transmission system was extended to 7 kHz, did syllable intelligibility equal that for direct communication. The intelligibility of words and phrases in connected speech, of course, is higher because some words can be guessed from the context.

Furthermore, in extensive word recognition tests carried out in connection with the design of communication systems, it was found that 20 frequency bands could be specified that provide equal contributions to speech intelligibility. These 20 contiguous bands (for equal signal-to-noise ratios) are listed in Table 3.

These and other experiments have shown that the upper frequencies of speech are as important in providing speech intelligibility as are the upper harmonics of musical sounds in determining musical quality.

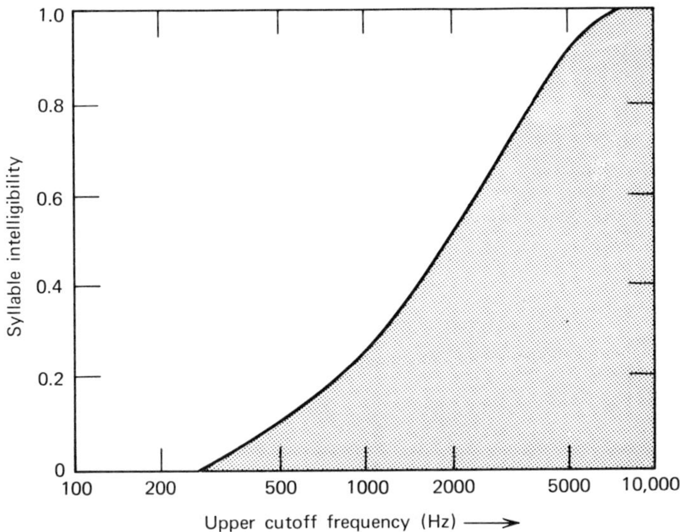

Figure 5   Syllable intelligibility as a function of the upper cutoff frequency. (From E. Meyer and E. G. Neumann, *Physical and Applied Acoustics,* Academic Press, 1972.)

A qualitative explanation of the importance of the higher frequencies relates to the fact that it is the *modulation* of "carrier" frequencies that provide speech intelligibility in the same sense that a radio signal conveys information only if it is modulated. In the case of a radio signal the carrier wave form is sinusoidal and the frequency is above the audible range. For voiced sounds the carrier wave form is complex with many frequencies lying within the audible range, and it is the *amplitude modulation* of these frequencies that provides the information transfer. The potential information transfer rate also increases with increasing bandwidth. In speech containing both vowels and consonants, both voiced and breath sounds comprise the sound carrier; in whispering, the steady emission of breath sound alone serves as the carrier. The *frequency modulation* of the carrier is the speech quality known as inflection, that is, the raising or lowering of the voice pitch.

The actual rate of information transfer in ordinary speech is quite limited in terms of information theory and in consideration of the bandwidth of the voice wave form. In normal conversation, about 10 phonemes are uttered per second, that is, one phoneme per 100 msec. Thus we see that the normal utterance rate of speech is not much different than is warranted by the average resolution time of the human ear (~50

Table 3   Twenty Frequency Bands of Equal Contribution to Speech Intelligibility[12]

Band	Frequency Limits (Hz)	Band	Frequency Limits (Hz)
1	200–330	11	1660–1830
2	330–430	12	1830–2020
3	430–560	13	2020–2240
4	560–700	14	2240–2500
5	700–840	15	2500–2820
6	840–1000	16	2820–3200
7	1000–1150	17	3200–3650
8	1150–1310	18	3650–4250
9	1310–1480	19	4250–5050
10	1480–1660	20	5050–6100

msec). This, however, is a written equivalent of an information transfer of less than 50 bits*/sec.[13]

A comparison of this rate with the potential rate of information transfer of the human voice channel is dramatic. Shannon[14,15] has shown that if a channel has a bandwidth of BW Hz, and signal and noise powers S and N, respectively, a method of coding is possible such that information can be signaled with little error, at a rate

$$C = BW \log_2 \left[1 + \frac{S}{N} \right] \text{ bits/sec} \tag{1}$$

Now a human voice channel has a bandwidth of about 3000 Hz and a signal-to-noise ratio of about 30 dB. According to the above formula the voice channel has an inherent capacity to transmit information at rates of the order of 30,000 bits/sec.

This immense discrepancy between the written equivalent of voiced phonemes and the voice channel potential has generated substantial research. It has been suggested that one reason for this discrepancy is that the waveform of speech is essentially a continuous one and its information transfer potential cannot be properly estimated by a written equivalent code. A speech, for example, conveys many moods and points of emphasis that are never evidenced in a printed script. In addition, there

---

* A unit of information equivalent to the result of a choice between two equally probable alternatives, and a shortened form of the term "binary digit" (0 or 1).

appears to be an upper limit on the rate at which the human brain can actually process information; a value of 50 bits/sec has been suggested as such a limit.[16]

ACOUSTICAL CORRELATES OF EMOTION IN SPEECH

Recent research has focused on correlating speech formants and other speech parameters with the emotional state of a speaker.[17] Results of these studies have shown that the fundamental frequency (F_0) and the range of the fundamental frequency band are substantially altered. Tape recordings of historical events having high emotional content have been of special interest. On May 6, 1937, the lighter-than-air dirigible *Hindenburg* was approaching Lakehurst, New Jersey after an historic trans-Atlantic voyage, and a radio announcer was providing a graphic description of its arrival—"here it comes, ladies and gentlemen." Suddenly, this hydrogen-filled zeppelin burst into flame and the announcer continued—"it's crashing, it's crashing terrible—." The pre- and post-moments of this disaster have been analyzed by voice spectrographs, and Table 4 shows the substantial increase in the median fundamental frequency, and in the extent of the frequency band.

Table 4   Change in Frequency with Emotional State[18]

	Fundamental Frequency F_0 (Hz)	F_0 Range (Hz)
Taped radio description:		
Before crash	166	124–196
After crash	196	152–260

Other controlled experiments with professional actors have shown a marked shift of the median fundamental frequency, F_0, for voices speaking in sorrow, neutral, fear, and anger situations, as shown in Fig. 6. The mean rate of articulation (syllables per second) for utterances in sorrow, fear, and anger was reported to be 1.9, 3.8, and 4.1, respectively.[19]

Such experiments do not always follow a consistent trend, and allowance must be made for pitch changes that accompany speaking condi-

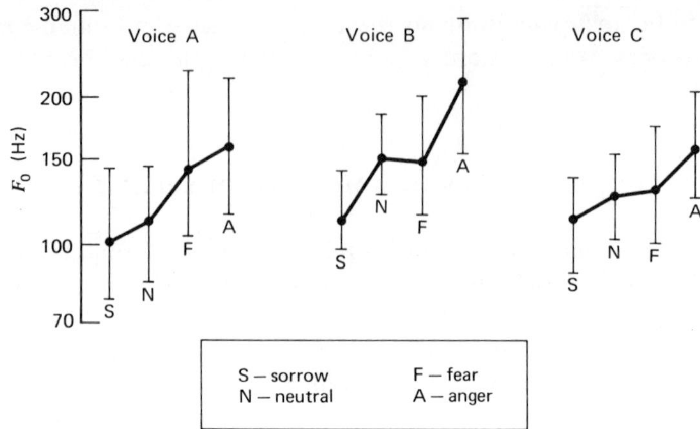

Figure 6   Median fundamental frequency F_0 for each of three voices speaking in sorrow, neutral, fear, and anger. [From C. E. Williams and K. N. Stevens, *J. Acoust. Soc. Am.*, **52**, 1238 (1972).]

tions from "normal" to "loud" to "soft." For example, the fundamental frequency may increase about 40% for loud speech, and decrease about 12% for soft speech—with respect to normal.[20] However, most experiments suggest that data can be obtained that reflects an emotional condition, and it is believed that such data will be of value if the physiological and emotional state of an individual needs to be determined.

PARAMETERS RELATING TO THE INTELLIGIBILITY OF SPEECH SOUNDS

Signal-to-Noise Ratio

Over a wide range of ambient noise levels, the intelligibility of speech will depend on the *relative* sound pressure level of speech signals to that of the ambient noise. This ratio is termed the *signal-to-noise* ratio (S/N). In this definition it is assumed that the noise is broad band and does not include strong pure tone components. The fact that it is the ratio that is important can be observed from common experience. In a quiet room, speech signals need only exceed the ambient by 10 to 20 dB, and in a noisy auditorium speech can still be heard clearly, provided the electronic amplification system boosts the speaker's voice above the general noise level. Speech is even quite intelligible in a football stadium if sufficient amplification is given to an announcer's voice. Quantitatively, it has been found that thresholds of intelligibility and the perception of speech, expressed in terms of a signal-to-noise ratio, are independent of the noise level in environments of 30 to 110 dB.[21] Only at very high noise levels ($>$ 110 dB) does a deterioration in intelligibility occur,* when the same S/N ratio is maintained.

The intelligibility of speech, of course, increases perceptibly as the S/N ratio increases. Figure 7 shows this increase, and it also shows that a reduction in the number of possible choices, that is, digits or the use of words in sentences, provide improved intelligibility scores for the same S/N ratio.[22] Intelligibility scores are also conditioned by speech rates.[23]

Talker-Listener Distance

In some instances where it is not feasible to reduce the background noise level, communication only becomes possible by (1) increased speech effort and (2) a decrease in the talker-listener distance. Figure 8 is a graphical representation of how these two parameters can be expected to change as a speaker attempts to communicate in an increasingly noisy environment.[24] The curves suggest the vocal effort for an average male voice that is required to provide good intelligibility of continuous speech, for increasingly noisy environments and for various talker-listener distances. In a background of noise of about 50 dB(A), communication at an average voice power is adequate for a distance of 16 ft. At succeedingly higher background noise levels, adjustments in either talker-listener distance or voice power are required.

In an interesting analysis of the acoustical dynamics of cocktail party noise,[25] it is observed that each talker, in trying to override the background noise will increase his voice power in small increments, after which the exponentially increasing voice power (of all guests) will be to no avail. A signal-to-noise ratio, which allows some information transfer, is then achieved only by all members of the group adjusting their talker-listener distances to distances that are much less than occur in any other dialogue.

* Because of this effect, the use of ear plugs actually improves speech intelligibility at very high noise levels.

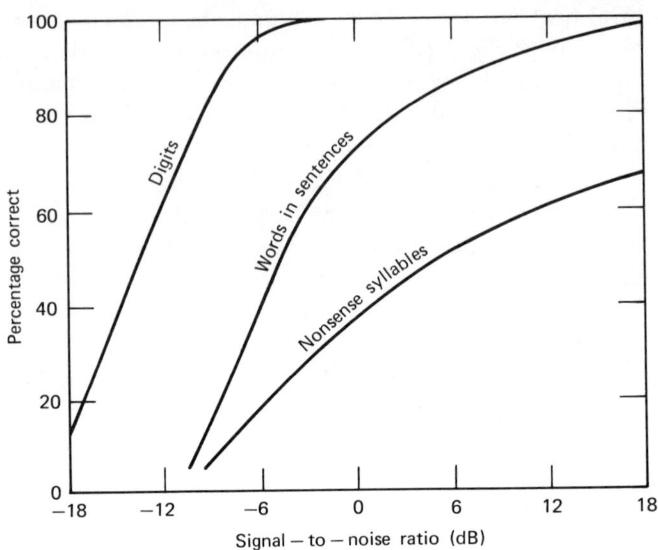

Figure 7   Relationship between intelligibility scores and signal-to-noise ratios. (From G. A. Miller et al., Ref. 22.)

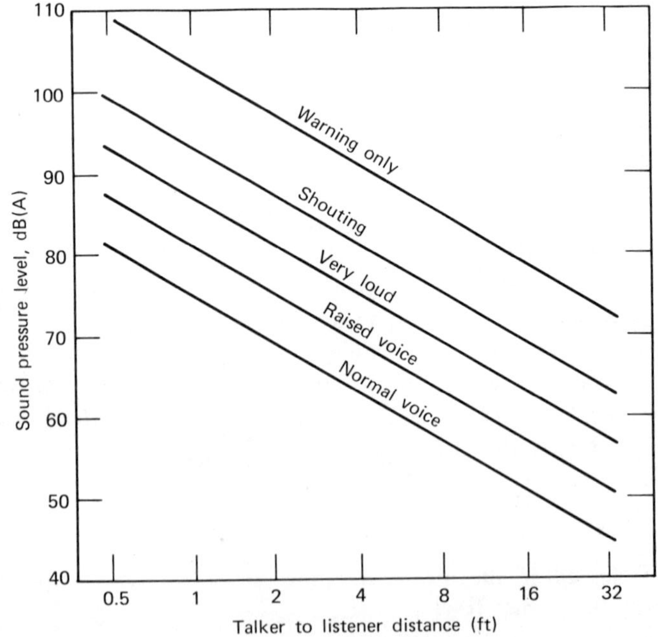

Figure 8   Quality of speech communication as a function of background noise and talker-listener distance. (Courtesy of Am. Petroleum Institute, Ref. 24.)

378

Articulation Index

In order to identify the purely physical parameters that relate to intelligibility of speech, the 20 frequency bands of Table 2 are assumed as comprising the entire speech spectrum. Each band is also assumed to contribute a *maximum* of 0.05 to the total *articulation index (AI)*; the actual contribution of each band will depend upon its own signal-to-noise ratio in real speech.[26] Experiments have shown that a 30 dB *change* in signal-to-noise will span the entire range of intelligibility scores from 0 to 100%, so that each decibel in each band contributes 1/30th of the 0.05 maximum.[27] With adequate instrumentation it is then possible to make acoustic measurements that permit the prediction of word or sentence intelligibility. This correlation of intelligibility with articulation index, *AI*, is shown in Fig. 9. It should be noted that an *AI* of 1.0 corresponds to ideal *S/N* ratios in each of the 20 frequency bands, but this ideal speech spectrum is rarely needed to recognize most speech sounds. However, the index is a useful scale by which physical changes in room acoustics can be monitored, and by which corresponding listening conditions or speech quality can be assessed. The changes in the articulation

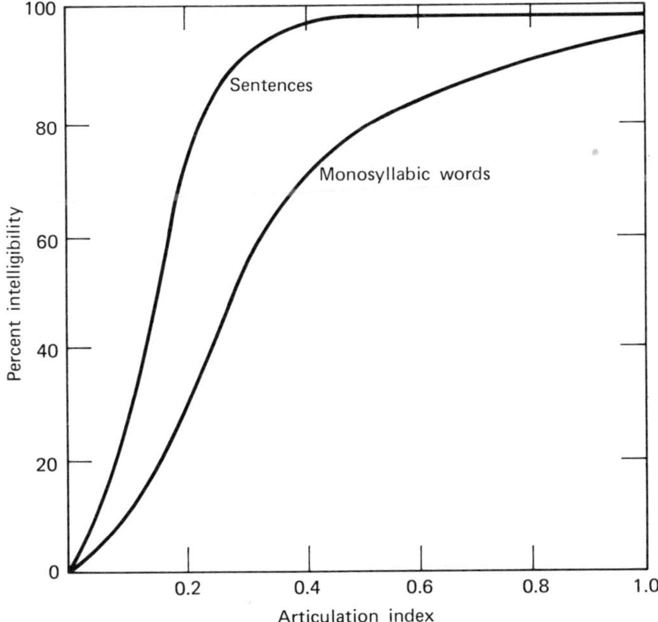

Figure 9   Intelligibility of sentences and words as a function of articulation index. (Ref. 26.)

index, ΔAI, due to changes in room acoustics are noted in subsequent sections.

Effect of Reverberation

In large rooms with highly reflecting walls, speech intelligibility will be decreased because of the long reverberation time. At normal speech rates, words will overlap and the signal-to-noise ratio of all words will decrease. The decrease in articulation index, ΔAI, as a function of reverberation time, is given in Table 5.

Table 5   Decrease   in Articulation Index With Increasing   Reverberation Time[28]

Reverberation Time (sec)	ΔAI
0	0
1	0.10
2	0.23
3	0.38
4	0.50
5	0.61

Speech Interference Level

A special noise index has been devised to assess the presence of frequencies that are most likely to interfere with speech communication. This index is called the *speech interference level (SIL)*. It is determined by taking noise measurements at three octave bands centered at 500, 1000 and 2000 Hz. The sound pressure levels of these octave bands are then added and divided by three, and the result is defined as the speech interference level of that environment. This is a particular instance in which decibel readings are added as linear units and averaged.

The index is a simple one to determine, and it provides a more meaningful measure of speech-interfering noise than A-weighted decibel readings. If speech is to be understood at close to 100% sentence intelligibility, the *SIL* measured at the listener's ear should be at least 12 dB below the overall sound pressure levels of speech at the same position. Maximum acceptable levels in various speaking situations are listed in Table 6.

Table 6   Speech Interference Levels in Rooms[29]

Type of Room	Maximum Acceptable SIL (dB)
Classroom	30
Conference room	30–35
Small private office	45
Typing room	60
Telephone booth	60

COMMUNICATION IN "OPEN PLAN" ROOMS AND LANDSCAPED OFFICES

The concept of "open plan" space utilization dates back to the advent of architecture itself. The current trend to the landscaped office or the large "open plan" schoolroom, however, poses new problems. Specifically, such rooms are supposed to afford conditions appropriate for speech and some speech privacy as well as to accommodate other functions. Such multiple requirements often lead to design compromises that are acoustically unacceptable.

In the past, open plan areas were rarely intended to be areas good for speech. In large industrial shops, the building had to be "open" so that overhead cranes could transport heavy materials from one work station to another; a foreman then walked from one building area to another to advise and instruct. Large office spaces were designed for compatible functions, that is, accounting, purchasing, drafting, and so on, where a primary function could be achieved without great consideration to speech or speech privacy. Large schoolroom areas were also restricted to gymnasiums, libraries, or group activity rooms where speech criteria played a minor role in their design.

In general, the "open plan" can provide neither the acoustical privacy nor the lower noise levels commensurate with that of multiple classrooms or the individual office. Nevertheless, flexibility in room use, economics, and other factors sometimes point up the desirability of open plan space utilization. Hence it is appropriate to inquire as to the acoustic compromise that results in an open plan situation. Most important, what acoustic factors affect the articulation index (*AI*) in environments that must accommodate speech communication?

The five variables that contribute to a change of the articulation index or speech privacy in open planning have been identified and assessed by Pirn[30] in Table 7.

Table 7   Factors Relating to a Change in Articulation Index in Open Planning[30]

	Approximate maximum *AI* change
Background noise	0.90
Speaker-to-listener distance	0.75
Speech effort	0.60
Attenuation by barriers	0.40
Speaker orientation	0.30

The change of *AI* as a function of background noise is presented in Fig. 10.

Each 6-dB increment in the ambient-noise level generally results in an *AI* decrease of 0.2. In a large open plan area, a 6 dB differential in noise level from a "high noise" to "average noise" area is not uncommon. Therefore, speech intelligibility may vary throughout the space—assuming that other factors are invariant. Nevertheless, there is really no such thing as a "quiet zone" in an open plan as long as there is a single noisy source—whether the source is contained within the room or whether noise is transmitted through a wall, window, or duct. Also, the "quiet groups" must always share in the noise that is generated by those engaged in noisy activities.

The *AI* will also vary with the speaker-listener distance. Because most open plan spaces are acoustically treated so as to be nonreverberant, the drop-off in speech level may approximate the 6 dB decrement that occurs outdoors with each doubling of the distance. A decrease of 5 dB with doubling of the speaker-listener distance is reasonable. In practice this

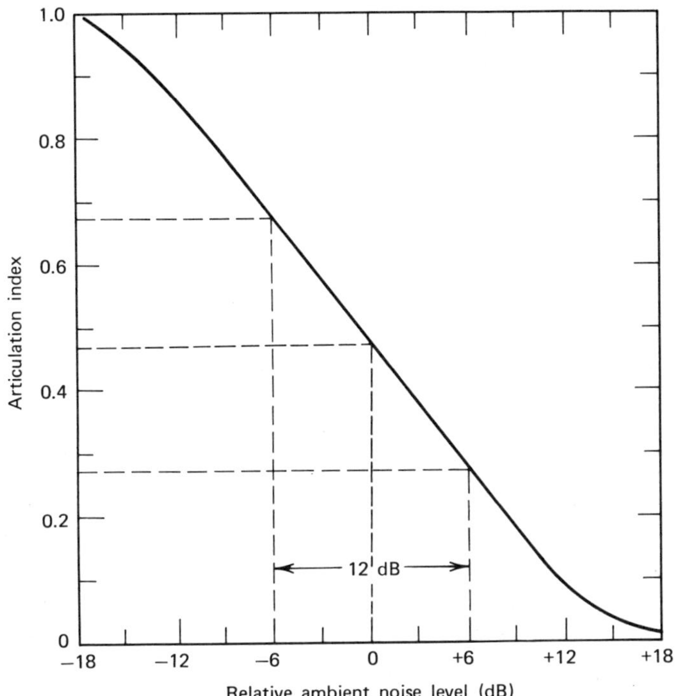

Figure 10   Articulation index as a function of background noise. A 6dB incre-ment in the ambient noise level results in an articulation index decrease of about 0.2 over much of the *AI* range. [From R. Pirn, *J. Acoust. Soc. Am.*, **49**, 133 (1971).]

means that persons in close proximity (i.e., adjacent desks) can com-municate easily without being overheard by persons in remote areas of the room.

The open plan environment also leads to a wider range of speech effort. In a small office or classroom only one speaker will be talking at a time, and wall reflections will permit conversation with a very low level of vocal effort. In the open plan, a speaker's voice level may vary from "low," "normal," "raised," to "loud" with a 6 dB increment between each level and a total differential range of 18 dB. The actual range will depend on whether the speaker is willing to converse only with persons a few feet away or include people across the room. The *AI* will be raised about 0.2 per 6 dB increase in vocal effort (at normal talking levels), but if many speakers do this simultaneously the overall noise level is in-creased substantially. Clearly, the open plan situation requires considerable self-discipline, and perhaps even a change in speech habits if acoustic privacy is to be realized.

The efficacy of sound barriers in open plan designs is often overestimated by nonacousticians. Barriers do not extend to the ceiling, and usually there are other "flanking' paths around the barrier (see Fig. 11). For a barrier to serve its acoustic function, sound transmission through the barrier must be much lower than transmission around it. Even when the flanking surfaces (walls, ceilings, and floors) are highly absorbent, low frequency sound will diffract easily around the barrier.

Figure 12 shows the effectiveness of typical partial height barriers as a function of frequency.[31] It will be noted that even a "good" barrier will provide less than a 6 dB loss at 500 Hz in a typical installation. At very low frequencies, a barrier does little more than to provide visual shielding.

Speaker orientation is an additional acoustical parameter, but as noted in Table 7, it plays a lesser role than other factors. Orientation relates to the speaker-listener angle. A "head-on" orientation is referred to as a 0° angle; a 180° orientation is when the speaker faces away from the listener. There is a gradual loss in the signal-to-noise ratio as a speaker turns away from a listener, and in a typical situation the AI will change 0.12 for every 60°. However, this directional effect has been studied extensively in connection with theater acoustics, and in the frequency range

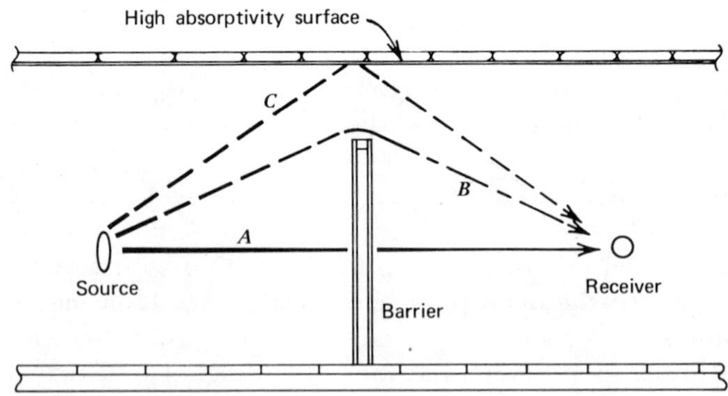

Figure 11   Schematic diagram of open-plan barrier. Transmission of sound from source to receiver is governed by the barrier's transmission loss (via path A), the diffraction of sound past the barrier (via path B), and the absorptivity of the flanking surfaces, that is, walls or ceiling (via path C).

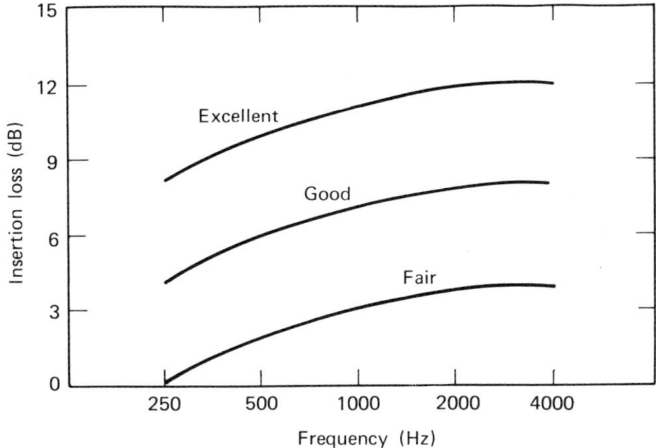

Figure 12   Various barrier losses obtained for partial height barriers utilized in open-plan spaces. Little sound attenuation is achieved at low frequencies. [From R. Pirn, *J. Acoust. Soc. Am.*, **49**, 1339 (1971.]

where much speech intelligibility lies (the *s*'s and *t*'s and *f*'s), 2000 to 4000 Hz levels are roughly 10 dB lower behind the speaker as compared to in front of the speaker.[32] This directional effect relating to speech intelligibility again points up the desirability of occupants of open plan offices and classrooms adopting speech habits that provide the maximum in signal-to-noise ratios for others as well as themselves.

Both objective and subjective assessments of noise in a number of landscaped offices have been conducted, and average noise levels in one study[33] are shown in Fig. 13. The room was designed for an occupancy of 47 persons, and included functions of management, accounting, typing, and duplicating. Note that within the entire space, the maximum noise differential is only 6 dB (51 to 57 dB). At the very least this example points up the desirability of utilizing the concept of wall-less spaces in a selective manner. The open plan relinquishes acoustic privacy, and a higher overall level of noise must be tolerated than in separate offices.

In educational situations, open planning appears to be gaining acceptance for instruction relating to arts and crafts, and activities that require true group participation. However, the acoustic privacy and the lower noise levels that can be realized within the isolated class room will continue to be preferred in many formal learning situations.

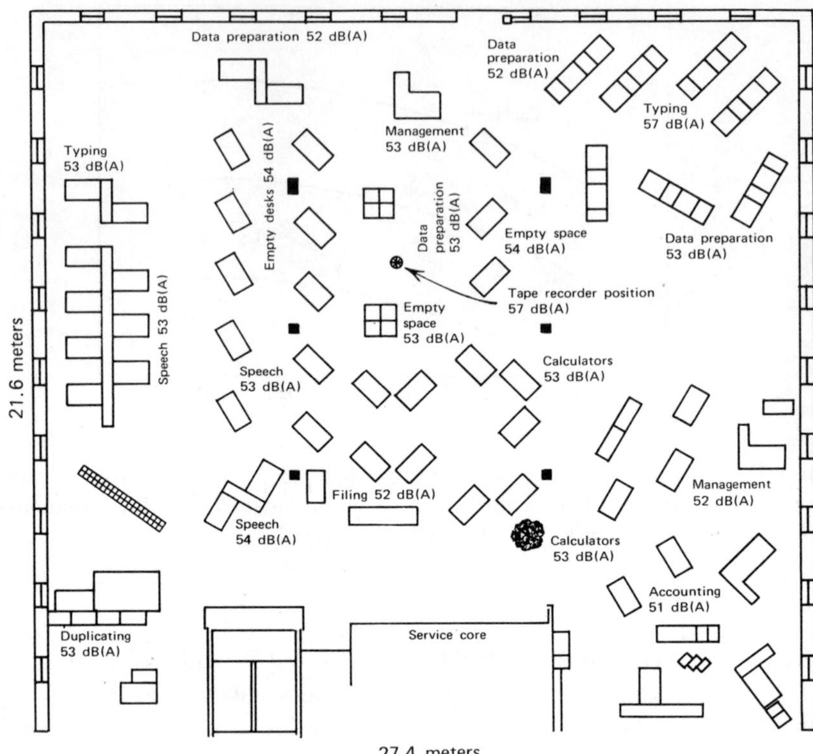

Figure 13   Layout of an open-plan office, having an occupancy of 47 people and including a variety of functions. Individual dB(A) readings are indicated for an average noise level of 54 dB(A). [From B. Hay and M. F. Kemp, *J. Sound Vib.*, **23**, 363 (1972).]

THE VOICE CONTOUR SPECTROGRAPH

Voice recognition apparatus is usually not associated with the acoustical studies of the environment; nevertheless, this type of instrumentation has the potential for providing information on the acoustical properties of auditoria as well as providing a "voiceprint" by which some speech utterance may be ascribed to a single individual. Of course, the ear itself is a highly sensitive voice analyzer and it often provides an instant identification of an individual on the basis of only a few syllables of speech. The voice spectrograph is designed to accomplish this same task by displaying (1) time, (2) frequency, and (3) sound pressure level. The interrelationship of these three parameters is so distinctive from one voice to another that a comparison of speech spectrograms will often provide a positive identification of the speaker from a very large group. In certain legal proceedings, it has also been accepted as proof for demonstrating that a voice recording was definitely *not* that of a suspect.

The basic elements of a voice spectrograph include a multiple band pass filter whose output is recorded on voltage-sensitive paper wrapped around a rotating drum, a reproducing head that turns simultaneously with the drum, and a mechanism that shifts the frequency range in small increments after each drum revolution. The prerecorded speech is carried on magnetic tape that is stretched over a stationary split drum with an aperture just wide enough so that the reproduce head can scan the tape at a fast rate ($\sim$90 in./sec). The three parameter display on a two dimensional graph can be achieved provided the variations of one variable is not registered continuously but in small steps. This technique is employed for topographic maps on which altitude variations are presented by contours; the highest altitude being represented by a dot, lower altitudes are displayed by successively expanding contours of equal altitude. The contour spectrograph is recorded in an analogous fashion. Frequency is the ordinate, time is the abscissa, and the sound pressure level variations are indicated in quantized steps by equal pressure contours.* Figure 14 shows a voice contour spectrogram covering a frequency range of 50 to 7000 Hz.

The essential features of a high speed spectrograph of speaker identification has been reviewed by Presti,[34] and an analytical treatment regarding the information content of a sound spectrogram has been reported by Olson[35] who developed a printout system for a trillion different syllable combinations.

POTENTIAL USE OF THE VOICE SPECTROGRAPH
FOR LEGAL PURPOSES

Because the voice spectrograph has been so useful in speech research, it is appropriate to ask what potential a *voiceprint* may have ultimately in forensic applications. As of this date, *voiceprints,* without other evidence, are given little weight. Several court cases have used voice identification to show that an accused person could *not* be the guilty party, and in furnishing this negative evidence the spectrograph is already an important tool. But can a voiceprint really be applied so as to unequivocally dis-

---

* In simpler versions of such instrumentation, sound pressure level variations are portrayed by the darkness of the tracing.

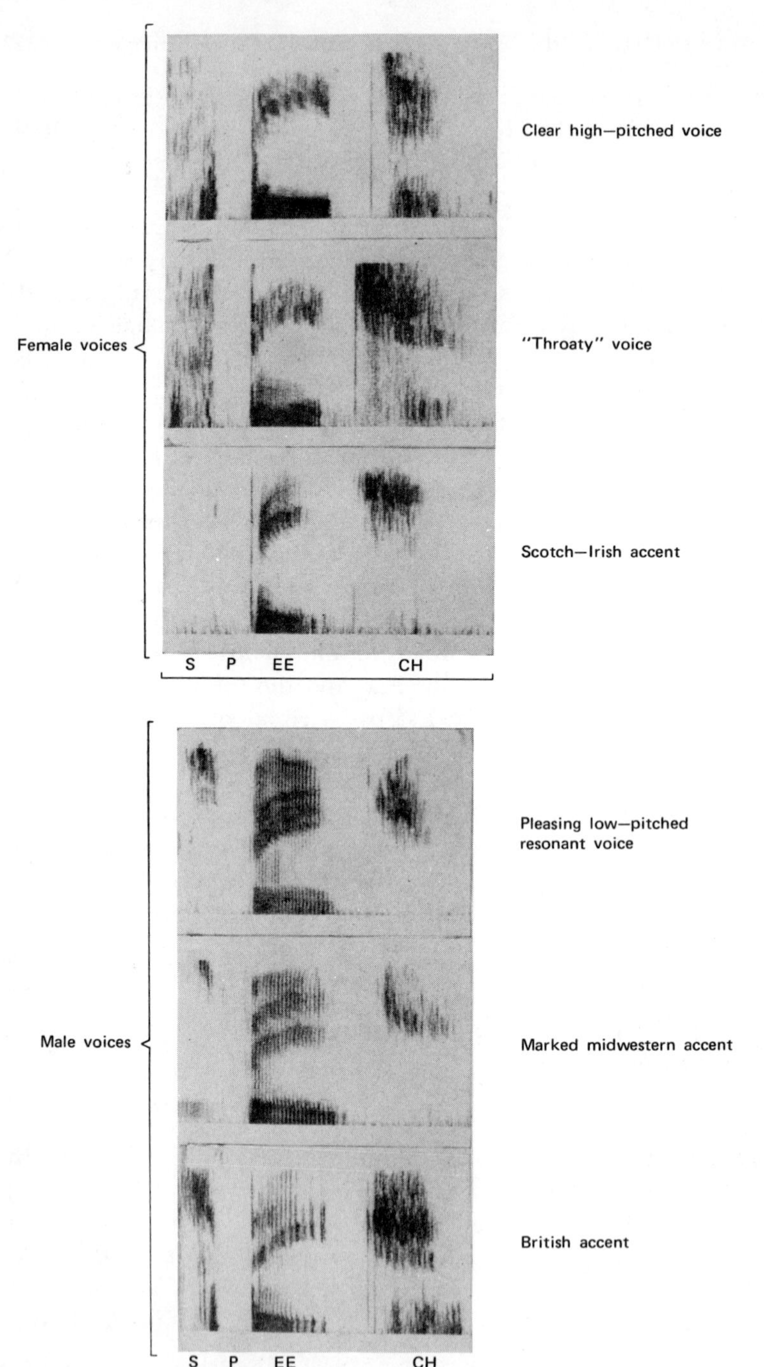

Figure 14   A spectrogram of the word "speech" by six different speakers including male and female voices and a wide range of pronunciations. (Courtesy of Bell Laboratories.)

tinguish one person from another? Probably not, certainly not with the reliability of a fingerprint, with existing instrumentation. An assessment of the spectrograph and its present inherent limitations has been reviewed by a committee acting on behalf of the Acoustical Society of America.[36] The committee posed such questions as: "When two voice spectrograms look alike, do similarities mean *same speaker* or merely *same word* spoken? Are the irrelevant similarities likely to mislead a jury in assessing conflicting testimony from opposing experts? How permanent are voice patterns? How distinctive are they for the individual? Can they be successfully disguised or faked?"

A partial answer to these questions relates to the significant difference between a *fingerprint* and a *voiceprint*. First, a fingerprint represents a discrete physical pattern whose features can be compared in great detail, whereas spectrographic patterns bear no simple relationship to a person's vocal anatomy. Second, the details of a fingerprint are determined by random processes in prenatal skin development and a sizable number of these anatomical details remain unchanged for life; voice features are known to change, even though many general features remain invariant with age. Third, speech of extended length, represents a highly complex encoding of syllables, words, and phrases, and a voiceprint has ambiguities based upon the speaker's mood and other variables. Formant frequencies, for example, are known to be affected by the speaker's dialect, the precision with which speech is uttered, as well as by the size and shape of the oral and nasal cavities or structure of the larynx, which would tend to clearly distinguish one speaker from another.

Thus, there are substantial uncertainties involved in voice pattern recognition, and these difficulties have been made apparent in a number of experimental tests relating to speaker identity. Recognition scores appear to be dependent upon (1) the duration of the speech sample—increasing sharply up to 1.2 sec, (2) the interference of noise, and (3) the frequency band. It has also been found that actual listening sometimes yields better talker identification than does a visual examination of spectrograms of the same utterances[37] and, in general, auditory comparisons yield smaller errors than do spectrographic matches. Both subjective and objective spectrographic methods have been used; in the latter case a computer makes points of comparison and renders a decision.

In its present state of development, a voiceprint can hardly be recognized as an "expert witness." Nevertheless, voicecoding does appear to be a very useful corroborative tool, joining other types of analytical instruments that have aided in proving either the innocence or the guilt of a defendant. The recording of a voice can be accomplished remotely, and voiceprinting has already been helpful in discouraging the use of the telephone for obscene calls. Some type of voice recognition apparatus would also appear to be useful to security guards, who could compare (via a computer) voice patterns of employees entering a plant with employee voiceprints stored in a memory bank.

GENERATION OF SPEECH SOUNDS BY COMPUTER

Although it will be some time before computers possess voices (and a large vocabulary) that truly resemble human speech, substantial progress toward this goal has already been made. There are many uses for computers that would provide an audio output: reading machines could be made available to the blind; weather and aircraft arrival reports could be revised at frequent intervals and be available by telephone; and the inventory of specific items in a warehouse could be provided through a speaker system. Reading machines might eventually be important tools in a reference library: a person could dial a computer, use a reference code, and have an excerpt of a book or journal remotely "read" to him. Interoffice memoranda could also be replaced by "verbal" instructions.

A major hurdle in providing a computer with a "human voice" begins with the exceedingly large number of speech sounds. In the English language some 130,000 words appear in a typical abridged dictionary; if we include the many derived words, modifying prefixes, or endings to denote plural nouns, adjectives, or adverbs, the number will approach 700,000. Two basic methods[38] have been proposed for providing computers with an output of "natural speech" for these words even though it has been admitted that such an output would probably have a "machine accent." The first method is termed *formant synthesis*. As indicated in a previous section, speech formants reflect the natural resonances of the human vocal tract. These resonances change in amplitude and center frequency with speech output, but typically three resonances occur in the 0 to 3 kHz range during normal talking. Word "libraries" are analyzed and stored in terms of these voiced formant frequencies. The speech synthesizer must have the capability of generating a continuous message with rules of timing, pitch, and duration, that are characteristic of the English language. Thus, segmented phonemes and words must be blended with smooth transition sounds that characterize real speech. Collectively, these requirements call for a computer with a large memory and elaborate software, but this formant synthesis scheme appears feasible.

The second method, called *text synthesis,* produces speech solely from a printed English text, that is, no element of human speech is involved in the input. The computer must read the text and make a syntax analysis of it. Sound, pitch, and duration times are derived from a vast array of stored rules that apply to the English language. A rather complex computational articulatory model is also utilized to simulate a voice tract, achieve phoneme synthesis, and provide other dynamic variations and constraints. Regardless of the technical approach to an automatic recognition of continuous speech, it seems clear that a new generation of "intelligence" machines will play an increasing role in man's future interactive environment, and that these machines will tend to augment "communications" in the broadest sense of the term.[39]

One may also speculate that if computer speech fulfills many of the promises of its advocates, "machine speech" will ultimately condition the further evolution of our natural spoken language.

REFERENCES

1. J. L. Flanagan, *Speech Analysis Synthesis and Perception*, Springer-Verlag, New York, 1972, p. 11.

2. E. Meyer and E. G. Neumann, *Physical and Applied Acoustics*, Academic Press, New York, 1972, p. 257.

3. H. K. Dunn and S. D. White, *J. Acoust. Soc. Am.*, **11**, 278 (1940).

4. L. E. Kinsler and A. R. Frey, *Fundamentals of Acoustics*, Wiley, New York, 1962, p. 383.

5. T. S. Littler, *The Physics of the Ear*, Pergamon Press, London, 1965, p. 175.

6. Dunn and White, *op. cit.*, p. 278.

7. P. Denes and E. Pinson, *The Speech Chain*, Waverly, Bell Telephone Laboratories, 1963, p. 118.

8. R. K. Potter and J. C. Steinberg, *J. Acoust. Soc. Am.*, **22**, 812 (1950).

9. H. M. Truby, "Acoustics-Cineradiographic Analysis Considerations," *Acta Radio.*, Suppl. 182 (1959).

10. E. Peterson, *J. Speech & Hear. Res.*, **2**, 173 (1959); also E. Peterson and H. Barney, *J. Acoust. Soc. Am*, **24**, 175 (1952).

11. Meyer and Neumann, *op. cit.*, p. 271.

12. L. L. Beranek, *Proc. Inst Radio Engr.*, **35**, 880 (1947).

13. Flanagan, *op. cit.* p. 4.

14. *Ibid.*, p. 4.

15. C. E. Shannon, *Bell Syst. Tech. J.*, **30**, 50 (1951).

16. Flanagan, *op. cit.*, p. 7.

17. C. E. Williams and K. N. Stevens, *J. Acoust. Soc. Am.*, **52**, 1238 (1972).

18. *Ibid.*, p. 1248.

19. *Ibid.*, p. 1245.

20. C. M. Harris, *J. Acoust. Soc. Am.*, **36**, 933 (1964).

21. J. Pollack and J. M. Pickett, *J. Acoust. Soc. Am.*, **30**, 127 (1958).

22. G. A. Miller, G. A. Heise, and W. Lichten, *J. Exp. Psychol.*, **41**, 329 (1951).

23. E. L. R. Corliss, *J. Acoust. Soc. Am.*, **50**, 672 (1971).

24. Committee on Medicine and Environmental Health, American Petroleum Institute, Report EA7301, *Guidelines on Noise*, p. 20.

25. W. R. McLean, *J. Acoust. Soc. Am.*, **31**, 79 (1959).

26. M. E. Hawley and K D. Kryter, *Effects of Noise on Speech*, Handbook of Noise Control, edited by C. M. Harris, **9**, McGraw-Hill, New York (1957).

27. N. R. French and J. C. Steinberg, *J. Acoust. Soc. Am.*, **19**, 90 (1947).

28. K. D. Kryter, *J. Acoust. Soc. Am.*, **34**, 1689 (1962).

29. C. Duerden, *Noise Abatement*, Butterworths, London, 1970, p. 22.

30. R. Pirn, *J. Acoust. Soc. Am.*, **49**, 1339 (1971).

31. *Ibid.*, p. 1343.

32. D. L. Klepper, *J. Aud. Eng. Soc.*, **22**, 15 (1974).

33. B. Hay and M. F. Kemp, *J. Sound & Vib.*, **23**, 363 (1972).

34. A. J. Presti, *J. Acoust. Soc. Am.*, **40**, 628 (1966).

35. H. F. Olson and H. Belar, *J. Acoust. Soc. Am.*, **34**, 166 (1962).

36. R. H. Bolt, F. S. Cooper, E. E. David, Jr., P. B. Denes, J. M. Pickett, and K. S. Stevens, *J. Acoust. Soc. Am.*, **47**, 597-612 (1970).

37. K. S. Stevens, *J. Acoust. Soc. Am.*, **44**, 1596 (1968).

38. J. L. Flanagan, *IEEE Spectrum*, 22, (October 1970); also J. L. Flanagan, *J. Acoust. Soc. Am.*, **51**, 1375 (1972).

39. N. R. Dixon and C. C. Tappert, *J. Cybern.*, **1**, 67 (1971).

15

MUSICAL ACOUSTICS

INTERVALS

Single tones mean little to the musician. Music begins only when one tone bears a pleasing relationship to other tones. In other words, it is the change or difference in pitch, rather than any specific pitch, that yields either an interesting or unacceptable sound pattern. We call this ratio between the frequencies of any two tones a musical *interval*. When the tones forming the interval are played consecutively, the interval is said to be *melodic*; when the tones are played simultaneously (i.e., in a chord), the interval is termed *harmonic*.

Prior to the advent of contemporary music, intervals were classified into two groups: *consonant* and *dissonant*. The consonant intervals were those generally judged to be "pleasing," "smooth," and "melodic," while dissonant intervals were judged to be "rough" and "harsh." Modern music has made this distinction somewhat vague, but there is still substantial agreement that certain combinations of tones give rise to an aesthetically satisfying sensation.

Figure 1 indicates the notes of the piano, to which we must refer in indicating the intervals of Table 1. Figure 2 shows the location of such notes in terms of standard musical notation.

Table 1 identifies some standard intervals and the corresponding piano keyboard notes, together with the frequency ratios of the higher to the reference or lower note.

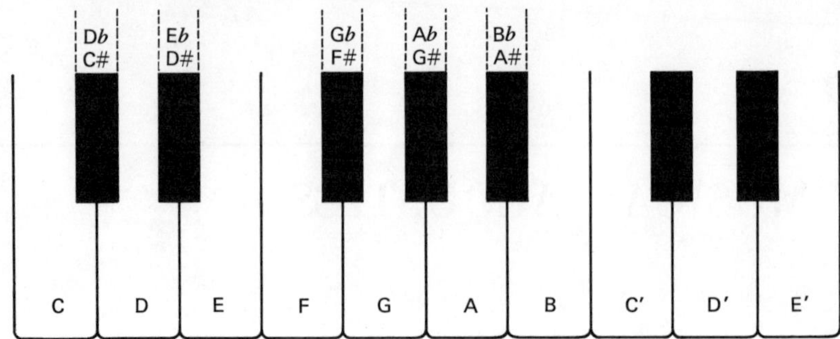

Figure 1   The notes of the piano keyboard.

The interval having the ratio of 2:1 is termed an *octave,* and it is the only interval common to *all* musical scales. It is deemed the most consonant of all intervals. Another interval that is generally judged to be exceedingly smooth is the *fifth;* it has a ratio of 3:2 with respect to the reference note. Other pleasing intervals are those having the ratios 4:3, 5:4, 6:5, and 5:3. No one can explain completely why these ratios are universally preferred, but it is true that the most pleasing combinations

Table 1

Interval[a]	Keyboard Notes	Frequency Ratio	Largest Integer Occurring in Ratio
Unison	C–C	1:1	1
Octave	C–C'	2:1	2
Fifth	C–G	3:2	3
Fourth	C–F	4:3	4
Major third	C–E	5:4	5
Major sixth	C–A	5:3	5
Minor third	C–E^b	6:5	6
Minor sixth	C–A^b	8:5	8
Second	C–D	9:8	9

[a] The terminology is based on numbering the white notes of the piano, calling C = 1, and identifying the successive notes D, E, F, G, etc. as 2, 3, 4, 5, etc. Intervals, of course, can be referenced to other starting notes, so that not only C to G, but D to A and E to B are intervals of a fifth.

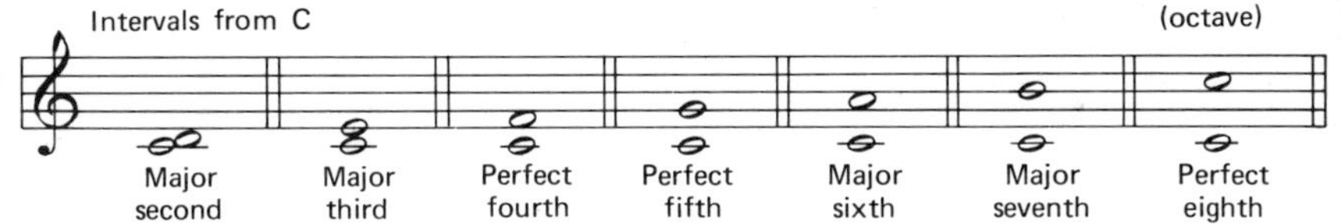

Figure 2 Location of notes and intervals in terms of standard musical notation.

of tones involve ratios expressible by two integers, neither of which is large.

MUSICAL SCALES

In the previous section we have identified a limited number of intervals and suggested that some have a greater degree of aesthetic acceptance than others. We have also specified these intervals in terms of both frequency ratios and the notes of the piano scale, that is, 12 notes per octave. Actually, many scales have been developed in the past and our present scale represents a compromise from so-called perfect intervals.

First, we should recognize that virtually any succession of notes or intervals could, in principle, be used as a scale. The historical development of scales, however, suggests that the nature of music is such that *appreciable* changes in pitch are required to produce a melody. A scale utilizing the hundreds of barely discernible pitches has, to date, been neither appealing nor practical.

One of the earliest contributions to the theoretical evolution of the musical scale came from the school of Greek philosophers founded by Pythagoras. Pythagoras discovered that simple musical intervals were related to lengths of segments of stretched vibrating strings. He stated that "intervals in music are rather to be judged intellectually through numbers than sensibly through the ear" and that "the simpler the ratio of the two parts into which the vibrating strings is divided the more perfect is the consonance of the two sounds." Pythagoras proceeded to develop a scale based on octaves (2:1 ratio), fourths (4:3 ratio), and fifths (3:2 ratio).

The Pythagorean scale is shown in Table 2. Calculation shows that the interval $(256/243) = 1.053$ is a slightly smaller one than the $(9/8) = 1.125$ interval. This scale is also an approximation to our modern *tempered* scale.

Other scales evolved from other cultures. Homophonic music (i.e., one part compositions) still exists among the Chinese, Indians, Arabs, and

Table 2   The Pythagorean Scale[1]

Note	C	D	E	F	G	A	B	C'
Frequency	1	$\dfrac{9}{8}$	$\dfrac{81}{64}$	$\dfrac{4}{3}$	$\dfrac{3}{2}$	$\dfrac{27}{16}$	$\dfrac{243}{128}$	2
Interval		$\dfrac{9}{8}$	$\dfrac{9}{8}$	$\dfrac{256}{243}$	$\dfrac{9}{8}$	$\dfrac{9}{8}$	$\dfrac{9}{8}$	$\dfrac{256}{243}$

Turks. When music is homophonic, a much greater variety of tones is possible than in polyphonic music, where several voices or parts must be consonant. The Arabs divided the octave into 16 unequal intervals so that some of their music is quite unlike that of the Western world.[2] The Hindus divided their scale into 22 steps although only seven are used, and the Persians once employed as many as 24 tones. The Chinese originally had a pentatonic scale, approximating the five black notes of the piano:

$$\text{F}^{\#} \quad \text{G}^{\#} \quad \text{A}^{\#} \quad \text{C}'^{\#} \quad \text{D}'^{\#}$$

This pentatonic scale is utilized in American Indian music, and a large number of Scottish melodies can be played employing only the black notes of the piano.

Occasionally, even modern composers will make use of special scales. In Debussy's compositions, we frequently find the whole tone scale represented by

$$\text{C} \quad \text{D} \quad \text{E} \quad \text{F}^{\#} \quad \text{G}^{\#} \quad \text{A}^{\#}$$

This sequence, which is also used in oriental music, seems to have an unusual character because it differs from the scale to which we are accustomed.

We will mention only one additional scale that is antecedent to our present one. This was termed the major *diatonic scale*,[3] or more specifically, the *major diatonic just scale in C*. This scale was probably of Greek origin, although its present form was first published in the sixteenth century. It is based upon three major chords, or *triads*. These triads are C-E-G, F-A-C', and G-B-D', as indicated below:

Note	C	D	E	F	G	A	B	C'	D'
Relative Frequency	4.........5.........6								
				4.........5.........6					
						4.........5.........6			

Each of these major triads were found to be quite pleasing; they satisfied the small number rule for intervals, and so were used to produce the scale that appears in Table 3.

Table 3   The Diatonic Scale

Note	C	D	E	F	G	A	B	C'
Relative frequency	1	$\frac{9}{8}$	$\frac{5}{4}$	$\frac{4}{3}$	$\frac{3}{2}$	$\frac{5}{3}$	$\frac{15}{8}$	$\frac{2}{1}$
Decimal	1.000	1.125	1.250	1.333	1.500	1.667	1.875	2.000
Interval ratio		$\frac{9}{8}$	$\frac{10}{9}$	$\frac{16}{15}$	$\frac{9}{8}$	$\frac{10}{9}$	$\frac{9}{8}$	$\frac{16}{15}$

An examination of this diatonic scale reveals that there are only three intervals: $9/8 = 1.125$; $10/9 = 1.111$; and $16/15 = 1.067$. The first two intervals are called *whole tones,* and the last is a *half tone* or *semitone,* although the latter interval, 16/15, is slightly greater than half of a major tone.

THE TEMPERED SCALE

About the time of Bach (1685–1750) there were several developments that led to the adoption of a new scale which represented a compromise to the "perfect" intervals of some previous scales. Music began to become more complex and composers chose to write in different keys. In transposing, a number of new notes had to be added to the scale. In a variable or nonfixed pitched instrument such as the human voice or violin, such transpositions posed no difficulty. But with fixed pitch instruments such as the piano or trumpet, the number of new notes or keys that were required to maintain the exact intervals of the diatonic C scale was impractical. Thus a few musicians adopted a compromise that made two semitones equal to one major tone, and all major tones of the diatonic scale were equalized.

Other factors leading to the adoption of our modern tempered scale included (1) the introduction of the Bach chorales during the Protestant Reformation, and (2) the influence of Bach through his many organ compositions. His first collection of preludes and fugues in all keys, *The Well-Tempered Clavichord,* is dated about 1722. The pipe organ itself, comprising thousands of individual pipes, was also a factor, because this vast harmonic ensemble had to be playable in many keys and still maintain chords that were harmonious.

The new tempered scale required the introduction of five tones to each octave corresponding to the black notes on the piano, and identical steps between each note. It will be seen that if there are to be 12 equal steps within each octave (i.e., doubling of the frequency), then the octave will be the factor n^{12}, where n is the ratio between successive notes. Thus

$$n^{12} = 2$$

or

$$n = (2)^{\frac{1}{12}} = 1.05946$$

The tempered scale thus has values for the octave in the middle of the piano keyboard as listed in Table 4.

Table 4   The Tempered Scale

Note	Frequency Ratio	Frequency (Hz)
C	1.000	262
C$\#$, D^b	1.059	277
D	1.122	294
D$\#$, E^b	1.189	311
E	1.260	330
F	1.335	349
F$\#$, G^b	1.414	370
G	1.498	392
G$\#$, A^b	1.587	415
A	1.682	440
A$\#$, B^b	1.782	466
B	1.888	494
C$'$	2.000	523

The actual frequencies, based on A $=$ 440, are listed to the nearest integer only.

This "equal temperament" scale lacks exactness with respect to the major intervals (fifth, fourth, etc.), and there are critics who state that music rendered on this scale is inferior in quality to that played on the diatonic scale. But for practical purposes and to most listeners the tempered scale is quite acceptable. Perhaps electronically engineered systems can provide true intervals in any keys, but the mainstream of music will undoubtedly utilize the tempered scale until there is a substantial change in the design of musical instruments.

THE STANDARD OF PITCH

Any musical scale requires only one standard frequency or pitch as a reference note. The history of several centuries of effort to establish a standard pitch has been an interesting one. Even in this century there has been a prolonged conflict between physicists and musicians. The musicians won.

The desire of the physicists was to establish a standard pitch, based on octaves being successive powers of two. On this basis "middle" C, designated as $C_4 = 2^8 = 256$; $C_5 = 2^9 = 512$; $C_6 = 2^{10} = 1024$; etc. The system was simple and many sets of tuning forks were manufactured to serve as standards. With this scheme, the A above middle C has a value of 426.6, which is a pitch very close to the A adopted by Handel.

The standard pitch in musical literature has varied greatly.[4] Mersenne, the French mathematician and so-called father of acoustics, first determined the pitch of a musical note. In his time (circa 1648) there were different pitches for different types of music; the lowest "church" pitch was A = 373.7 and the pitch for "chamber" music was A = 402.9.[5] Handel's standard pitch in 1751 was A = 422.5. A contemporary of Handel, John Shore, invented the tuning fork and some of the first of these devices were tuned to this reference.[6] A pitch of 440 was suggested in Stuttgart in 1834, and an orchestral A of 435 was adopted in France in 1859. This latter pitch was soon adopted by several symphony orchestras, including the Boston Symphony Orchestra (1883). In 1812 the Piano Manufacturer's Association adopted the French pitch of A = 435 and designated this as "international pitch."

Nevertheless, many attempts were made by musicians to achieve "brightness" by continually increasing the frequency. Music was sung and played at substantially higher pitches than were intended by many classic composers. Some musical performances at the end of the nineteenth century were played at A =445 in England and A = 461 in the United States.[7] In 1939 an international conference on pitch was held in London to resolve this dilemma, and an agreement was reached to recommend that A = 440 be adopted as the standard of orchestral pitch. Finally, in 1953 the International Standards Organization (ISO) recommended the adoption of 440 as a world standard. This frequency is now radio broadcast by the Bureau of Standards from Washington, D. C. at frequent intervals as a public service.

HARMONICS AND TRANSIENTS

It is the harmonic content of a sound by which we are able to distinguish one sound from another sound of like pitch and loudness. Indeed, when we say that a musical sound has a desirable *quality*, we are implying that we are pleased with the number, intensity, and distribution of the harmonics or overtones that accompany the fundamental tone. Various musical terms that are sometimes employed to denote harmonic content include the words *timbre* (French) and *Klangfarbe* (German); the latter loosely translates into "tone color." Thus, the tone color of a violin is different from that of a clarinet, an oboe, or a trumpet, for example, because each of these instruments possesses a different harmonic spectrum.

It is conventional to call the fundamental frequency the first harmonic. Musicians often refer to the harmonic tones above the fundamental as *upper partials* (to denote that many partial tones accompany a musical tone) or *overtones*. Physicists and engineers, however, generally refer to the overtones simply as *harmonics*; by definition, all harmonics must have frequencies that are integral multiples of the fundamental frequency. Thus, if n is the frequency of the fundamental tone, the general form of the harmonic series will be n, $2n$ (second harmonic), $3n$ (third harmonic), $4n$ (fourth harmonic), . . . etc. In musical notation these harmonics are given the names octave $(2n)$, twelfth $(3n)$, fifteenth $(4n)$, . . . etc., corresponding to the piano keyboard location of the harmonic tone, with respect to the fundamental.

Figure 3 indicates 16 terms of a harmonic series, based on the fundamental note $C_2 = 65.4$ Hz (the second C of the piano keyboard). In the figure the notes are indicated that are the closest to the true "harmonic" frequencies, and it can be observed that above 392 Hz, only the octaves correspond exactly to terms of the harmonic series (frequencies of notes are listed to the nearest integer only). Students of music will recognize the first six harmonics as pitches that are indeed "harmonic" in the subjective sense. Musical tones that include these lower harmonics in reasonable strength are usually considered "harmonious" and of interesting tone quality. In comparison with pure tones of a single frequency, such tones are "rich" and, if the very high harmonics are absent, there is a fullness of tone without harshness.

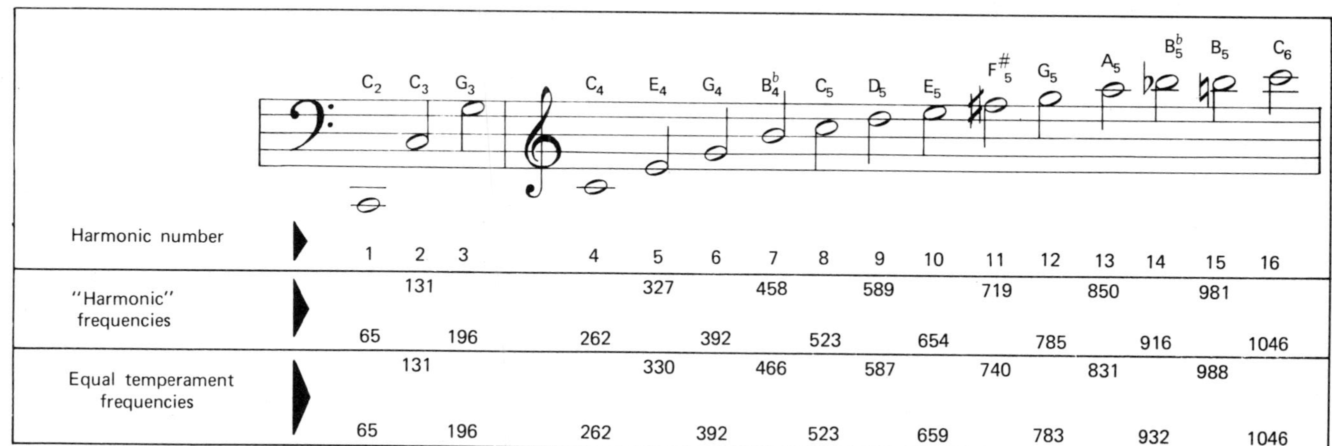

Harmonic number	1	2	3	4	5	6	7	8	9	10	11	12	13	14	15	16
"Harmonic" frequencies	65	131	196	262	327	392	458	523	589	654	719	785	850	916	981	1046
Equal temperament frequencies	65	131	196	262	330	392	466	523	587	659	740	783	831	932	988	1046

Figure 3   A harmonic series based on C_2 (the second C of the piano keyboard). Frequencies are listed to the nearest integer.

Another factor that is important to musical tones is the starting "transient" that is associated with the initial buildup of sound from a violin or horn, for example. A finite time is required to generate a steady-state tone, and the transient will be different from one instrument to another. The short term variation in sound pressure level will also vary with the particular note or fundamental frequency that is being bowed, blown, or produced by other means. Figures 4a and 4b show the relative amplitude

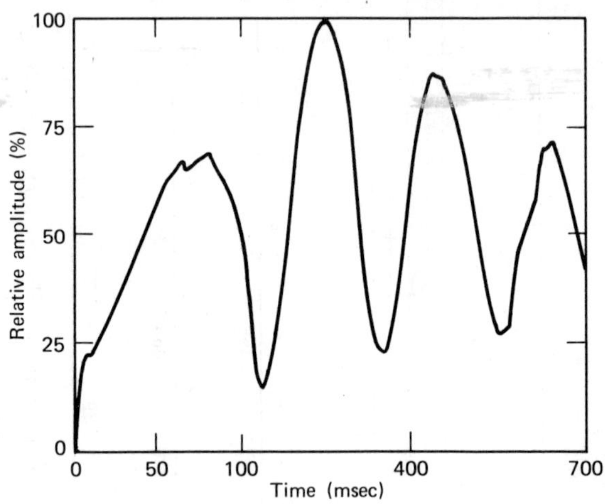

Figure 4a   Example of a flute "transient." [From W. Strong and M. Clark, *J. Acoust. Soc. Am.*, **41**, 39 (1967).]

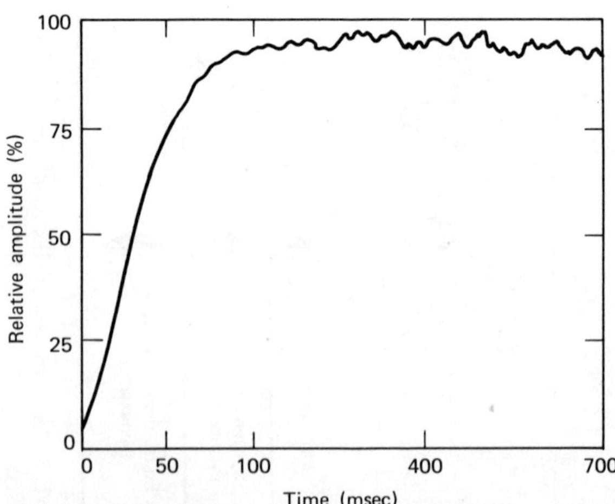

Figure 4b   Example of a clarinet "transient." [From W. Strong and M. Clark, *J. Acoust. Soc. Am.*, **41**, 39 (1967).]

of flute and clarinet transients, during the first 0.7 sec period.[8] A flute tone will actually generate an oscillating pressure wave indefinitely.

It has also been shown that transients are an important factor in the correct identification of an instrument. In an experiment where taped sounds were cut to delete the transient, the correct identification of musical instruments by trained musicians decreased from 47% to 32%;[9] that is, adding the transient enhances discrimination about 50% over a steady-state tone.

FUNDAMENTAL FREQUENCY RANGE OF MUSICAL INSTRUMENTS

The frequency range of musical instruments varies widely. A pipe organ will emit fundamental tones to the lower threshold of hearing and upper notes that approach 10,000 Hz. The bass viol has the lowest frequency of the most common orchestral instruments (41 Hz); the piccolo at 3729 Hz has the highest. The approximate frequency range of many orchestral instruments and the human voice are presented in Fig. 5.

These are fundamental frequency ranges, that is, the actual playable notes only; the prolific harmonic structure is not included. Of course, the harmonic content of a particular note may be contained within the frequency range shown, but most "high" notes of any musical instrument will generate harmonic frequencies that exceed greatly the range of all its fundamental notes. Overtones of some instruments exceed the upper limit of hearing (20 kHz). Some instruments generate noises and subharmonics having frequency components below their primary notes. As previously mentioned, it is the rich harmonic content of any instrument that makes it an interesting sound source; and without harmonics or overtones, many instruments would be indistinguishable from one another, except on the basis of loudness.

ACOUSTIC POWER OF INSTRUMENTS

The acoustic power of even the loudest musical instruments is relatively small. A large pipe organ may require a 10,000 W blower motor to supply air to its wind chests, but of this energy no more than about 13 W will be transformed into sound. A concert pianist may also be expending a substantial amount of energy in a recital (perhaps 200 W), yet only approximately 0.4 W will be broadcast from the piano as sound waves. Table 5 lists the acoustic power of some instruments; unless otherwise noted the values correspond to "loudest" or maximum sounds.

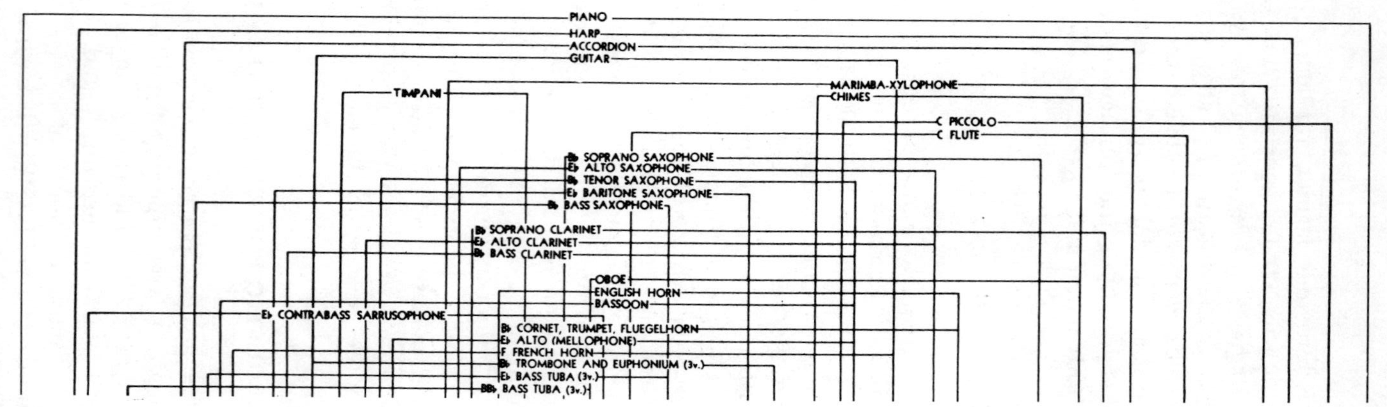

PIANO
HARP
ACCORDION
GUITAR
MARIMBA-XYLOPHONE
CHIMES
TIMPANI
C PICCOLO
C FLUTE
Bb SOPRANO SAXOPHONE
Eb ALTO SAXOPHONE
Bb TENOR SAXOPHONE
Eb BARITONE SAXOPHONE
Bb BASS SAXOPHONE
Bb SOPRANO CLARINET
Eb ALTO CLARINET
Bb BASS CLARINET
OBOE
ENGLISH HORN
BASSOON
Eb CONTRABASS SARRUSOPHONE
Bb CORNET, TRUMPET, FLUEGELHORN
Eb ALTO (MELLOPHONE)
F FRENCH HORN
Bb TROMBONE AND EUPHONIUM (3v.)
Eb BASS TUBA (3v.)
BBb BASS TUBA (3v.)

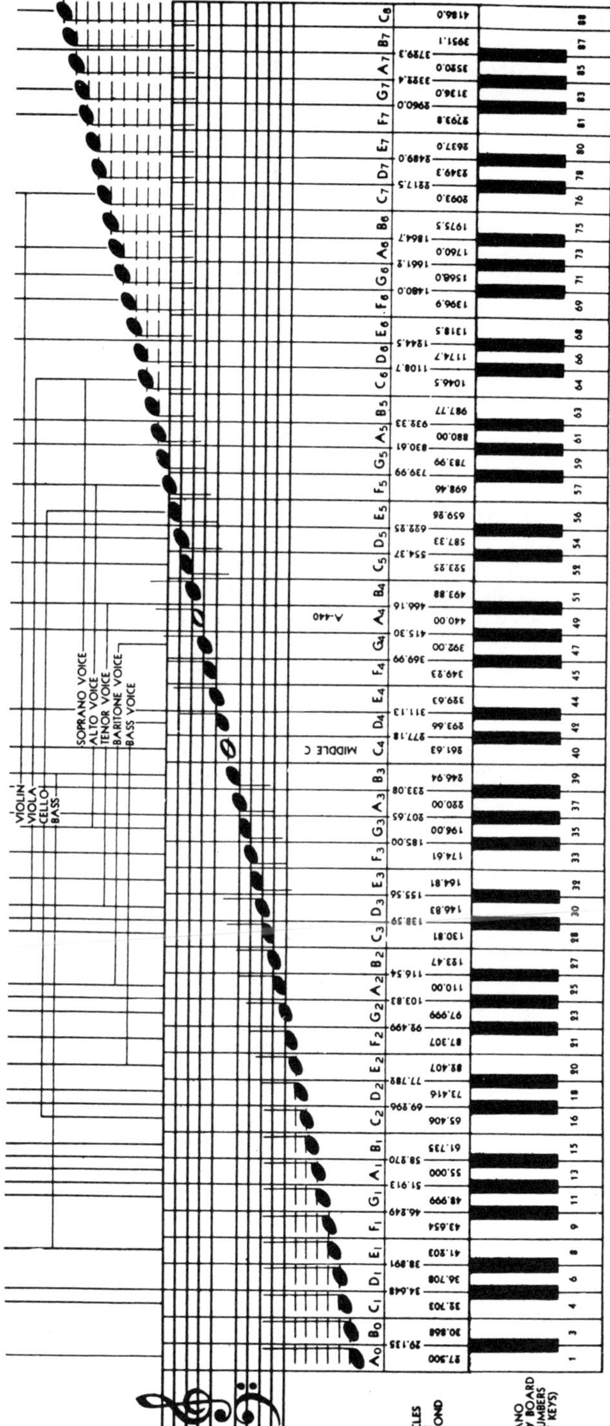

Figure 5 Chart indicating the frequency range of the human voice and various musical instruments. (Courtesy of C. C. Conn, Ltd.)

405

Table 5   Acoustic Power of Musical Instruments

Instrument	Acoustic Power (Watts)
Orchestra of 75 performers	70
Bass drum	25
Pipe organ	13
Snare drum	12
Trombone	6
Piano	0.4
Trumpet	0.3
Tuba	0.2
Double bass	0.16
Piccolo	0.08
Clarinet	0.05
Flute	0.05
Human voice — Bass singing ff	0.03
Human voice — Alto singing pp	0.001
Violin at softest used in a concert	0.0000038

HARMONIC SPECTRA OF INSTRUMENTS

Stringed Instruments

The violin as we now know it, made its appearance in the middle of the sixteenth century, and it is the only important orchestral instrument that has survived several centuries without significant change. An advantage of the violin compared to the piano or harp is its capacity to produce a sustained tone. In the bowing process the string is displaced from its equilibrium position by friction. An increasing displacement is accompanied by increased tension in the string. When the friction of the bow can no longer sustain the displacement, the string slips back to its original position and beyond. Subsequently the hairs of the bow again catch the string and cause another displacement until slippage occurs again. Figure 6 shows a qualitative curve for the displacement of a bowed string; the curve is somewhat saw-toothed; the displacement with the bow and "slipback" occurring with different velocities. String vibration is thus initiated by a complex frictional force, rather than by the hammerlike action in the piano or plucking of the harp.

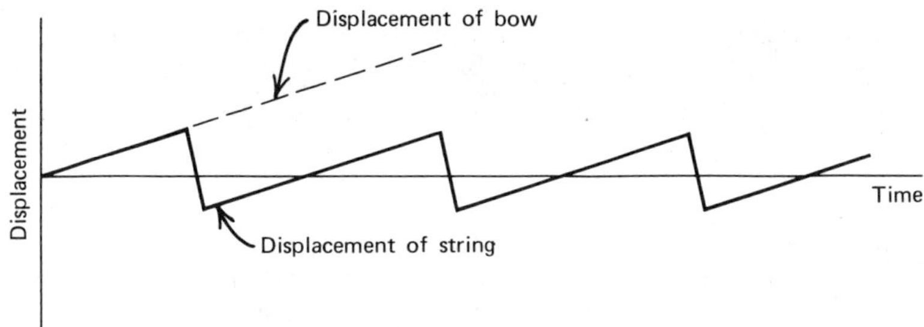

Figure 6   Displacement vs. time for a bowed string: the string first moves with a uniform velocity (of the bow) but then slips back, generally at a different velocity, and the process is repeated periodically.

The quality of any given note is affected both by the speed of the bowing and by the bowing pressure. An increase in the pressure applied to the bow tends to enhance the higher harmonics. The precise point of bow contact on the string is also important. The intensity of sound is primarily but not exclusively ·dependent upon bowing speed, and the *range* of intensity for violin produced sound is normally 25 to 30 dB. The fundamental frequency of each of the four violin strings (open position) G, D, A, and E is given by the formula for the strings fixed at both ends:

$$ n = \frac{1}{2L} \sqrt{\frac{T}{\rho}} \tag{1} $$

where L is the length, T is the tension and, ρ is the density per unit length. Other modes of vibration will be integral multiples ($2n$, $3n$, $4n$, . . .) of the fundamental frequency. Figure 7 shows the first three modes of vibration of a stretched violin string.

As noted in the formula, for any given length, the frequency can be varied by adjusting the tension or changing the string density. In practice both of these parameters are varied. The lowest violin note G, is "tuned" with a lesser tension than the higher frequency E string; also the density per unit length, ρ, is substantially greater in order that a lower frequency can be produced.

The harmonic content of four "open" violin strings is shown in Fig. 8. It will be noted that the harmonic content of the strings is exceedingly rich, and that each string has its own individual timbre or sound spectrum.

Violins of high quality have substantial harmonic content to about the twelfth harmonic; poor violins will generate even higher harmonics of

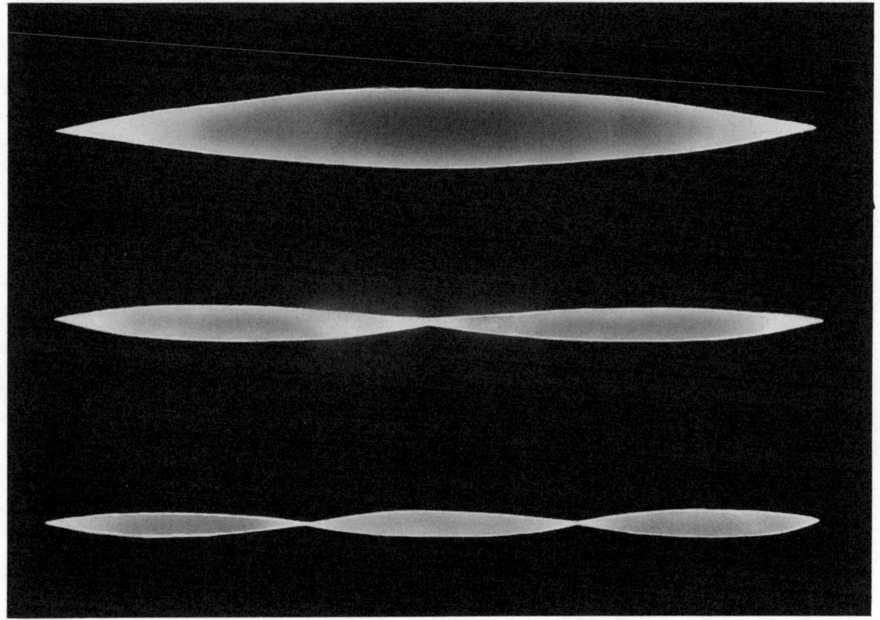

Figure 7   First three modes of vibration of a stretched string.

considerable energy, resulting in harshness of tone. It is also important to note that most of the sound energy that is emitted, either fundamental or harmonics, is radiated via the body of the violin rather than by the strings themselves.

The modern grand piano consists of approximately 240 separate strings playable on 88 individual keys. For the upper notes, three wires are tuned in unison for each key and struck simultaneously by the hammer in order that both the quality and intensity of the sound may be enhanced. Some of the low frequency notes have single strings, and their effective density is increased by having them wound with a helical overlay of copper or alloy wire. By this means, low frequencies can be attained with reasonable tension. The entire string assembly is stretched upon a massive steel frame, built to withstand a tensile force that may be as much as 30 tons. One end of each wire is permanently fixed; the other is wound around a peg that can be turned so as to increase or decrease string tension for tuning.

Steel strings are set into vibration by a felt-covered wooden hammer that is controlled by the key action. The hammer mechanism is designed

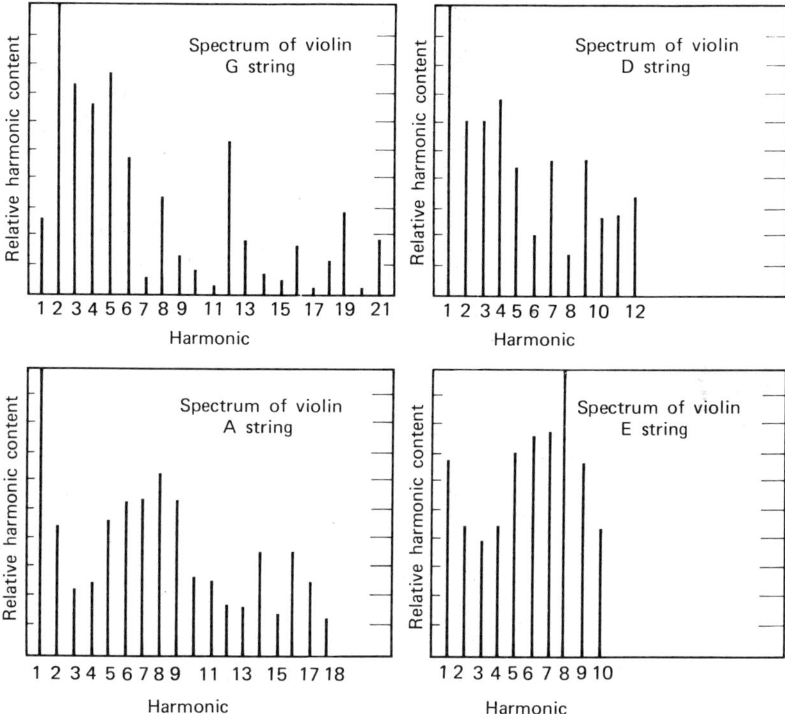

Figure 8   Typical harmonic spectra of violin strings (From *Musical Acoustics,* C. A. Culver, copyright 1956 McGraw-Hill, used with permission of McGraw-Hill Book Co.)

(in the middle part of the instrument) so that it strikes the string at a point that is about one-seventh to one-ninth of the length of the string. Such a design tends to suppress harmonics above the seventh, to prevent harshness of tone.

A spectrum of piano tone is shown in Fig. 9. The same key action provides a damper to attenuate the string vibration when a key or note is released. Basically the piano is a percussive instrument; there is a violent transient followed by an exponential decay. The sounding board provides amplification for the vibrating strings, and its shape and choice of wood are important in determining the sound quality heard by a listener.

The actual spectrum of tone emitted from a piano is known to vary with time. Under certain conditions the fundamental frequency predominates, but shortly after a key is struck, the amplitude of the fundamental frequency may decay to a value below that of the octave (2nd harmonic).

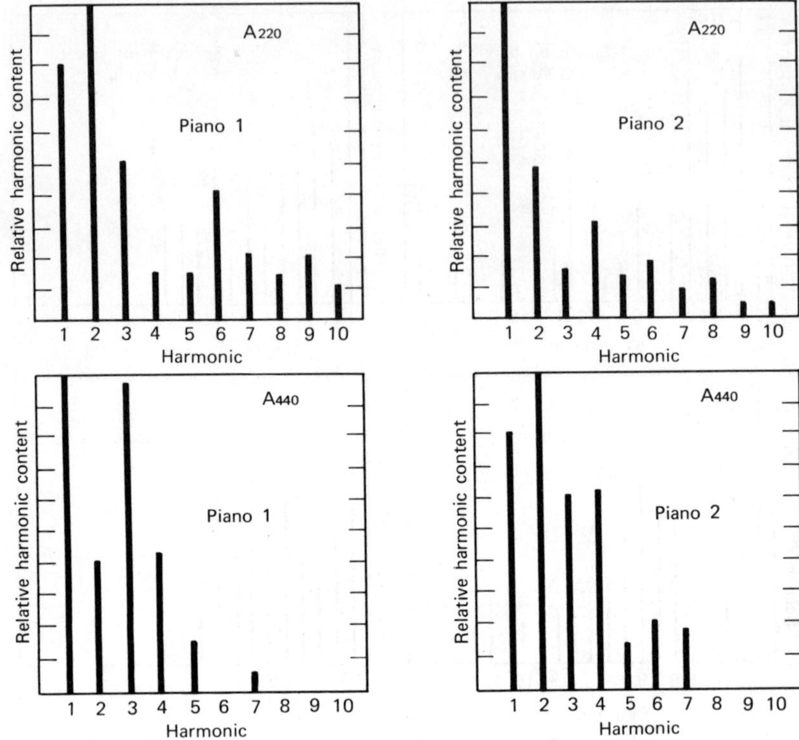

Figure 9   Harmonic frequencies of piano tone. (From *Musical Acoustics*, C. A. Culver, copyright 1956 McGraw-Hill, used with permission of McGraw-Hill Book Co.)

Some pianists claim that they can produce different tones depending upon the "touch" or the manner in which they strike the key. There is no question but that the harmonic spectrum as well as the intensity of sound vary greatly with the velocity at which the hammer strikes a set of piano strings. However, extensive modern experiments indicate that musical quality is not affected by "touch" or the particular manner in which a key is struck. Provided that the hammer mechanism provides an equal amplitude of vibration, there is no detectable difference in the acoustical spectrum of a single note struck by an artist or triggered by a mechanical device. It is only when a succession of notes or chords are played that an artist can introduce tonal variations that are indeed real.

Brass Instruments

This group is comprised of the trumpet, the cornet, the trombone, the tuba, the French horn, and others. Common features of these instruments include (1) a cup-shaped mouthpiece (2) a coil-shaped tubulation, (3) a system of valves, (a slider in the case of the trombone) and (4) a bell. Figure 10 is a schematic diagram of a brass instrument.

The mouthpiece is pressed against the lips of the performer and air is forced through the lips in a manner such that the lips act as a vibrating reed coupled to the pipe. The length of the pipe is effectively changed by manipulating the valves that can cause any one or any combination of additional lengths to be added to the main tubulation. The tubing is flared into the shape of a bell at the output end.

The tonal quality of a brass instrument, in addition to the all important technique of the musician, is dependent upon

    (a) the tube shape.
    (b) the *scale* of the tube.
    (c) the bell.

In some brass instruments the tube is designed to be approximately "exponential" over some portion of its length. This implies that for equal increments along the axis of the tube, the diameter of the tube in-

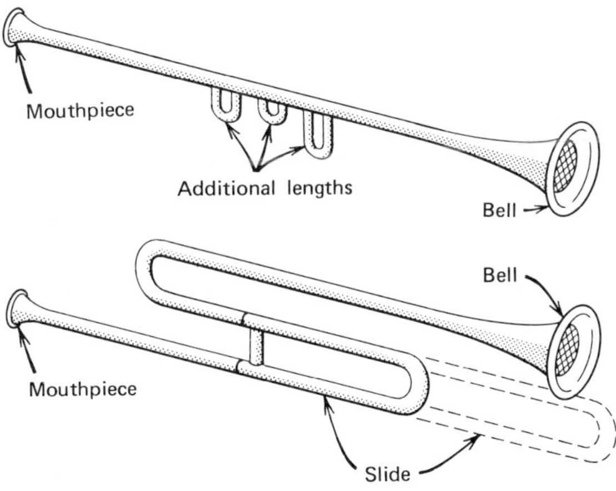

Figure 10 Schematic diagram of a brass instrument, including a cup-shaped mouthpiece, coil-shaped tubulation, valves, and bell. Different resonant frequencies are achieved by adding discrete tube lengths or by using a telescopic slide.

creases in a constant ratio and, hence, the cross-sectional area increases as the square of this increase in diameter. Many early horns were conical in shape, but it was found that the exponential configuration was more efficient in projecting sound. Also, in order to generate adequate bass tones, it was found that the exponential flare must not be too rapid. The *scale* of the tube is defined as the ratio of diameter to length. In general, the narrower the tube for a given length, the more brilliant will be the quality of tone. Most brass instruments terminate with a bell in which the flare expands rapidly. A large bell tends to produce mellowness of tone, while small bells tend to retain brilliance.

The harmonic content of a brass instrument will depend upon the particular note that is sounded, and the loudness of the note that is blown, as well as upon the instrument's inherent design. An increase in the radiated power is achieved largely by an increase in the amplitudes of the high frequency harmonics. Table 6 is an analysis of a French horn.

The tone of the French horn is often characterized as having a soft plaintive quality that makes it attractive for certain solo passages. The rich harmonic structure also allows it to blend well with other woodwind and brass instruments. It will be noted from Table 6 that the fundamental tone (when played loudly) is a minor contributor to the spectrum at frequencies of about 130 Hz and below.

The trumpet and the cornet are quite similar. The cornet tube is shorter and the bore is larger; as a consequence of this larger bore, harmonics above the seventh are difficult to produce. Spectra of cornet and trumpet tones are shown in Fig. 11*a* and *b*.

The trombone is another of the brass group. Played loudly, it can generate, momentarily, up to 6 W of acoustic power. A common characteristic of these brass instruments relates to their directional characteristics. At the high harmonic frequencies where the wavelength of sound is less than the bell diameter, diffraction effects decrease and the sound tends to be focused along the axis of the instrument. For this reason a brass instrument will appear to have a greater brilliance when pointed directly toward a listener than at some oblique angle. Several trumpets pointed directly toward an audience will cause most of the higher harmonics to reach it; a more mellow tone is experienced when these instruments are pointed away from the audience because of the absorption of these harmonics by diffuse reflection.

Woodwinds

The class of instruments called *woodwinds* includes the flute, piccolo, clarinet, oboe, bassoon, and so on. The term *woodwind* is derived from

Table 6   Relative Amplitude of Harmonics of the French Horn[13]

Note/Frequency		Harmonic Number									
	a	1	2	3	4	5	6	7	8	9	10
B^b_4/466	f	90	9	1							
	p	86	12	2							
A_4/440	f	99	1								
	p	26	73	1							
F_4/349	f	66	29	4	1						
	p	94	6								
A_3/220	f	26	31	26	5	9	2				
	p	77	6	14	2						
F_3/175	f	14	32	46	7	1					
	p	10	43	36	9						
C_3/130	f	1	19	21	48	4	5	2			
	p	9	30	25	30	5	1				
A_2/110	f	2	22	34	6	21	3	1			
	p	11	34	4	25	11	9	4	1	1	
F_2/87	f	1	43	22	19	3	6	4	1		
	p	0	12	7	10	15	16	27	8	3	2

[a] The relative amplitudes of the harmonics will depend on whether the instrument is played *forte* (loud) or *piano* (soft).

the fact that originally all these instruments were made of wood. Many modern flutes are made of metal; silver, gold, and even platinum have been used. Despite personal preferences by musicians, it has been shown that the material has a negligible influence on the internal standing wave, and wall vibrations have little to do with the actual sound radiated by the instrument.[10]

In the case of the flute, air is blown across a hole at one end of the tube, rather than down its length. This air stream then generates a vibrating air column, with resonant frequencies being produced according to the covering or uncovering of holes along the length of the instrument. With all holes covered, the fundamental or lowest tone of the flute is produced (262 Hz). The piccolo, which is about one half the length of the flute, has a lowest fundamental frequency an octave higher. Flute tones have little sound energy in their higher harmonics, and the upper

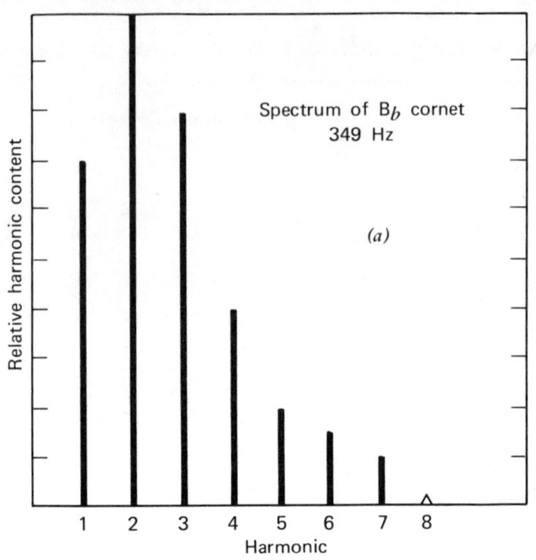

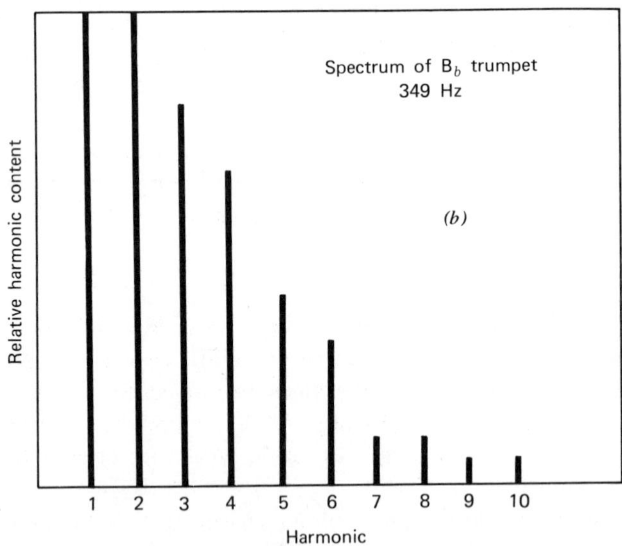

Figure 11   Spectra of (a) cornet and (b) trumpet tone. (From *Musical Acoustics,* C. A. Culver, copyright 1956 McGraw Hill, used with permission of McGraw-Hill Book Co.)

414

notes of the flute are close to pure tones (i.e., almost no higher harmonics). The harmonic structure of flute tone is shown in Fig. 12a.

The clarinet has a vibrating reed for producing the oscillations within a cylindrical tube. The tube is basically closed at one end and open at the other so that its fundamental frequency is half that of a tube open at both ends. For this reason the fundamental note of a clarinet is approximately an octave lower than the flute, although of almost equal length. Clarinet tone has a characteristic "reedy" harmonic structure, approximated by the overtone spectrum of Fig. 12b. When used in military and concert bands, clarinets substitute for the violins of an orchestral ensemble.[11]

Organ Pipes

The large number of individual pipes contained in a modern pipe organ permit musical compositions to be rendered with a great dynamic range and with many distinctive tones.[12] A single set or *rank* of pipes, usually consists of 61 pipes if these pipes are playable on the five-octave manual keyboard. A pedal *rank*, played on a 32-note pedal or foot keyboard, has 32 pipes. The pedal pipes are large, because they must provide the low frequency tones. The lowest note on most organs will have a pipe that is approximately 16 feet in length. The highest note will have a pipe whose

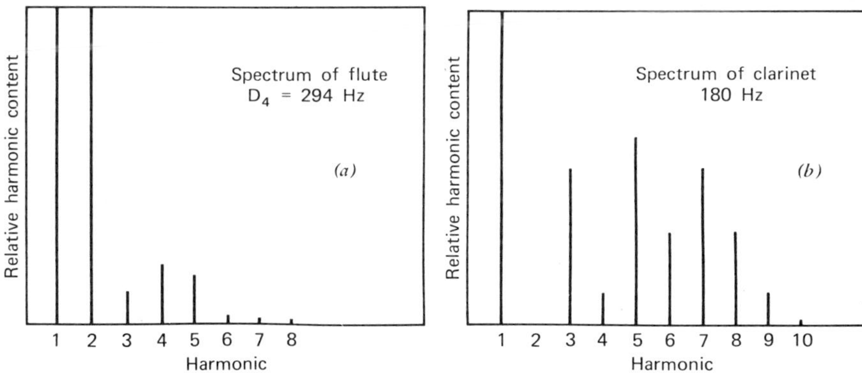

Figure 12   Harmonic structure of (a) flute tone and (b) clarinet tone. (From *Musical Acoustics,* C. A. Culver, copyright 1956 McGraw-Hill, used with permission of McGraw-Hill Book Co.)

effective pipe length is a fraction of an inch. A section of a modern organ having many manual ranks of pipes is shown in Fig. 13.

The various ranks of pipes are designed to produce diverse tone colors, including those of some orchestral instruments. However, all pipes can usually be divided into the following groups:

1. Diapasons
2. Flutes
3. Strings
4. Reeds

Diapason tone serves as the foundation tone for the organ, and such pipes do not attempt to simulate any other instrument. Diapason pipes may be made of either wood or metal; the metal is usually pure zinc or

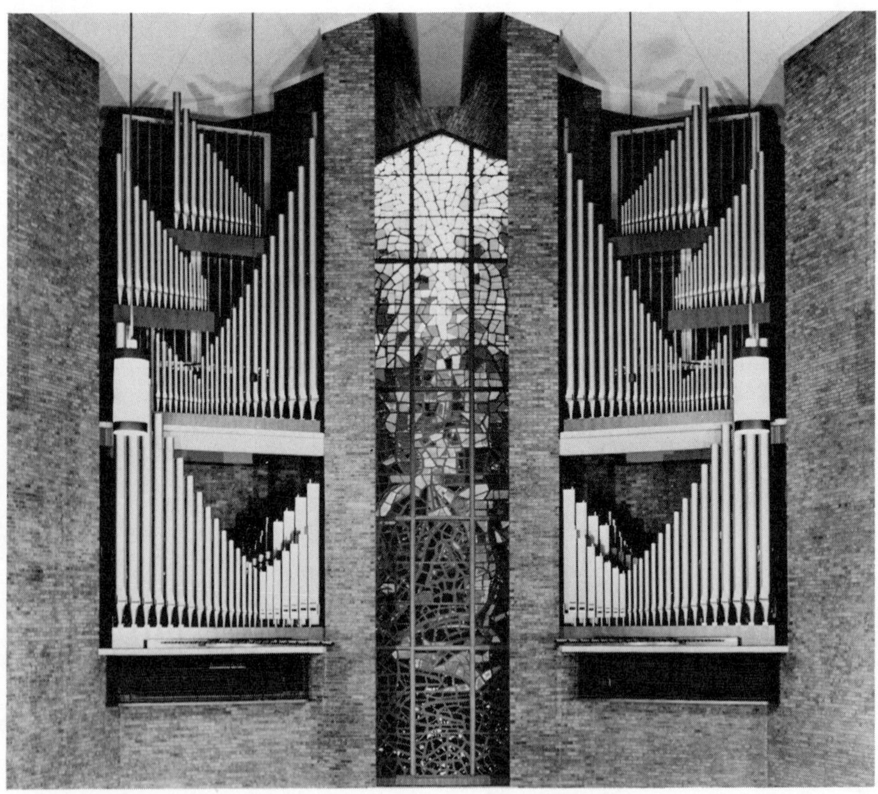

Figure 13   Section of a modern pipe organ comprised of various ranks of pipes. (Courtesy of Schantz Organ Co.)

an alloy of tin and lead. A *diapason chorus* is comprised of several diapason pipe ranks being played together. For example, a diapason chorus might consist of five pipe ranks, connected to a single keyboard. One pipe will emit the fundamental frequency, the second pipe the octave or second harmonic, the third the fifth above the octave or third harmonic, etc. In practice all diapason pipes generate harmonics as well as the fundamental, but a unique buildup of tone color is possible when individual tone sources are designed to generate specific harmonics.

Flute pipes of an organ emit sounds that are remarkably similar to their orchestral counterparts. Some flute tones are reasonably "pure," that is, the sound approximates a sinusoidal vibration, but most contain many harmonics. If a flute pipe is *closed* or capped at one end, it can be shown that the relationship of the pipe length, L, to the wavelength of the fundamental tone will be $L = \lambda/4$, whereas an *open* pipe will have the relation $L = \lambda/2$. The series of standing waves that can be contained in open or closed pipes is shown in Fig. 14. In principle, open pipes per-

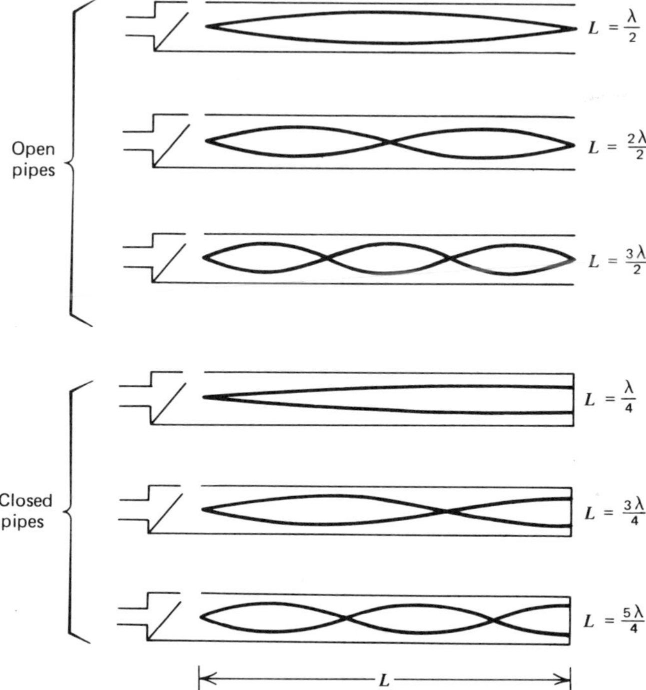

Figure 14   Standing waves generated in organ pipes. Open pipes generate all harmonics; closed pipes generate standing waves of only the odd harmonics.

mit the generation of all harmonics $n = 1, 2, 3, 4, \ldots$, whereas closed pipes generate standing waves containing only the odd harmonics $n = 1, 3, 5, \ldots$. A flute generating only odd harmonics has a distinctive musical sound.

The so-called *string* tones of a pipe organ come from pipes that are small in diameter and *open*; thus all harmonics are produced. Organ string tones are somewhat similar to the violin or cello, although the sustained tone of an organ pipe derived from a wind chest can never include the variety of transient sounds that are produced by a single performer on an orchestral instrument. One artifice that is employed with string organ pipes is to *voice* one rank either slightly sharp or flat with respect to another rank. When played together, there results a fullness of tone that cannot be achieved with pipes that are perfectly in tune.

Reed pipes use an actual vibrating metal reed at the bottom of the pipe to generate the vibrating air column in the pipe. The resonance frequency of such a pipe thus depends upon both the natural frequency of the reed and the dimensions of the resonating tube. Reed pipes are often conical in shape, and if so, they emit all harmonic frequencies. Principal reed pipes on a modern organ include orchestral sounds of trumpet, clarinet, oboe, and special tones that are indigenous to the organ.

Figure 15 is a photo of pipes representing the several families of pipes discussed above.

Bells and Carillons

Bells are acoustically related to vibrating plates, and they can be considered to be derived from a disk that is curved in shape and loaded at its center. The classical shape of a bell is shown in Fig. 16.

The optimum shape for a bell has been the result of centuries of craftsmanship. The bell resembles slightly an inverted cup with a flared mouth and it contains a *clapper* that activates the bell. The clapper is hung free within the bell so that when the bell is rotated the clapper strikes the bell at the thick portion, which is termed the *soundbow*. The fundamental frequency is a complex function of several parameters, the bell diameter, wall thickness, and the density and elastic modulus of the metal. The metal is usually an alloy consisting of three to four parts of copper to one part of tin. The pitch or fundamental frequency of bells of the same shape and material varies inversely with their diameter; the pitch also varies inversely as the cube root of the weight, that is, the larger and heavier the bell the lower its pitch.

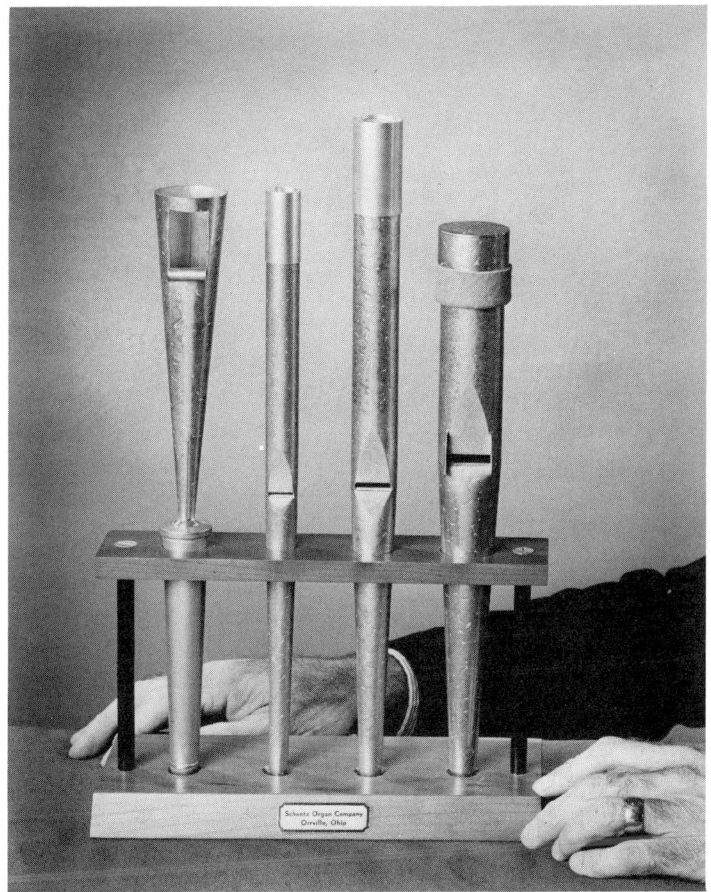

Figure 15   Photo of various organ pipes: (from right to left) an open diapason, flute, string, and trumpet reed. (Courtesy of Schantz Organ Co.)

The largest bell ever built in Europe was the Great Bell of the Kremlin.[13] Its weight was approximately 180 tons and its height over 19 ft. This bell was cast in 1734, and the clapper was swung by 25 men pulling from each side. The bell was broken in 1737, when it fell from its supporting structure. The largest ringable bells in Europe now include Big Ben ($13\frac{1}{2}$ tons) at Westminister and Great Paul ($16\frac{3}{4}$ tons) at St. Paul's Cathedral in England.

When a large number of bells is activated from a keyboard, the set of such bells is termed a carillon. The bells are tuned to a musical scale

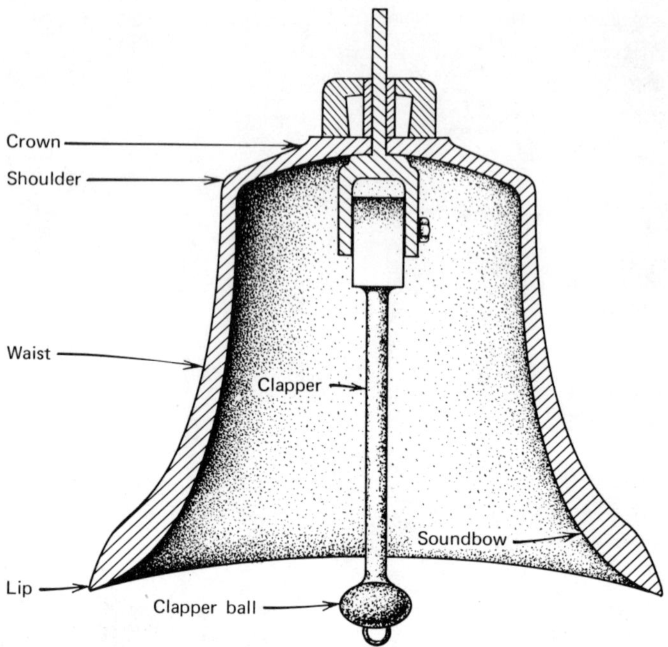

Figure 16   Schematic diagram of a bell.

and hung "dead" (i.e., the bells are immobile). The bells are then played by a carilloneur on a keyboard mechanism that causes the clapper to strike the bell. In electrified carillons, the clappers are activated by large solenoids, so that successive notes can be played easily.

The number of bells in a modern carillon may range from 23 to 72. The Riverside Carillon in New York City has 64 bells, the largest of which weighs $18\frac{1}{4}$ tons.

The harmonic spectrum of a bell is unlike that of any other musical instrument. For a typical bell of pleasing sound, the frequencies have been determined to be in the following ratios:[14]

$$1.00, \ 1.51, \ 2.02, \ 2.93, \ 3.43, \ 4.33, \ 4.60, \ 4.63$$

It will be noted that only two ratios in this series (2.02 and 2.93) are even close to being integral multiples of the lowest or fundamental tone, which is sometimes called the *hum note*. Musicians will also recognize that the series contains an interval approximating a minor-third (Table 1), which is generally considered to be a desirable interval for a bell, as it adds the plaintive or somber quality that is associated with bell tone. Higher inharmonic frequencies will also be present, and some of these will give rise to the production of beats. The actual pitch of the bell, that is, the tone discerned by a listener, is not the fundamental or *hum note,* nor any member of the above series.[15] Rather, several investigators have concluded that the *strike note,* which is defined as the perceived pitch of the bell, corresponds to the octave below the fifth term of the series. The *strike note* is thus half of 3.43, or 1.71 times the fundamental frequency. A partial explanation for this aural illusion is that this pitch is sensed because of the different decay rates of the many frequencies that comprise the bell spectrum.

The Human Voice

The human voice is, in one sense, a wind instrument whose harmonic spectrum is somewhat conditioned by a number of cavities, that is, the pharynx, oral and nasal cavities, and windpipe. But the voice also functions much like a stringed instrument in the sense that the many frequencies that are generated depend on the tension of the vocal cords, which are set in vibration by air forced through them from the lungs. This variable tension of the vocal cords combined with modulations of the shapes of the cavities in the vocal tract permit the production of sound having a harmonic structure of great diversity.

The maximum power of the singing voice is less than a watt, and the fundamental frequency range extends from about 60 Hz for a low bass to about 1300 Hz for a high soprano. The range of any single voice, however, rarely is greater than two octaves so that the fundamental frequency range is less than most musical instruments. The normal singer also cannot attain the pitch precision that is possible with instruments having "tuned" and sharply defined resonances. Uncertainties in pitch or frequency location, after hearing a reference tone, range to $\pm$ 0.8%; in the free singing of intervals, the uncertainty may be as much as 3%.[16]

What constitutes "good voice-quality" is somewhat subjective, but there is general agreement among voice critics that a good voice requires contol of pitch, intensity, and timbre (harmonic structure). Figure 17 shows the harmonic spectrum of a soprano after four months of training compared to that of the same singer after five years of training. The difference in the harmonics is apparent.[17]

Good male voices will also be rich in harmonics. A strong "low formant" will be in the range of 500 Hz, and a strong "high formant" will be in the range of 2400 to 3200 Hz. Although this latter formant (or fre-

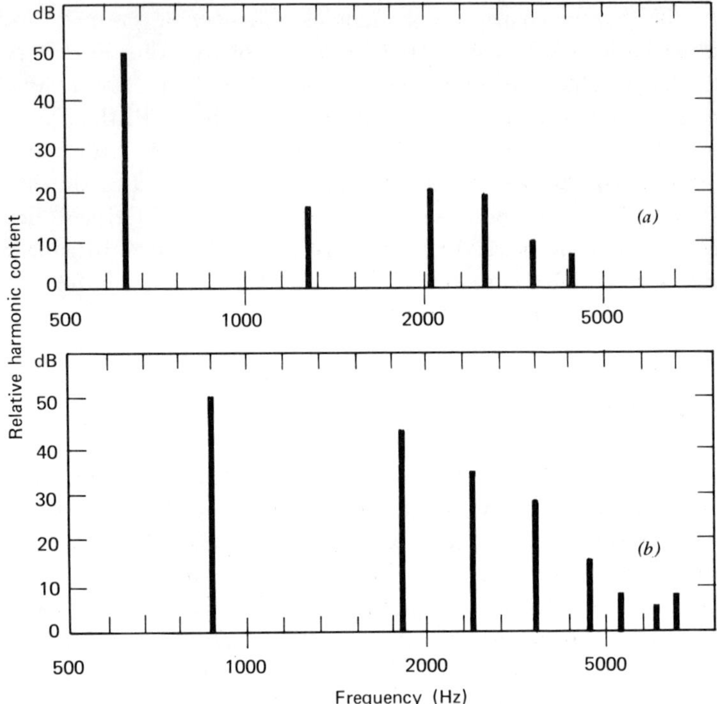

Figure 17   Harmonic spectra of a soprano voice. (*a*) After four months of train-ing. (*b*) After five years of training. [From A. Bjørklund, *J. Acoust. Soc. Am.,* **33,** 575 (1961).]

quency band) may vary with pitch and intensity, a good male voice will have many harmonics in this 2400 to 3200 Hz region regardless of whether the tone is produced by tenor or bass, and regardless of the pitch of the fundamental tone.[18]

CHORUS EFFECT OF AN INSTRUMENTAL ENSEMBLE

We have seen that for any single musical instrument the quality of sound is greatly dependent on the richness of its harmonic spectrum and on its starting transient. The sound spectrum from an ensemble of instruments is even more complex. From experience we know that two violins play-ing together do not sound the same as a single violin that is being played to yield an equivalent loudness. A partial answer is that no two violins can be played exactly in tune, even by experts. This slight "out-of-tuneness" gives rise to "beats," as discussed in Chapter 6, so that the ear senses a modulation of tone that cannot be produced by a single sound source. Not only are such beats produced by the fundamental frequencies of two instruments, but many additional beats are generated by the higher harmonics, which are also slightly separated in frequency. This blending of many slightly different frequencies is called the *chorus effect.* This chorus effect is an essential ingredient of symphonic music, that is, a large string section will provide a richer blend of tone than a small one, in addition to furnishing an adequate balance of loudness compared to other divisions of the orchestra.

The chorus effect is especially noticeable with pipe organs, which are sometimes comprised of thousands of pipes and which are often required by musical compositions to produce sustained tones. A detailed study conducted at the Smith Auditorium at Brigham Young University has identified the large number of frequencies that are emitted from such an instrument. For example, when three keys for a single major chord were depressed, some 229 individual frequencies were identified, whereas only 38 harmonics should have been present if all pipes of the organ had been "in tune." Table 7 contains a small number of these 229 frequencies, arising from a single note, when 12 organ "stops" (types of pipes) were combined.

TONALITY

Briefly, *tonality* refers to the characteristic key (C, F$^\#$, etc.) or musical scale (major or minor) in which a composition is written. Do different keys or scales evoke a different emotional or psychological response? Since the days of Aristotle many musicians have held an affirmative view. Composition, in minor scales have usually been associated with melancholy, darkness, lament, or gloom. The term "minor key" has even become an accepted figure of speech. The "character" is presumably associated with the order in which the semitones occur in the scale, and the distinction between the major and minor chords (minor chords involve different frequency ratios). Major scales have always been used for reflecting a joyous mood. Specific major keys were even chosen by the masters. F sharp major was considered by Handel to be a transcendental or visionary key, and F major was the key of the pastoral idyll through most of the eighteenth century. Beethoven also chose this key for his *Pastoral Symphony* and for his *Spring Sonata* for violin and piano. C

Table 7   Measured Frequencies Occurring from Various Organ Pipes, Sounding the Same Musical Note[19]

Name of Organ Pipe	Detected Frequency (Hz)
First diapason	994
Second diapason	992
Gedeckt	990
Principal	990
Octave	989
Horn diapason	976
Fagotto	974
Trompette	978
Clarion	975
Mixture	988

major was a distinctive key for Bach, and this key had special significance for other composers.

Today, ascribing a given psychological mood to different musical keys is difficult to justify. The tempered scale has identical intervals between each note of the scale so there should be no preferential brillance to chords and no relative frequency difference to a sequence of notes in various major keys. There will, however, be different natural resonances associated with musical instruments and there is an easily detectable change in the tone of violin notes for the "open" strings.

With persons having a very keen sense of pitch, it is indeed possible that a slightly different stimulus is experienced, dependent upon key composition. However, any psychological mood-key relationship that may exist is more likely due to associations that have been built up with familiar compositions.

A modern innovation has been the creation of keyless or *atonal* music, of the type composed by Arnold Schoenberg. All 12 tones of the tempered scale are given equal importance, and there are no "reference" tones. To date, however, all musical scales make use of a very small number of quantized pitches or tones, even though the average ear is capable of distinguishing a very large number of discrete frequencies. Specifically, in the equally tempered scale covering the audio range of 16 to 16,000 Hz, there are only about 120 different tones, even though a good ear can distinguish 1400 discrete intervals.

ABSOLUTE PITCH

Absolute pitch refers to a person's ability to determine pitch without comparison to any instrumental tone or reference frequency. Few people possess such an acute sense of pitch. Those who do, however, rarely make a mistake in naming a pitch, that is, an isolated piano note, even when they have not been exposed to any musical tones for hours or days. Generally, people either have this faculty or they do not, although there is some evidence that one's sense of pitch can be improved with training. At present, there is no scientific explanation for this phenomenon. One proposed theory suggests that to person's having *absolute* or *perfect* pitch, each note of the scale (C, D, . . . , etc.) has a certain "chroma" or distinctive tone color.[20,21] Another theory suggests a type of musical "memory" or mental imprinting, in the same sense that visual memory varies widely among individuals.

REFERENCES

1. J. Backus, *The Acoustical Foundations of Music*, Norton, New York, 1969, p. 119.
2. C. A. Culver, *Musical Acoustics*, McGraw-Hill, New York, 1956, p. 132.
3. J. J. Josephs, *The Physics of Musical Sounds*, Van Nostrand, Princeton, 1967, p. 81.
4. J. Jeans, *Science and Music*, University Press, Cambridge, 1953, p. 23.
5. Culver, *op. cit.*, p. 85.
6. Backus, *op. cit.*, p. 131.
7. *Ibid.*, p. 132.
8. W. Strong and M. Clark, *J. Acoust. Soc. Am.*, **41**, 39 (1967).
9. E. L. Saldanha and J. F. Corso, *J. Acoust. Soc. Am.*, **36**, 2021 (1964).
10. J. Backus, *J. Acoust. Soc. Am.*, **36**, 1881 (1964).
11. Culver, *op. cit.*, p. 205.
12. W. H. Barnes, *The Contemporary American Organist*, Fischer, Glen Rock, N. J., 1964.
13. A. Wood, *The Physics of Music*, Dover, New York, 1950, Chapter 9.
14. Josephs, *op. cit.*, p. 133.

15.   A. T. Jones, *Phys. Rev.,* **16,** 247-259 (1920).

16.   F. Winckel, *Music, Sound and Sensation,* Dover, New York, 1967, p. 127.

17.   A. Bjørklund, *J. Acoust. Soc. Am.,* **33,** 575 (1961).

18.   W. T. Bartholemew, *J. Acoust. Soc. Am.,* **6,** 25 (1934).

19.   H. Fletcher, E. D. Blackman, and D. A. Christensen, *J. Acoust. Soc. Am.,* **35,** 314 (1963).

20.   A. Bachem, *J. Acoust. Soc. Am.,* **27,** 1180 (1955).

21.   L. A. Jeffress, *J. Acoust. Soc. Am.,* **34,** 987 (1962).

16

AUDITORIUMS AND CONCERT HALLS

THE EVOLUTION OF ARCHITECTURAL ACOUSTICS

The word *auditorium* literally means "the hearing place," as opposed to the word *theater,* of Greek origin, which loosely translated means "the seeing place." The essential feature of the early Greek theaters was that seats were arranged in semicircles upon a steep embankment.* In some theaters, the seats were carved out of the native rock. Sound-reflecting surfaces were located at the stage, and a circular stone "orchestra" floor was at the center of the semicircular seating arrangement.[1] The net effect was to project the voices of the speakers through a decreased solid angle, thus increasing the intensity of sound reaching the audience. The Romans made a number of modifications to the Greek design. The "stage" area was widened and also lowered for the benefit of the front row audience. These changes caused sound to travel at less than an optimum angle, and the wider and sometimes deeper stage resulted in poorer reflection patterns and a loss of intensity. By modern standards neither the early Greek or Roman theaters would be acceptable for speech or music.[2]

The change from open air to enclosed spaces came with the advent of the *odeion.* The Greek odeion was a square structure of moderate size

* Figure 3, Chapter 1.

with stepped curved rows of seats and a wooden roof. Audience capacity ranged from a mere 200 seats to a large 2000 seat capacity theater located at Athens. These enclosed theaters provided shielding from extraneous noise, and both music and speech was heard with increased brilliance and clarity. Reverberation in the larger structures was controlled by heavy fabrics suspended from the ceiling.[3]

Several other architectural developments were antecedent to the modern auditorium and concert hall. One was the advent of Christian churches, often patterned along the lines of the Roman basilicas. The massive columns, piers, and sculpture aided in the diffusion of sound, and the building materials provided substantial reverberation for sound reinforcement. However, when the sanctuaries increased greatly in size,* as did the medieval Gothic structures, the acoustic quality diminished.

Then circa 1600, in Italy and Sicily, some of these massive edifices were supplemented by small chapels or *oratorios,* adjacent to the sanctuaries.[4] These *oratorios* were rectangular, high-ceilinged rooms of moderate size, which proved to be ideal for both choral and instrumental music. In fact, the oratorio (as a distinctive musical composition) was named after this type of room that was favorable to musical acoustics. A further salient development was the design of the Italian opera house (Fig. 1). In this horseshoe-shaped room, with multiple balconies and steeply tiered rows of seats, acceptable listening for speech and music was assured for virtually all persons in the audience. In the nineteenth century, the secular palace ballroom provided an excellent environment for musical performances. Here, classical music flourished, and at least a few ballrooms possessed acoustics comparable to the acoustics of the modern concert hall. Thus, the development of the classic music hall is an architectural outgrowth of the church, the oratorio, and the palace ballroom—the spaces first chosen by musicians to develop their art form.

The scientific foundation for auditorium acoustics, however, began in 1895 when President Eliot of Harvard University told a member of his faculty to "do something" about the poor acoustical properties of the large lecture hall in Harvard's new Fogg Art Museum,[5] (now called Hunt Hall). Professor Wallace Sabine responded by making some highly quantitative reverberation measurements using a Gemshorn organ pipe of 512 Hz, a stop watch, and a few trained listeners. On the basis of these measurements, Sabine defined reverberation time, established a formula

---

* St. Peter's Basilica, in Rome, has an air volume of approximately 20 million ft^3. In such a structure the reverberation time is long and the effect of air absorption of sound is important.

Figure 1 The opera house, La Scala, in Milan, Italy. This horseshoe-shaped theater was originally completed in 1778, and rebuilt after World War II. (Photograph by E. Piccaguani, and courtesy of Teatro alla Scala, Milan.)

for its measurement, and shortened the reverberation time of the lecture hall from its original 5.6 sec by means of absorbing materials. Shortly thereafter, Sabine became the acoustical consultant for Boston Symphony Hall—the first music hall to be designed with the aid of quantitative acoustical relationships.

THE SOUND FIELD IN AN AUDITORIUM

The sound field experienced by a listener in an auditorium depends markedly upon the architectural planning of the building and specific acoustical considerations. Isolation from external noise sources is essential, and the acoustical properties of the enclosure will, in a large measure, depend on the balance between direct and reflected sound.

In most of the old famous halls designed for music or the opera, the structures were sufficiently massive to attenuate most noise from outside sources. Today, special care must be exercised to eliminate not only traffic noise but the intense broad band noise of jet aircraft. For example, Philharmonic Hall at the Lincoln Center in New York was planned to control both earthborne and airborne vibrations.[6] Lead-asbestos isolation pads were placed under all columns of the building and a sound barrier (glass-fiber pads plus gravel) was placed around all exterior walls to isolate subway noise. The ceiling of the auditorium was floated from elastic hangers, exterior walls were made of 8-in. concrete block, and a massive interior glass wall isolates lobby areas and surrounds the auditorium proper. In addition, the air-conditioning system is equipped with low-noise fans, acoustically lined ducts, and special sound attenuators are placed between the air supply and the auditorium vents.

The ratio between the direct and indirect sound reaching a listener can be qualitatively understood with the aid of Fig. 2a. There is the direct sound that arrives at a given point in an auditorium, followed by an initial time-delay gap, after which multiple reflections arrive that comprise the reverberant field. Figure 2b is a diagram that suggests the reflection sound pattern incident upon the listener from stage, ceiling, and wall areas. Rarely will this type of discrete sound pattern be detected as there are many multiple paths and the ear cannot resolve sounds that are separated by less than 50 msec. In a practical case the sound pattern

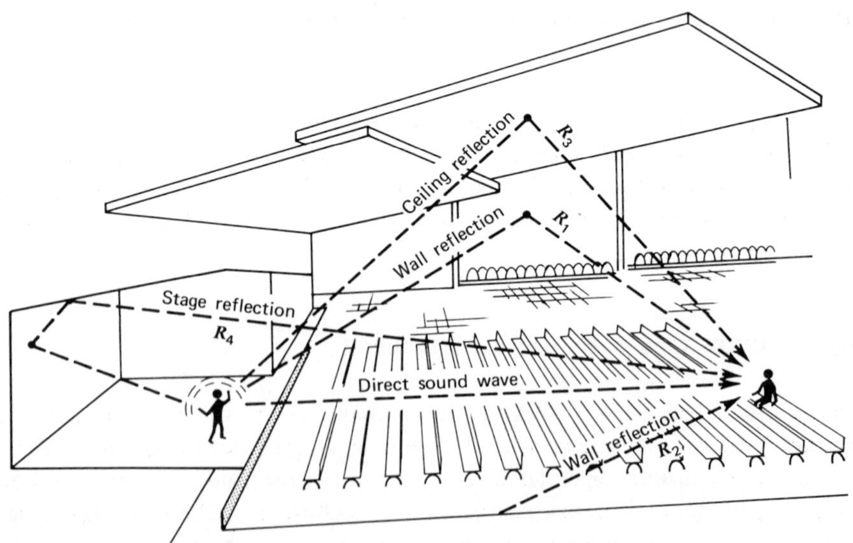

Figure 2a   Paths of direct and reflected sound in an auditorium.

will more closely approximate the schematic diagram of Fig. 2c. As seen in this figure, the listener will hear the direct sound, and if the dimensions of the hall are large, he may be able to detect a time delay. The reverberant field, however, will essentially be sensed as one of exponential decay. The apparent quality of sound will depend upon the reflection coefficients of sound of the several surfaces, as a function of frequency. The distribution of sound energy, that is, the spatial acoustic

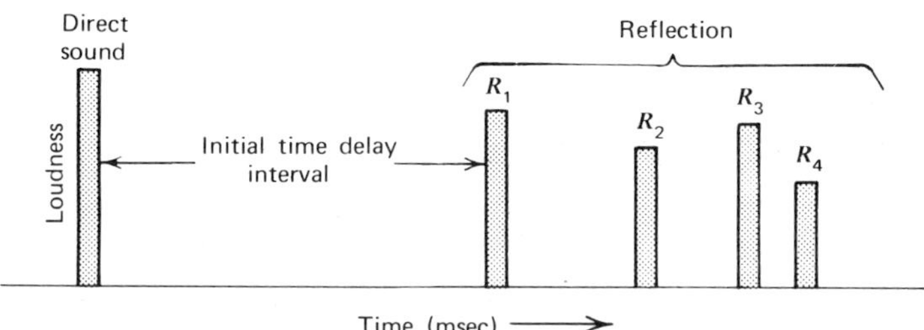

Figure 2b   For a short pulse of sound, the listener will receive direct sound, followed by reflection. The initial time delay gap will depend on room geometry and listener position. In a large auditorium, additional reflections may be incident upon the listener for an extended period.

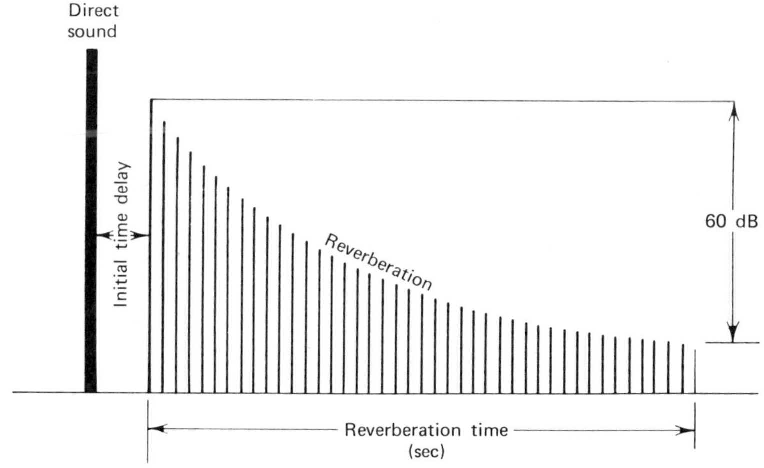

Figure 2c   Direct and reverberant sound. Direct sound, first sensed by a listener, is followed by many reflected sound waves. In most instances these multiple reflections will not be resolved, but blend into a general decay curve. The reverberation time is defined as the time required for reverberant sound to decay 60 decibels.

energy density, will depend upon room geometry, the extent of effective diffusing surfaces, and the extent of the primary sound source. The increase in the sound pressure level at a given point in an auditorium, can easily be 25 dB greater than the direct sound energy.

Echoes can sometimes be sensed in large empty halls with highly reflecting parallel walls—as distinct reflections *above* the level of the general reverberant field. A delay of about 70 msec from the direct sound can produce an echo (i.e., approximately an 80 ft path differential). In properly designed auditoria for the performance of music, distinct reflections as echoes, should not be apparent.

ACOUSTIC CRITERIA FOR SPEECH

Criteria for speech in an auditorium include the following:

1. The ambient auditorium noise level should be sufficiently low for good syllable recognition.
2. The reverberation time should be sufficiently short to provide good speech intelligibility.
3. The sound field should be highly directional (i.e., a direct speaker-listener communication link should exist).

Figure 3 shows the relationship between percentage articulation and the sound level of speech for two conditions of noise: (1) for a very quiet room having a noise level of 0 dB, and (2) for a noise level of 43 dB having a frequency spectrum that is typical of "average room noise."[7] These data assume that reverberation is minimal and that the threshold of the listener corresponds to the minimum audible threshold. An increase in syllable articulation accompanies an increase in the speech power to noise level ratio over a wide range.

A short reverberation time is also essential in order that each phoneme is distinguishable. There is, of course, some latitude in the rapidity with which speech can be understood, depending upon linguistic factors as well as upon acoustical conditions. An audience that knows the context of a specific topic will have a much greater comprehension of a speech than a general audience. Nevertheless, the rapidity with which speech components follow each other, places severe limits on reverberation time. As noted in Chapter 14, some high frequency consonants are exceedingly short and of low intensity, and they can be easily masked by stronger speech sounds or ambient noise. Thus one might conclude that the optimum conditions for speech would be an anechoic chamber having a reverberation time approaching zero. Such a condition is approximated

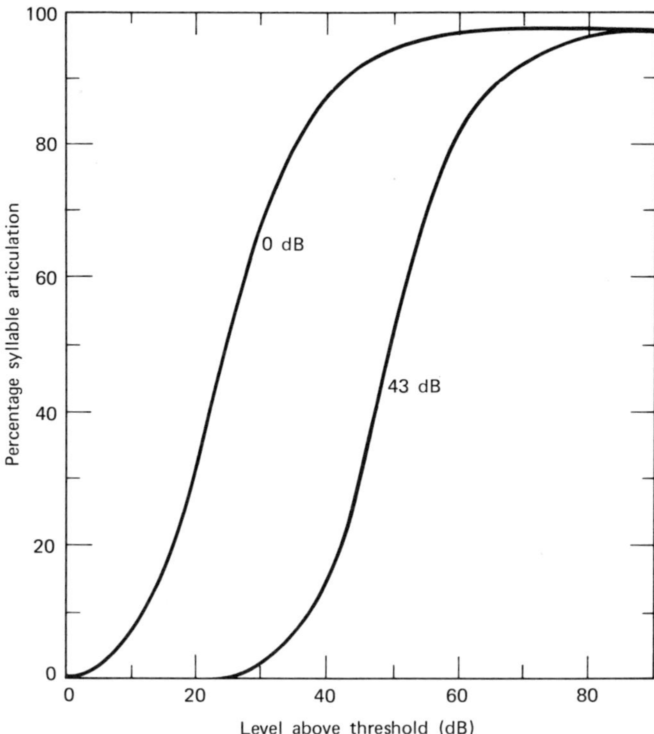

Figure 3   Percentage syllable articulation vs. speech level for (1) a very quiet environment having a noise level of 0 dB, and (2) a noise level of 43 dB having the frequency spectrum of typical room noise. (From V. O. Knudsen and C. M. Harris, *Acoustical Designing in Architecture*, Wiley, 1950.)

by small rooms such as broadcast studios, but in an auditorium the vocal effort required of the speaker would be unrealistic. Furthermore, to some degree, a speaker requires reflected sound in order to monitor his own delivery; that is, some acoustic feedback is essential to adjust speech rate and speech power.

Therefore a compromise is required in an auditorium designed primarily for speech. Speech power must not be attenuated unduly, but a short reverberation time is required for speech intelligibility. Acoustically acceptable theaters have been built, without electronic reinforcement, having midfrequency reverberation times between 0.3 and 1.7 sec. Clearly, the optimum reverberation time must be related to room volume.

Figure 4 shows the percentage syllable articulation in auditoria of different volumes as a function of reverberation time.[8] Reverberation tends

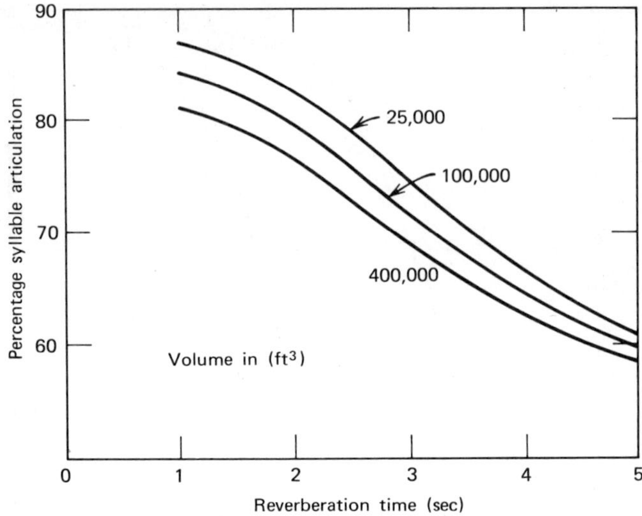

Figure 4   Percentage syllable articulation as a function of reverberation time in auditoriums. The volume of these auditoriums is also noted. (From V. O. Knudsen and C. M. Harris, *Acoustical Designing in Architecture*, Wiley, 1950.)

to reduce articulation scores since it extends speech sounds into the gaps between individual syllables or words and obscures the consonants. A very slow speech delivery rate is the only option available in a hall with very long reverberation times. If the reverberation time of an auditorium is markedly frequency dependent, it is clear that a speaker's voice will also tend to have an unnatural quality.

It is also a fact of common experience that intelligibility in a given room tends to decrease with increasing listener distance from the sound source. However, there exists a critical distance, D_c,* beyond which intelligibility tends to remain constant. In this "outer zone" ($>D_c$) intelligibility is only dependent upon room volume and reverberation time. For consonants, this critical distance is given by the simple formula:[9]

$$D_c = 0.20 \sqrt{V/T} \text{ meters} \tag{1}$$

where V is the volume of the room, in cubic meters, and T is the reverberation time (at about 1400 Hz), in seconds.

* D_c is related to the "room radius" (Chapter IV, equation 1), and for any given room there will be a speaker-listener distance beyond which the direct sound component will be less than the reverberant component.

An adequate *directional* sound field is very important for speech. In the case of music it is desirable to be surrounded by sound; with speech one desires speaker-listener directionality. Subjectively, it is very annoying to see a speaker but hear his voice arriving from a rear wall (either via pure reflected or electronically amplified sound). Thus an auditorium suitable for speech must provide a direct sound field that masks single reflections or echoes.

Consider the case of an auditorium about 100 ft long.[10] Assume the first row of listeners is 10 ft from the speaker, and the last row is 80 ft distant. The intensity of the direct sound reaching the last row will be 18 dB less, simply by applying the inverse square law and neglecting any reflections. If the auditorium floor is not steeply banked there will be a further attenuation, because of audience "shadowing," that is, because of the grazing incidence of the sound. Clearly, some type of sound reinforcement is desirable for the distant listener. In a wide fanshaped auditroium, the ceiling offers the best means for "early" reflections to reinforce the direct sound.

There are two interrelated considerations for sound reinforcement. The first relates to loudness; the second to the sense of direction of the sound. We have previously noted that about 50 msec is the resolving time of the ear. Hence, if identical speech sounds arrive at a rate substantially below this interval, they add and give rise to an increased loudness without a decrease in intelligibility.

If the individual sounds of speech have larger intervals between them than 50 msec, intelligibility will be lost and at about 100 msec the listener may be able to discern discrete echoes. In discussing the Haas effect (Chapter 5), we noted that it is the *first* sound that impinges upon a listener that determines direction. In a 100 ft long auditorium with a listener 55 ft from the speaker, the direct sound will take about 50 msec to travel from the speaker to listener. If "early" reflections arrive shortly after the onset of direct sound (i.e., 25 msec later), the apparent direction of the source will be determined *only* by the first sound; the promptly reflected sound—from whatever direction—will not introduce directional ambiguity. Figure 5 indicates auditorium dimensions and transit times that permit a retention of this directionality sense.

The Haas effect is always taken into account in the placement of loudspeakers as shown in Fig. 6. Loudspeakers can preserve the directionality of speech if electronic delays are introduced in the circuitry to compensate for the transit time of the direct sound field. Provided only that a secondary sound source is not much greater than the primary sound ($<$10 dB), a listener will not be conscious of it as a separate source, and the illusion of line-of-sight voice contact will be maintained.

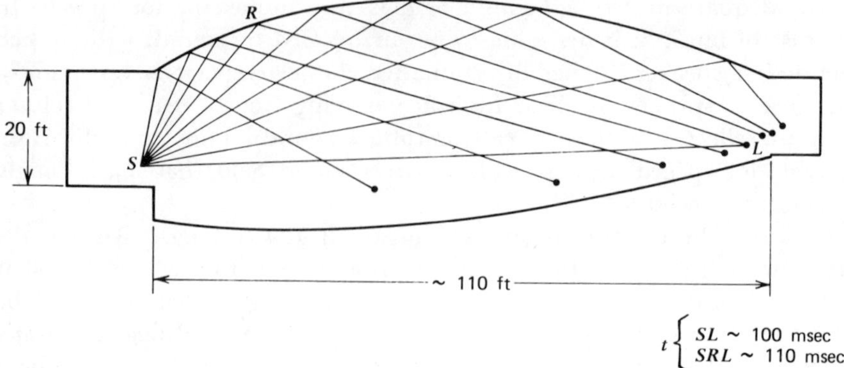

Figure 5   Ceiling reflections provide effective sound reinforcement. If the time delay between direct and reflected path is short, the reflected sound will reinforce the direct sound, and the directionality of sound will be preserved.

ACOUSTIC CRITERIA FOR MUSIC

Because there exists such a wide range in various types of music and because of the subjective nature of the listening experience, no rigid rules can be specified for the ideal music hall. Indeed, some classic composers wrote some of their works with a specific auditorium in mind and not a few compositions were conceived for specific state occasions. Bach, for example, wrote some of his major works for performance in the Leipzig cathedral* where he was organist and choirmaster, and Handel's compositions were often designed for special festivals. Thus, perhaps one of the first criteria for the performance of music is to have reasonable match between specific musical work and the environment in which it is to be rendered. For major symphonic works, a *live* or reverberant auditorium is essential, and Beranek[11] has cited the several qualities that characterize the ideal concert hall:

Uniform Loudness

Only in a reverberant hall can the loudness be reasonably uniform throughout the entire space. The intensity of direct sound will diminish with distance, so that the ratio between direct and reflected sound will

* The St. Thomas Church in Leipzig, with a volume of about 78,000 m³ and a reverberation time of 1.9 sec when fully occupied, was quite suitable for Bach's polyphonic compositions.

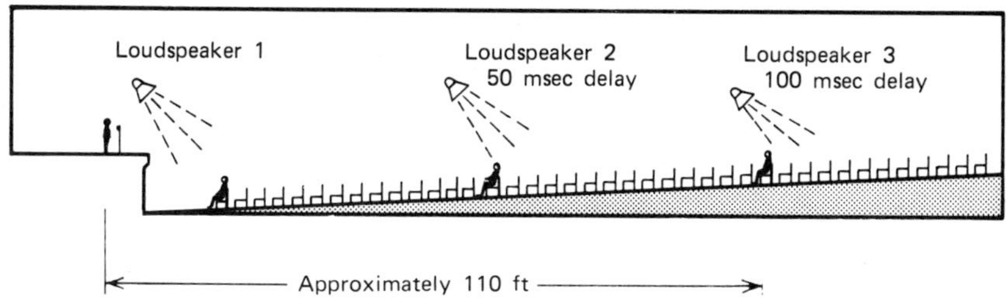

Figure 6 Schematic diagram of an auditorium with a time-delayed speech reinforcement system. The time delay circuitry permits the blending of sound from the first and successive loudspeakers, and the listener senses a single sound source.

change. However, in a good concert hall, the multiple reflections will cause the sound intensity to be sustained in the more remote sections of the enclosure, and help provide a proper ratio of direct to reflected sound close to the stage area.

Enhancement of Bass and Treble

Through proper choice of materials, a reverberant hall can sustain and sometimes provide an enhancement of the bass and of the treble. Augmentation of the bass can sometimes be achieved by using appropriate materials with very low absorption coefficients in this range.

It is interesting to note that the perception of warmth (or rich bass sound) is especially important in the reverberant sound spectrum. Even if the direct sound is deficient in low frequencies, the music will be sensed as satisfactory if there is sufficient low frequency energy in the reflected sound. In symphonic music, at least, "the ear seems to judge spectral balance in terms of an integration over several hundred milliseconds—and it is willing to wait for the reverberant energy before making its evaluation of musical warmth."[12] The brilliance of the treble can also be affected by the strategic placement of sound-reflecting panels. Figure 7 shows the effectiveness of suspended ceiling panels in reinforcing direct sound in Avery Fisher Hall, New York in the original design.[13] The lack of energy at low frequencies was attributed in part to the small panel sizes compared to sound wavelengths. Figure 8 is a photo of Avery Fisher Hall after extensive modifications had been made to the ceiling and other portions of the hall.

In general, ceiling reflectors will bring "early-reflected" sounds to a larger fraction of the audience, and transmit the "transients" that are so important in giving character and clarity to all instrumental sound.

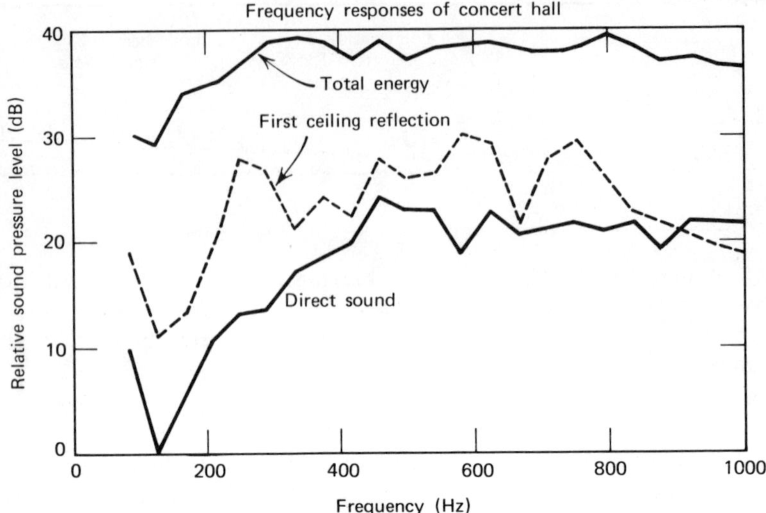

Figure 7   Energy received at a listening position near the center of the orchestra floor in Avery Fisher Hall, New York. [From M. R. Schroeder, *J. Acoust. Soc. Am.*, **45**, 1077 (1969).]

Fullness of Tone

The fullness of tone that is sensed in a reverberant hall results from the blending of a tone with preceding and subsequent tones. This fullness or breadth of tone is especially desirable in many musical passages, and this quality is quite distinct from the sensation of loudness.

Range of Crescendo

Only a reverberant hall can provide the maximum possible range of crescendos and diminuendos. The listener can, under ideal conditions, sense a single instrument at 30 dB and hear a full orchestra *double forte* that grows and swells to a 90 dB climax. The reverberation sensed after single chords, trumpet fanfares, or percussion effects also provides much of the musical emotional impact.

Diffusion of Sound

The diffusion of sound is another desirable consequence of the phenomenon of reverberation. Sound impinges upon the listener's ears from all directions, giving him the sensation of being immersed in a sea of sound. In the ideal concert hall, music literally fills the hall.

Figure 8   Avery Fisher Hall (formerly Philharmonic Hall). (Photograph by Sandor Acs, and courtesy of the Lincoln Center for the Performing Arts.)

All of the above mentioned qualities relate in some manner to reverberation time—either physically or subjectively. First, reverberation provides the "bridging" of the slight gaps between musical notes to provide fullness of tone. Second, reverberation provides the listener with sound from all directions. Third, the dynamic range and perceived loudness depends on the ear performing an integration over a finite time, in addition to the fact that the actual sound pressure level will be increased with increasing reverberation times.

A reverberation time of approximately 2 secs is acceptable for symphonic music in a large concert hall. Figure 9 indicates the range of reverberation times that seem to be appropriate for various types of music. In order to preserve the natural tone of musical instruments, the reverberation time should be nearly constant from 300 Hz to 1500 Hz; at higher frequencies the reverberation times in large halls will decrease because of air absorption.

Recent studies have also attempted to more clearly identify subjective responses to music with auditorium geometries. In one study it has been suggested that a "spatial impression" can be attributed to the lateral reflections (from the side walls) rather than ceiling reflections, and that the degree of this spatial impression is probably related to the ratio of lateral to nonlateral sound arriving within 80 msec of the direct sound.[14]

BASIC HALL SHAPES

There are three basic shapes for concert halls and opera houses (1) rectangular, (2) fanshaped, and (3) horseshoe. A fourth type might be called

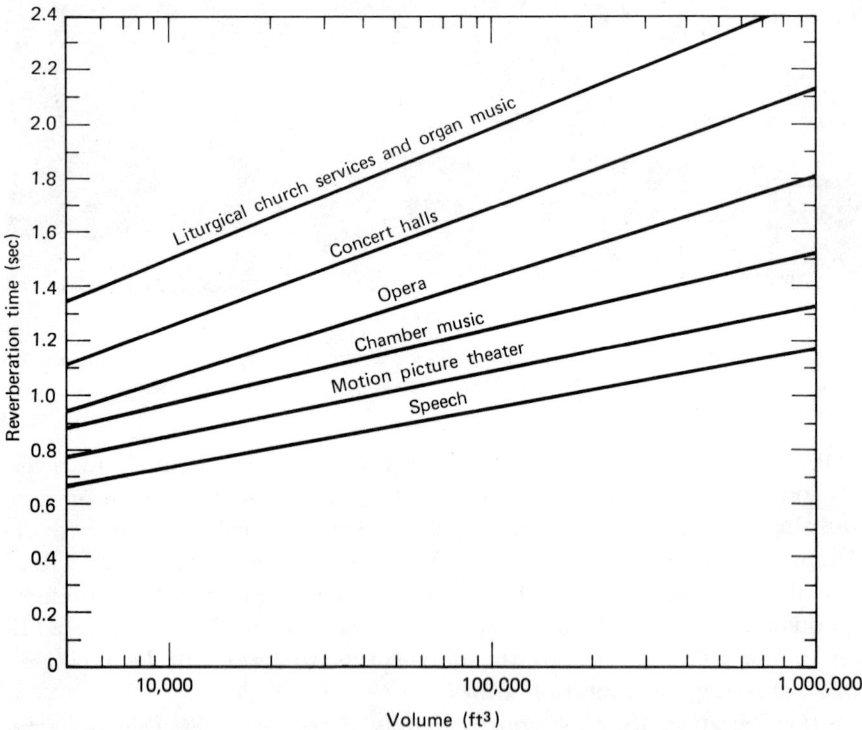

Figure 9   Typical reverberation times for various auditoria and functions.

a "modified arena"—such as the Royal Albert Hall (London) built in 1871, which is nearly elliptical in shape; the Concertgebouw in Amsterdam, built in 1887; the Sydney Opera House; and the new hall being designed in Denver. The first three basic hall shapes are shown in Fig. 10.

The rectangular hall is traditional, and it has been built to accommodate both small and large audiences. Acoustically, a rectangular hall will always generate cross-reflections between the parallel walls. There will also be some direct reflected sound from the rear wall back to the stage area, although the amount of this reflection will depend upon balcony layout and sound-absorbent treatment. For a rectangular hall of modest dimensions, these reflections aid in the build-up of sound and provide a reasonable degree of diffusion. If the dimensions of the hall are very large, however, reflections from the parallel wall can give rise to standing wave resonances and flutter echoes. The rectangular design is best exemplified by such great music halls as the Symphony Hall (Boston) built in 1900, and the Grösser Musikvereinssaal (Vienna), completed in 1870.

A fanshaped hall has the advantage of making reasonable the distance from the stage to the most remote listener, even for a hall with a large seating capacity. The nonparallel walls prevent objectionable resonances, and most listeners are able to obtain a good balance between direct and reflected sound. A disadvantage in terms of early time delay gap is the distance from the side walls. Therefore, it is often essential to add a series of inner reflectors or forestage canopies to maintain articulation and other desired acoustic characteristics. In any event, in the United States architects have been required to design concert halls for large audiences and a basic fan shape, with many modifications, provides a large seating capacity while retaining a reasonable degree of both visual and aural coupling to the stage area. Examples of this design include the Eastman Theatre, Rochester, New York; the Henry and Edsel Ford Auditorium, Detroit, Michigan; the Kleinhans Music Hall, Buffalo, New York; the Orchestra Hall in Chicago; and the Dorothy Chandler Pavilion in Los Angeles.

Horseshoe-shaped structures have, for centuries, been popular for opera houses and concert halls of modest seating capacity. The geometry provides for a greater feeling of intimacy, and the concave surfaces are sufficiently textured to create adequate diffusion of sound. The multiple balconies permit excellent line-of-sight and a short path for direct sound. The arrangement is also well suited for the opera where clarity of sound is of greater importance than fullness of tone. The most renowned example of the horseshoe design is La Scala Opera House in Milan, Italy

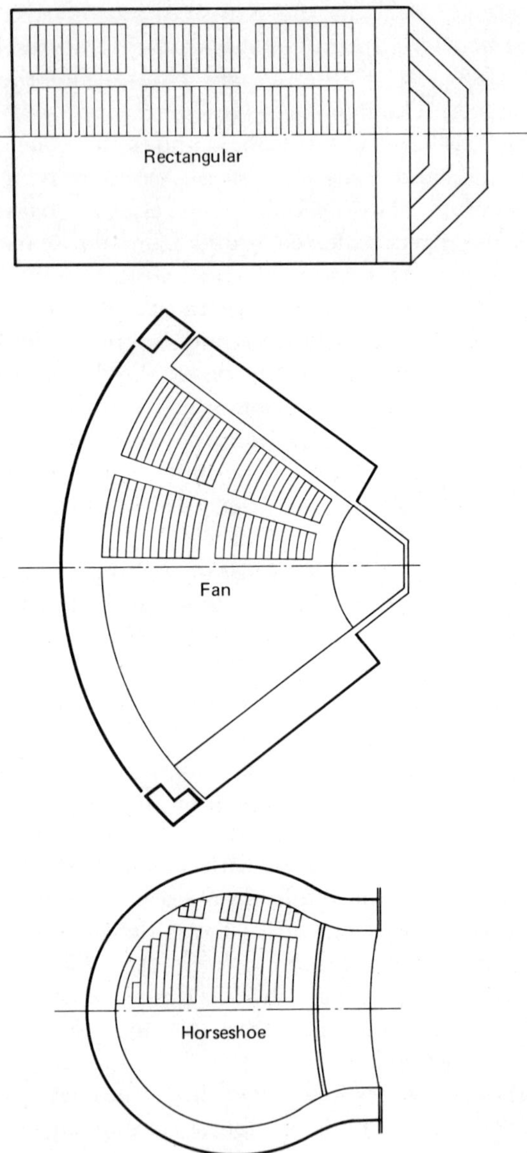

Figure 10   Basic hall shapes: rectangular, fan-shaped, and horseshoe-shaped.

(Fig. 1). In the United States, Carnegie Hall in New York and the Academy of Music in Philadelphia typify this Italian-style opera house, providing an intimate coupling between orchestra, singers, and audience.

Virtually all concert halls have balconies; they are desirable from an acoustic point of view in that for a given seating capacity, a structure can be built with a smaller volume. Balconies also place listeners in a more intimate relationship with the stage. The balcony depths are usually not greater than twice their vertical opening to the stage, and a smaller ratio is actually preferred to prevent undue sound attenuation at the rear wall. In contemporary architectural planning, a rule of thumb is that their depth should not exceed 1.4 times the opening at the front of the balcony.

Ceilings provide the best means for transporting sound energy from the stage to distant listeners in all designs, and if the ceiling is not too high, there will not be a great time difference between direct and ceiling-reflected sound (Fig. 11). The floor profile is also important in establishing a proper ratio of direct to indirect sound, and splays on side walls are effective in promoting diffusion and uniform loudness. Rear walls are usually absorbent to prevent echoes being sent back to the stage.

CLASSIC CONCERT HALLS

Among the hundreds of music halls throughout the world, some have been widely acclaimed as excellent and some as poor. All of the "excellent" halls have a few common characteristics. First, none are overly large with a capacity limited to about 2800 seats. Second, all permit the audi-

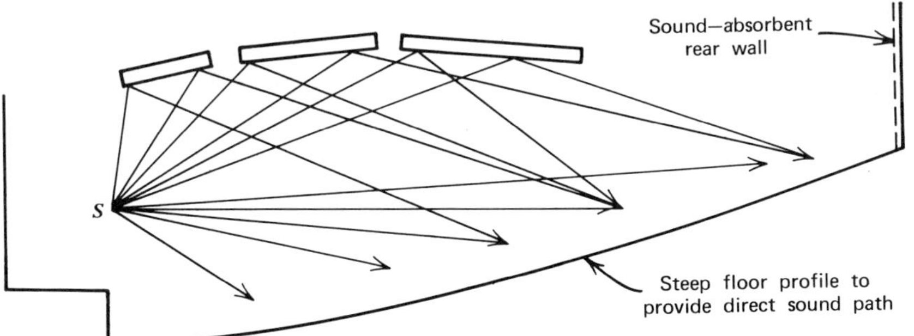

Figure 11 Ceiling and floor profiling aid in transmission of sound to all areas of an auditorium.

ence to distinguish clearly the individual instruments of the orchestra without losing the fullness or blending of tone associated with reverberant sound. In other words, there is a proper balance between the direct and reverberant sound field; initial transients can be detected before they are masked by reflected sound. Third, the "excellent" halls are the choice of both the orchestras and the audience; a good hall allows the orchestra to "hear itself."

There is general agreement that the Symphony Hall in Boston rates acoustically as one of great classic halls in the world. Designed by Wallace Sabine, often called the father of architectural acoustics, it is a rectangular structure with a high, textured ceiling and two balconies that extend along three walls. Figure 12 is a photo of this famous hall as viewed from the stage area, and Fig. 13 shows the stage area. Volume is 662,000 ft,^{3} seating capacity is 2631, and the reverberation time in the 500 to 1000 Hz range is 1.8 sec (occupied).[15]

Many of the world's great conductors have performed in this hall and all are unanimous in their praise. "It is the most noble of American con-

Figure 12   Boston Symphony Hall, viewed from the stage. (Courtesy of the Boston Symphony Orchestra.)

Figure 13   Boston Symphony Hall; the stage area. (Courtesy of the Boston Symphony Orchestra).

cert halls"—Bruno Walter; "—it is to the orchestra what a Stradivarius is to the great violinist in providing a sound box of the utmost brilliance and sensitivity"—Boston *Herald* music critic. The octave-band reverberation time measurements of this hall (unoccupied) are: 125 Hz (2.0 sec); 250 Hz (2.05 sec); 500 Hz (2.05 sec); 1000 Hz (2.0 sec); 2000 Hz (1.95 sec); 4000 Hz (1.6 sec); 6000 Hz (1.1 sec).

The Orchestra Hall in Minneapolis,* completed in 1974 is also patterned after the classic rectangular design. The slanted floor of the hall contains 1590 seats, and three-stepped balcony tiers hold an additional 983 seats for a total audience capacity of 2573. The ceiling is covered by a random pattern of plaster cubelike shapes that are effective in diffusing the sound throughout the hall. This overlay of cubes continues down the back wall behind the stage as shown in Fig. 14. Wood paneling partially covers the walls, and wood floor construction has been utilized in

* Cyril M. Harris, acoustical consultant.

Figure 14   Minnesota Orchestra Hall. (Courtesy of the Minnesota Orchestral Association.)

the stage and audience areas. Overall, this concert hall provides great dynamic range, balance, and clarity.

Table 1 is a list of a select number of other important halls throughout the world, together with seating capacity and reverberation times. Note the substantial variation among these halls with respect to reverberation time. Also, some halls have been constructed with seating and carpeting materials that yield almost equal reverberation times in the occupied and unoccupied state. Such a condition is usually an asset because the conductor does not have to be concerned with audience effects, and rehearsals can simulate an actual performance. It is generally agreed that the ideal concert hall cannot seat more than 3000 persons. A very large hall may have an acceptable reverberation time; nevertheless, there is a limit to the sound intensity that can be generated by an orchestra or soloist, and it is difficult to attain adequate contrast between pianissimo and fortissimo passages.

AIR ABSORPTION IN LARGE HALLS

In the preceding sections we have tacitly assumed that sound absorption occurs only at the boundaries. For small rooms and auditoria of up to about 200,000 ft³ this assumption is generally justified. However, for spaces of very large volume and when the reflecting surfaces have low absorption coefficients (i.e., in highly reverberant rooms), there can be a measurable absorption because of air at high frequencies. In fact, for auditoria or astrodomes of very large dimensions, air absorption will place an upper limit on the reverberation time even if the boundaries are highly reflecting.

Thus, it is necessary to make a modification to our simple reverberation equations of Chapter 4, by adding the term $4mV$ to the total absorption, that is,

$$(S\alpha)_{\text{total}} = S\alpha + 4mV \tag{2}$$

where $S\alpha$ is the absorption of the wall surfaces, V is the room volume in ft³ or m³, and m is the air attenuation constant in reciprocal feet or meters.

The reverberation time for the auditorium will then be

$$T = 0.05 \frac{V}{S\alpha + 4mV} \text{ sec} \tag{3}$$

when S and V are expressed in ft² and ft³, respectively, and

$$T = 0.16 \frac{V}{S\alpha + 4mV} \tag{4}$$

when S and V are expressed in m² and m³, respectively.

Values of the correction term, $4m$, at a temperature of 68°F (20°C) and for several relative humidities and frequencies, are listed in Table 2. The first entry is the value of $4m$ in ft⁻¹; the second is in m^{-1}.

Some simple calculations will show that, at the higher frequencies, air absorption will substantially decrease the reverberation time in auditoria

Table 1   Concert Halls and Auditoriums[16]

	Volume (ft³)	Seating Capacity	Reverberation Time†	
			Occupied	Unoccupied
United States				
Baltimore, Lyric Theatre	744,000	2616	1.47	2.02
Boston Symphony Hall	662,000	2631	1.8	2.77
Buffalo, Kleinhans Music Hall	644,000	2839	1.32	1.65
Cambridge, Kresge Auditorium	354,000	1238	1.47	1.7
Chicago, Arie Crown Theatre	1,291,000	5081	1.7	2.45
Cleveland, Severance Hall	554,000	1890	1.7	1.9
Detroit, Ford Auditorium	676,000	2926	1.55	1.95
New York, Carnegie Hall	857,000	2760	1.7	2.15
Philadelphia Academy of Music	555,000	2984	1.4	1.55
Purdue University Hall of Music	1,320,000	6107	1.45	1.6
Rochester, New York, Eastman Theatre	900,000	3347	1.65	1.82
Austria				
Vienna, Grosser Musikvereinssaal	530,000	1680	2.05	3.6
Belgium				
Brussels, Palais des Beaux-Arts	442,000	2150	1.42	1.95
Canada				
Edmonton and Calgary, Alberta Jubilee Halls	759,000	2731	1.42	1.8
Vancouver, Queen Elizabeth Theatre	592,000	2800	1.5	1.9
Denmark				
Tivoli Koncertsal	450,000	1789	1.3	2.25
Finland				
Turku, Konserttisali	340,000	1002	1.6	1.95
Germany				
Berlin, Musikhochschule Konzertsaal	340,000	1340	1.65	1.95
Bonn, Beethovenhalle	555,340	1407	1.7	1.95

(*Continued*)

Table 1   Concert Halls and Auditoriums (continued)

	Volume (ft³)	Seating Capacity	Reverberation Time[†] Occupied	Unoccupied
Great Britain				
Edinburgh, Usher Hall	565,000	2760	1.65	2.52
Liverpool Philharmonic Hall	479,000	1955	1.5	1.65
London, Royal Albert Hall	3,060,000	5080	2.5	3.7
London, Royal Festival Hall	755,000	3000	1.47	1.77
Israel				
Tel Aviv, Frederic R. Mann Auditorium	750,000	2715	1.55	1.97
Italy				
Milan, Teatro Alla Scala	397,300	2289	1.2	1.35
Netherlands				
Amsterdam, Concertgebouw	663,000	2206	2.0	2.4
Sweden				
Gothenburg, Konserthus	420,000	1371	1.7	2.0
Switzerland				
Zurich, Grosser Tonhallesaal	402,500	1546	1.6	3.85
Venezuela				
Caracas, Aula Magna	880,000	2660	1.35	1.8

† At 500-1000 Hz.

where surface absorption is low. Indeed, in very large auditoria with highly reflecting surfaces, air absorption at high frequencies may be the dominant phenomenon. An examination of equations 2 and 3 and Table 2 will also explain why a musical performance by the same orchestra can seem more or less "brilliant" or "live," depending on the auditorium characteristics and the humidity.

It should also be noted that in a large concert hall, air absorption occasionally plays a positive role, in that an adequate ratio of bass to high frequency sound is maintained via the late arriving reverberant field. In a small hall, where air absorption is negligible, some high fre-

Table 2   Air Absorption Values of 4m For Reverberation Time Correction, at 68°F (20°C)[17]

Relative Humidity	2000 Hz	3000 Hz	6300 Hz	8000 Hz
30%	0.0036 ft^{-1}	0.0116 ft^{-1}	0.0256 ft^{-1}	0.041 ft^{-1}
	0.0119 m^{-1}	0.0379 m^{-1}	0.0840 m^{-1}	0.136 m^{-1}
50%	0.0029 ft^{-1}	0.0074 ft^{-1}	0.0153 ft^{-1}	0.026 ft^{-1}
	0.0096 m^{-1}	0.0244 m^{-1}	0.0503 m^{-1}	0.086 m^{-1}
70%	0.0026 ft^{-1}	0.0065 ft^{-1}	0.0122 ft^{-1}	0.0184 ft^{-1}
	0.0085 m^{-1}	0.0213 m^{-1}	0.0399 m^{-1}	0.060 m^{-1}

quency absorption may have to be introduced to provide the proper balance of bass to treble.

ELECTRONIC ENHANCEMENT OF REVERBERATION

Because the reverberation time is the most important acoustical parameter of an auditorium, various schemes have been devised to control it by electronic means. Of particular interest is the problem of enhancing the reverberation time of an auditorium within a limited frequency range, in order to obtain a better "balance" of orchestral sounds.

Such a change was successfully achieved with an electronic system installed in the Royal Festival Hall, London,[18] where it was desired to lengthen the reverberation time for frequencies below 700 Hz. A number of "channels," each consisting of a microphone, amplifier, and loudspeaker, were used to cover a 58 to 700 Hz range, with each channel covering a narrow frequency band. All microphones and loudspeakers were located in the ceiling, and microphone positions were carefully selected to correspond with reasonably invariant sound pressure levels for a given frequency, and for a fixed power output from a sound source on the stage. Precautions were taken also to prevent excessive interaction between channels. Amplifier gains of the various channels were then set so as to increase the sound power of reflected sound and lengthen the reverberation time. Figure 15 shows the reverberation times in this Hall with the system on and off (solid and broken lines, respectively), when the orchestra was the sound source. The increase in the reverberation at

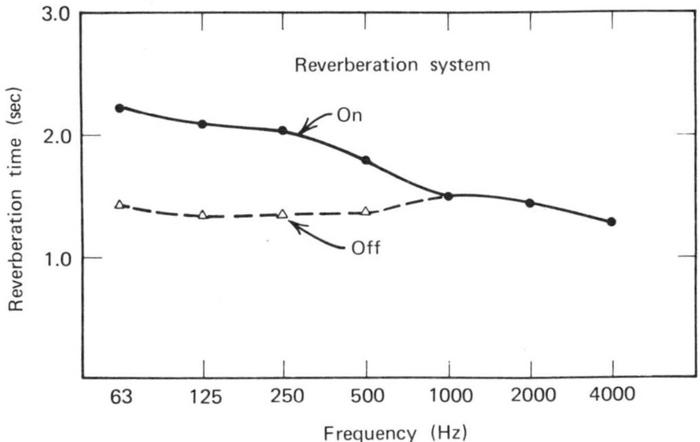

Figure 15   Electronic enhancement of reverberation time for low frequencies at the Royal Festival Hall, London. [From P. H. Parkin and K. Morgan, *J. Acoust. Soc. Am.*, **48**, 1025 (1970).]

low frequencies is reported to result in a substantial acoustical improvement. All the extra acoustical energy is radiated from the ceiling-mounted loudspeakers, but this secondary source of sound is not distinguishable, because of the Haas effect.

Changes in "effective reverberation" can also be achieved through the use of electronic circuitry at the sound source. Electronic organs, for example, can be made to sound somewhat like pipe organs, even when played in very small rooms. One method for achieving this simulation is to record the sound from the tone generators on magnetic tape, and have not one but several pickup heads replay the sound. These pickup heads are spatially separated so that when a tape passes through them, the sound will be reproduced several times, that is, at times Δt_1, Δt_2, Δt_3, Δt_4 . . . subsequent to the recording time, t_0. The net result is that when a single note or chord of short duration is played, (and simultaneously taped), the reproduced sound will bear a remarkable resemblance to the time-delayed multiple reflections in a large reverberant room. With such devices, and with recently developed digital time delay circuitry, signal decay rates corresponding to reverberation times of several seconds can be produced.

The theatrical uses of reverberation control have been important for several decades[19] and special "electro-acoustic environments" now represent an integral part of the performing arts.

J. F. KENNEDY CENTER FOR THE PERFORMING ARTS

This Center, opened in Washington, D. C. in 1971, was originally established by an act of Congress in 1958 as a National Cultural Center. However, subsequent to the death of President Kennedy in 1963, Congress renamed it as a memorial. The $67,000,000 Center consists of a single structure within which is housed a 2759-seat concert hall, a 2319-seat opera house, and 1142-seat capacity theater.

The environmental acoustic problems associated with its design were especially severe.[20] The Center is situated on a site facing the Potomac River in close proximity to Washington's National Airport. Aircraft, both private and commercial, sometimes fly as low as a few hundred feet directly over the roof, and on occasion helicopters fly along the Potomac River at close to rooftop levels. In addition, vehicular traffic runs along the river and directly beneath the plaza of the Center. These exterior noise sources are attenuated by utilizing a box-within-a-box design, such that the three auditoriums (Fig. 16) are wholly contained within an exterior shell, and the columns supporting each auditorium are built to isolate both airborne sound and mechanical vibrations from interior ceilings, walls, and floors. Much of the double-wall construction consists of 6-in. solid high-density concrete blocks separated by a 2-in. air gap. The massive windows of the Grand Foyer overlooking the river consist of 1/2-in. and 1/4-in. thick glass sheets separated by a 4-in. air space. Noise control of interior noise sources—for example, transformers and air-conditioning units—is accomplished by means of resilient mounts, flexible connectors, and acoustically lined ducts. Special doors are utilized at all auditorium entrances, together with "sound locks" between foyer and the auditorium interiors.

The Concert Hall has a volume of 682,000 ft^3 and a seating capacity of 2759. It is rectangular in basic design, and large contoured wall surfaces and a coffered ceiling enhance the diffusion of sound at both low and high frequencies. Eleven massive crystal chandeliers (2700 lb each), gifts of the Norwegian government, also contribute effectively to diffusion. Balconies are relatively shallow to avoid a reduction in sound level under the balcony overhang. The reverberation characteristics of the hall are given in Table 3.

The Opera House is shown in Figure 17, being designed to bring the audience as close to the stage as possible, and having a seating capacity of 2319.

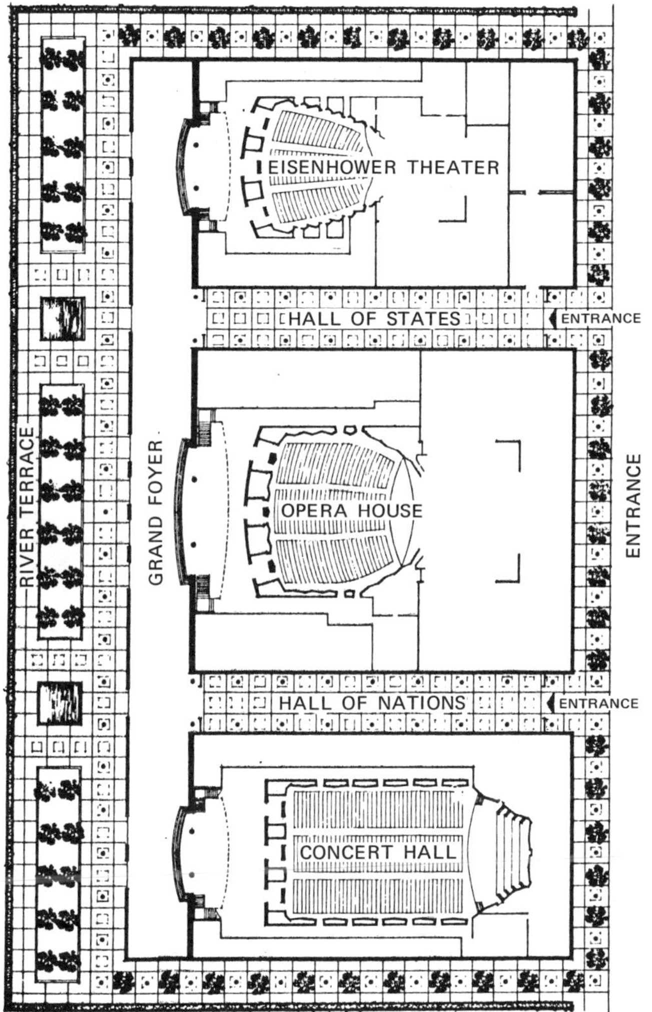

Figure 16   Schematic of the three auditoriums of the J. F. Kennedy Center for the Performing Arts. [From C. M. Harris, *J. Acoust. Soc. Am.*, **51**, 1114 (1972).]

Table 3   Reverberation Time vs. Frequency in the J. F. Kennedy Concert Hall[20]

Frequency	125 Hz	250 Hz	500 Hz	1000 Hz	2000 Hz	4000 Hz	6300 Hz
Occupied	2.5	2.2	2.0	1.8	1.7	1.4	1.4
Unoccupied	2.8	2.4	2.2	2.0	1.9	1.6	1.6

Figure 17   The interior of the J. F. Kennedy Center Opera House. (Photograph by Fletcher Drake, and courtesy of the J. F. Kennedy Center.)

BAND SHELLS AND "OPEN" AUDITORIUMS

Within the last few decades, there has been a rebirth of "outdoor" musical performances. In the nineteenth century band concerts were a common occurrence throughout the United States, and in many cities and villages there were small permanent band stands erected in the parks. Today, we have modern band shells for open air concerts and semi-open structures have been designed specifically to provide excellent musical acoustics for large audiences.[21]

There are several reasons for these new structures. First, building costs of music halls are becoming prohibitive, except for the large metropolitan areas. Second, even the finest music halls can accommodate an audience of very limited size. Third, there is a growing tendency for people to seek entertainment in an informal setting and, if possible, in a rural area that has an environment quite distinct from the city.

The band shell has two functions. It directs the sound to the audience area, and the shell configuration assists the members of the musical group to hear each other. It is almost impossible for a large orchestral group to play effectively as an ensemble in the open air.

Figure 18   The Hollywood Bowl. (Courtesy of the Southern California Symphony—Hollywood Bowl Assoc.)

The site of the band shell is important. Ideally, it should be completely isolated from the noise of traffic or overhead aircraft. The topography of the land is also a salient feature. If the shell can be located on a hillside, there will be substantially less attenuation of sound by both ground and audience than if the shell is located on a plateau. Visual contact is also measurably improved. A final contouring of the land is, of course, achieved by earth-moving machinery. Ambient noise levels should be no greater than 35 dB(A) except that there will probably be some intermittent natural sounds.

Figure 18 is a photo of a modern band or orchestra shell, the Hollywood Bowl.* Clearly, in such an outdoor "concert hall" there is no reverberant sound field. The only reflected sound will be that of the early reflection from the shell. This sound will reinforce the direct orchestra-listener sound power, but the stage distances are sufficiently short so that

* The Hollywood Bowl has the distinction of being able to somewhat control its environmental noise through an arrangement with commercial airlines. During important concerts, air traffic is routed away from the Bowl area via radio messages to aircraft pilots.

Figure 19   The Tanglewood Music Shed, Lenox, Massachusetts. A massive ceiling canopy provides reflected sound and adds brilliance and dynamic range. (Courtesy of the Boston Symphony Orchestra.)

the sound is heard without any distinguishable time delay gap. Shells can never provide the dynamic range of sound power that is realizable in an enclosed reverberant concert hall. For full symphony orchestra or band ensembles, however, they provide sound fields sufficient to be acceptable to an audience of many thousands. Amplification systems of high quality are also sometimes employed.

The semi-enclosed structure or "music shed" specifically designed for music and the performing arts is a fairly modern development. The Tanglewood Music Shed in the Berkshire Hills of Massachusetts is one of the nation's foremost such structures. It is the summer home of the Boston Symphony Orchestra, and its acoustics provide an excellent environment for an audience of 6000. An additional 6000 persons can also hear through the partly open sides of this pavilion. Modifications to the original building were completed in 1959, providing a new orchestra enclosure and an acoustic canopy that covers the front third of the audi-

Figure 20   The Center for Performing Arts, Saratoga Springs, New York. (Courtesy of the Saratoga Performing Arts Center.)

torium.[22] Figure 19 is a photo showing the extensive canopy that projects and diffuses sound throughout a volume of 1,500,000 ft.[3] When occupied it has a reverberation time of 2 sec in the frequency range of 500 to 1000 Hz, which is excellent considering the fact that the floor consists only of packed earth. The ceiling is constructed of 2-in. thick wooden planks; side and rear walls are 3/4-in. fiberboard.

Another such modern "music shed" is the Performing Arts Center in Saratoga Springs, New York, which now serves as the summer abode for the Philadelphia Symphony Orchestra. Figure 20 is a photo of this new semi-open concert hall. Mention should also be made of the Blossom Music Center outside of Cleveland, Ohio, which provides concert hall acoustics for 4500 people. At both sites the music can be heard by a large lawn audience. In order to maintain the sound at an adequate level throughout the lawn area of the Blossom Center, a sound reinforcement system was employed. High fidelity speakers were placed at the perimeter of the building and connected to the microphone through a "delay line."

Figure 21   The 360° Music Center at Toronto, Canada. (Photograph by Craig, Zeidler & Strong, Architects; acoustical consultant, Christopher Jaffe.)

This electronic delay was made to match the transit time of sound coming directly from the stage to the lawn edge. Thus the direct sound field and amplified sound field are essentially in phase. The installation is so successful that one cannot sense the transition region where the direct and amplified sound fields are matched.[23]

Finally, the Forum at Ontario Place, Canada, should be cited, since it is the first 360-degree symphonic facility in North America. Located on the Toronto shorefront on Lake Ontario, it permits a 2800 seat audience to be within 50-ft of the performers, and lightweight turnable ceiling panels can be adjusted over the center orchestral sections to achieve an optimum distribution of sound. Audiences of up to 12,000 have listened to symphonic music from this symmetrical musical pavilion.[24] Figure 21 is a photo of this new center.

REFERENCES

1.   R. S. Shankland, *Phys. Today*, 30 (October 1973).
2.   R. S. Shankland, *Am. Sci.*, **60**, 201-209 (1972).
3.   *Ibid.*

4. R. S. Shankland and H. K. Shankland, *J. Acoust. Soc. Am.,* **50,** 389 (1971).

5. W. C. Sabine, *Collected Papers on Acoustics,* Dover, New York, 1964, p. 6.

6. L. L. Beranek, F. R. Johnson, T. J. Schultz, and B. G. Watters, *J. Acoust. Soc. Am.,* **36,** 1247 (1964).

7. V. O. Knudsen and C. M. Harris, *Acoustical Designing in Architecture,* Wiley, New York, 1950, p. 178.

8. *Ibid.,* p. 179.

9. J. M. A. Pentz, *J. Aud. Eng. Soc.,* **19,** 915 (1971).

10. P. H. Parkin and H. R. Humphreys, *Acoustics Noise and Buildings,* Praeger, New York, 1958, p. 140.

11. L. L. Beranek, *Music, Acoustics and Architecture,* Wiley, New York, 1962.

12. T. J. Shultz, *IEEE Spectrum,* June 1965.

13. M. R. Schroeder, *J. Acoust. Soc. Am.,* **45,** 1077 (1969).

14. M. Barron, *J. Sound & Vib.,* **15,** 475 (1971).

15. Beranek, *op. cit.,* p. 97.

16. *Ibid.,* p. 555-567.

17. C. M. Harris, *J. Acoust. Soc. Am.,* **40,** 148 (1966).

18. P. H. Parkin and K. Morgan, *J. Acoust. Soc. Am.,* **48,** 1025 (1970).

19. H. Burris-Meyer, *J Acoust. Soc. Am.,* **13,** 16 (1941).

20. C. M. Harris, *J. Acoust. Soc. Am.,* **51,** 1114 (1972).

21. C. Jaffe, *J. Audio Eng. Soc.* **22,** 163 (1974).

22. F. R. Johnson, L. L. Beranek, R. B. Newman, R. H. Bolt, and D. L. Klepper, *J. Acoust. Soc. Am.,* **33,** 475 (1961).

23. Shankland, *op. cit.,* p. 203.

24. C. Jaffe, *The Design of a 360 Degree Symphonic Facility,* 84th Meeting of the Acoustical Society of America, Miami Beach, November 1972.

17

PHYSIOLOGICAL AND PSYCHOLOGICAL EFFECTS OF SOUND

GENERAL PHYSIOLOGICAL EFFECTS

It is now known that high sound levels can induce responses in the human body that are not specifically related to the auditory system.[1] These physiological responses represent essentially a failure of the human organism to remain a passive system. In most instances a physiological change is evident only during, or for a short period after the noise exposure. Specific studies have focused on a (1) the effect of noise on blood circulation, (2) the resistance of the skin to electrical potentials, (3) skeletal-muscle tension, (4) breathing amplitude, and (5) the effect of noise on sleep. There is also some evidence that noise is a contributing factor to changes in the motility of the gastrointestinal tract, changes in the size of the pupils of the eyes, and changes in the rate of saliva and gastric secretions.[2,3]

Blood Circulation

In connection with reasonably short, intense sounds, a general constriction of the peripheral blood vessels with a reduction in peripheral blood flow has been observed.[4] Generally, the sound levels must be greater than

460

70 dB(A), and the response may diminish or disappear with a predictable repetition of the signals. Figure 1 shows results reported from a controlled experiment in which subjects were exposed to intermittent wide band noise during rest periods after quiet manual work. The data was obtained in a series of 45 tests on 7 subjects.

A statistical study of 1005 German industrial workers[5] also revealed that a larger number of workers encountered blood circulation problems in noisy environments than in environments of average sound levels (Fig. 2). Additional data from noisy industries in Russia (e.g., foundries) have indicated that workers undergo an unusually high percentage of circulatory, digestive, and metabolic difficulties.[6]

Skin Resistance

Noise has been shown to induce a galvanic skin response (GSR). A relative measure of this response has been reported by Davis et al.[2] The relative reduction in the resistance of the skin to electric current was observed as a function of repeated signals of a 98 dB(A) tone near 800 Hz. The change is an indication of an activation of the peripheral visceral nervous system.

Muscle Tension

Limited data is also available on the magnitude of the muscle-tension reflex in response to increasing sound levels. For a steady noise level of 90 dB(A), increased tension was detected in muscles, and the sound also

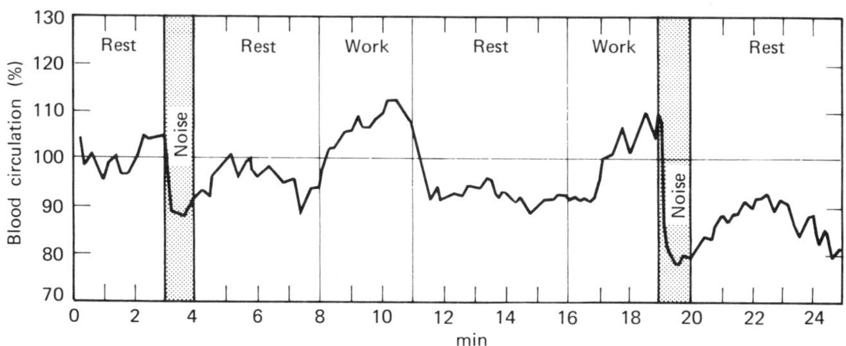

Figure 1   The effect on the blood circulatory system, attributed to periods of rest, work, and noise. A reduced blood circulation suggests vasoconstriction of the peripheral blood vessels. (From G. Jansen, Ref. 3, and K. D. Kryter, *The Effects of Noise on Man*, Academic Press, 1970.)

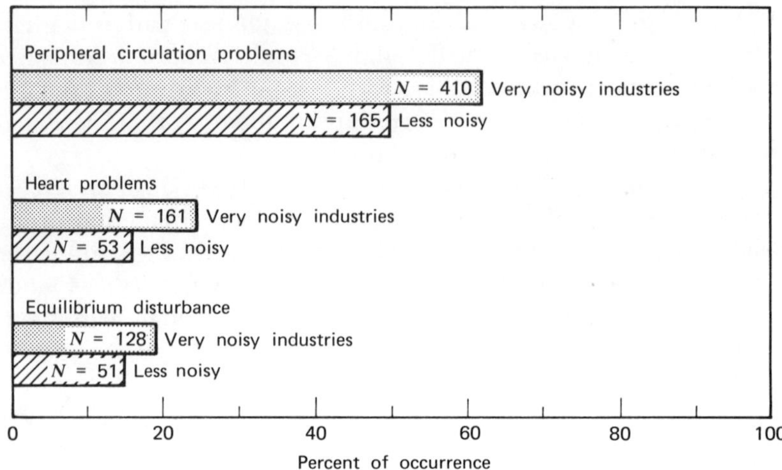

Figure 2 Differences in peripheral circulation and other physiological responses among German industrial workers. (From G. Jansen, Ref. 5, and K. D. Kryter, *The Effects of Noise on Man,* Academic Press, 1970.)

influenced the response time of subjects in reacting to a simple task choice.[4] The degree of muscular activity has also been linked to the increase and decrease of sound intensity.

Breathing

Changes in the breathing pattern have been noted for intermittent sound levels of 98 dB (800 Hz).[2] The response was found to be highly variable with stimulus occurrence, as in related tests for galvanic skin response.

EFFECT OF NOISE ON SLEEP

Disturbance of sleep by excessive or unfamiliar noise has been studied extensively, although so many variables are present that testing is difficult. Sleep is a succession of states or stages, which will vary with age, sex, sleep deprivation, mental and physical states, and general stress, as well as external environmental conditions.

Nevertheless, sleep deprivation is one of the most serious noise-related problems. Even if people are not awakened by noise, they are deprived of the deeper stages of sleep that are essential to health. The several sleep stages are sometimes categorized as I, I-REM, II, III, and IV. A typical night of sleep follows an irregular pattern, but there is a sequence of states from being awake to a deep sleep, and a return to an awakened state, that occurs several times. For young adults, the following fractions of the sleep cycle are:[7] I-5%; I-REM- 20 to 25%; II-50%; III and IV total about 20 to 25%. Stage I-REM is the so-called *rapid eye movement* portion of sleep. The sleeper is in a light stage of sleep (as in Stage I), but there is evidence of rapid eye motion, and it is in this stage that most dreams occur. Stages II, III, and IV are successively "deeper" sleep stages that have been identified by an examination of brain waves (electro-encephalograms, or EEG patterns) and other physiological criteria.

From adolescence to about age 40, approximately 5% of the "sleep" period is really in an awake state; this period increases to nearly 20% for older persons. Also the number of awakenings that occur after initially falling asleep increases from an average of two at age six, to six at age 90.[8]

There exist no quantitative criteria that predict when a sleeper can be stimulated to awaken. The following factors, however, are usually relevant.

1. *Environmental Familiarity.* An urban resident may be disturbed by the sounds of wildlife when in the country, and the sounds of traffic are usually sufficient to awaken rural persons.

2. *Motivation.* A mother may respond to the cry of her child, but fail to be disturbed by sounds of much greater intensities (lightning storm, aircraft, etc.).

3. *Sleep Stage.* Arousal from sleep will depend upon the sleep stage, for noises that have no special or familiar meaning.

4. *Age.* Older persons are more sensitive to intrusive sounds (simulated sonic booms, aircraft noise, etc.) than younger persons.

5. *Sleep Deprivation.* There is an increased insensitivity to arousal with prior sleep deprivation.

6. *Noise-Induced Sleep.* There is some evidence that certain rhythmic sounds are conducive to sleep, such as ocean surf. The masking noise of a fan may also induce or maintain sleep. A claim has also been made that 320 to 350 Hz tones calms infants.[9]

A salient question is whether the sense organs remain sensitive during the sleep experience or whether the peripheral nervous system is "blocked." Research to date suggests that the sensory pathways are not blocked[10] and that the senses remain fully sensitive, and that signals from the sense organs indeed reach the highest centers of the brain even during deepest sleep. Electrical measurements of anesthetized men and animals support this thesis. Hence, it is hypothesized that the apparent insensitivity to signals during sleep is due to a complex coding or reorganization of the brain process,[11] which differs substantially from the awakened to sleeping states.

HEARING LOSS

Of all the physiological effects induced by excessively loud sounds, the most notable is hearing loss. As early as the 1880s attention was being drawn to the effect of railroad noise on hearing, and occupational hearing loss among boilermakers was of concern as early as 1890. In recent years attention has been focused not only upon hearing loss attributable to occupational activities, but also to entertainment and recreational pursuits. Loss in hearing sensitivity does not appear to have a threshold; increased exposure to noise of all types and even moderate intensities probably has some cumulative effect over many years.

Severe hearing impairment is often due to illness, infection, and other factors quite unrelated to noise. Regardless of cause, hearing disabilities are identified in terms of (1) *conductive,* (2) *nerve,* or (3) *cortical* losses in sensitivity. Conductive losses in hearing are due to defects in the external canal (meatus), the eardrum, and ossicular chain. A blocking of the canal, thickening of the eardrum, or otosclerosis (immobilization of the ossicular chain) all result in conductive hearing losses. Fortunately, such defects are sometimes amenable to surgical treatment. Nerve deafness is due to loss of sensitivity in the hair cells of the inner ear; such a hearing loss is usually different for different frequencies and there is no medical remedy. Cortical losses relate to damage or degeneration of neurons in the brain, and such a degeneration is irreversible.

Presbycusis

Virtually all persons experience a general hearing loss with age. This loss in hearing sensitivity is termed *presbycusis.* A portion of this loss may be due to changes in the middle or inner ear and a portion may be due to the degeneration of brain neurons. Even among people who have suffered no serious illness nor were exposed to high occupational noise levels, this "normal loss of hearing with age" is substantial.[12] Figure 3*a* shows the average shift with age of the threshold of hearing for pure tones, for men with "normal" hearing. Figure 3*b* shows a similar shift for women.

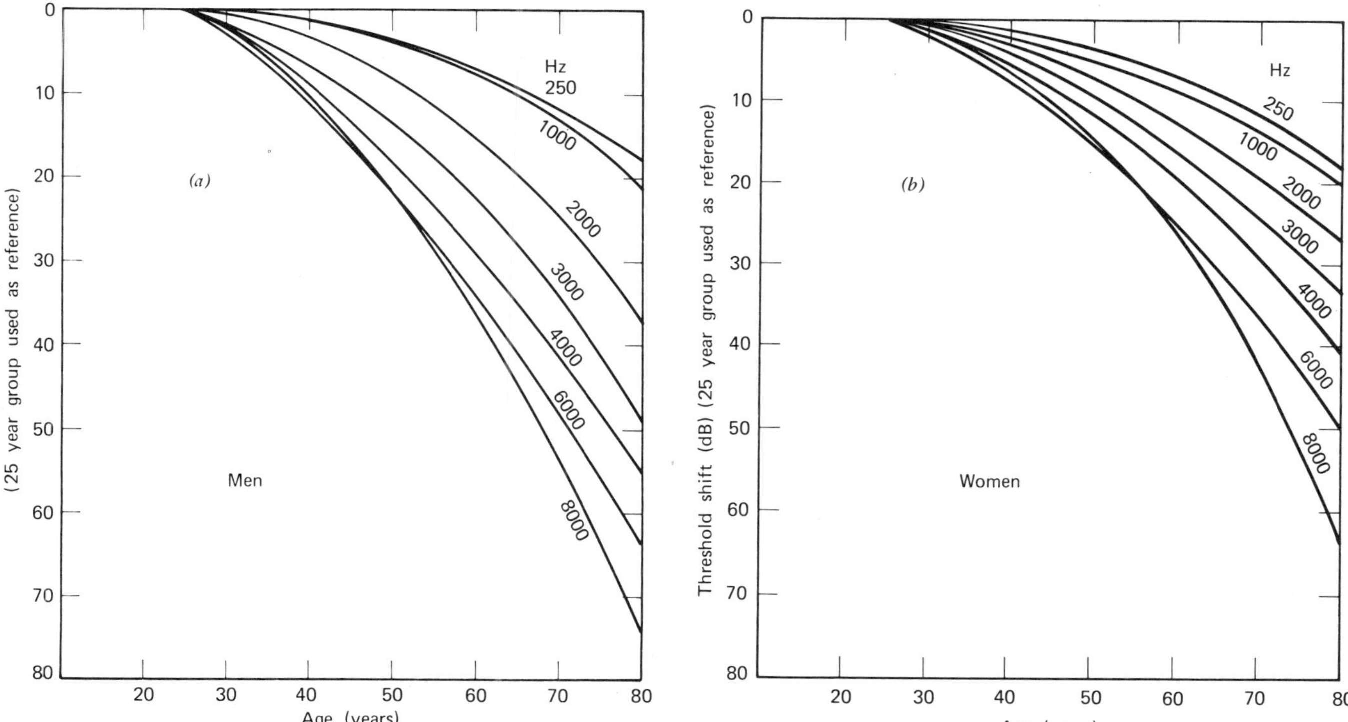

Figure 3*a* Presbycusis: the average shifts with age of the threshold of pure tones of men with normal hearing. Figure 3*b* Presbycusis: the average shifts with age of the threshold of pure tones of women with normal hearing. (From A. Spoor, Ref. 12, and published in *Handbook of Noise Measurement,* by A. P. G. Peterson and E. E. Gross, Jr., General Radio Co., 1972.)

These curves are termed "presbycusis curves," and they are normalized to the hearing threshold (i.e., threshold = 0 dB) for persons 25 years of age. A general hearing loss is somewhat greater for men than for women at high frequencies.

There are substantial departures from these "normal" hearing loss curves among special groups. Rosen *et al.*[13] has reported finding little evidence of presbycusis among a primitive, tribal people (a Mabuan tribe) located in the Sudan in equatorial Africa. This group of about 20,000 exists in an unusually quiet environment—no guns, no drums, and they still hunt and fish with spears. Dr. Rosen also determined that in this group there was an ample blood supply to the ear, indicating a generally healthy cardiovascular system. He stated that "hearing is significantly more acute in all Mabuans aged 10 to 70 than in people of the same age who live in industrial areas of the United States," and he found that the average Mabuan man of 75 could hear as well as the average American male of 25 years.

Glorig and Nixon[14] have also found that the general environment does affect the degree of presbycusis experienced by a group, and suggest that presbycusis is probably a combination of aging, plus hearing losses due to *sociocusis*. Sociocusis is simply a term used to include all nonoccupational noises and stresses to which members of our modern society are exposed.

Temporary Threshold Shift

A temporary threshold shift (TTS) in hearing sensitivity is usually noise-induced. A temporary threshold shift can vary in magnitude from a few decibels over a narrow frequency band, to a hearing loss of such extent that a person is, for all practical purposes, temporarily deaf. After cessation of the noise, the time required for a return to normal hearing may be only a few minutes (for low noise levels) or two to three weeks (for very high noise exposure levels).[15] The amount of threshold shift will be dependent on the level of the noise, the duration of exposure, and the susceptibility of a given individual.

There is a very large variation with individuals who have undergone equal exposures to the same noise levels, and only by continual audiometric testing is it possible to detect those persons most susceptible to noise. In industry, the greatest rate of change in hearing sensitivity resulting from noise exposure occurs in the 3000 to 6000 Hz range. Hearing losses below and above this frequency band are somewhat less. Thus, if the audiogram of a group exposed to the same noise over an extended period shows a marked hearing loss at 4000 Hz, it is likely that the auditory loss was noise-induced. Furthermore, sounds limited to the high frequency band do not produce hearing losses at low frequencies. It should not be concluded, however, that an audiogram of noise-exposed employees mirrors the octave-band spectrum of the noise to which they have been exposed.

Because temporary threshold shift (TTS) in hearing sensitivity almost always precedes permanent shift, studies have been reported on this phenomenon. The following general findings have been reported by a Subcommittee of Noise in Industry:[16]

1. Exposure to steady-state noise with octave-band sound pressure levels of 85 to 95 dB for a full workday produces an average temporary threshold shift of about 10 dB for frequencies above 1000 Hz.

2. Exposure to intermittent or nonsteady noise with octave-band sound pressure levels of 80 to 120 dB for a full workday produces an average temporary threshold shift of about 5 dB for frequencies above 1000 Hz.

3. The extent of the temporary threshold shift is a function of the hearing loss already sustained (resting threshold). Normal thresholds may be shifted as much as 35 dB at frequencies above 1000 Hz. Impaired thresholds (45 dB or more) may be shifted 5 dB or less.

4. In steady-state exposures, a substantial percentage of the temporary threshold shift has occurred after half a working day.

There appears to be no one-to-one correspondence between temporary and permanent threshold shifts, but temporary impairment, repeatedly sustained, leads to permanent hearing loss. Temporary threshold shifts have been measured for industrial workers, airline pilots, and military personnel. The TTS for a group of 120 industrial workers,[17] exposed for one day to broad band noise (i.e., a flat spectrum) with an overall sound pressure level of 105 dB(A), is shown in Fig. 4. The top curve shows the median permanent threshold of the group. The dashed line shows the TTS measured at the end of the working day.

Dramatic temporary threshold shifts have also been observed in "rock" musicians and their audiences. Such groups have been studied by the U.S. Public Health Service and the TTS of one band (Fig. 5) is not atypical of the very marked loss of hearing sensitivity experienced by band musicians using high power amplifier-speaker systems. Figure 5 shows the substantial TTS in the 3000 to 6000 Hz range immediately following a 3-hr exposure at about 112 dB(A). The exposure graph also suggests that mem-

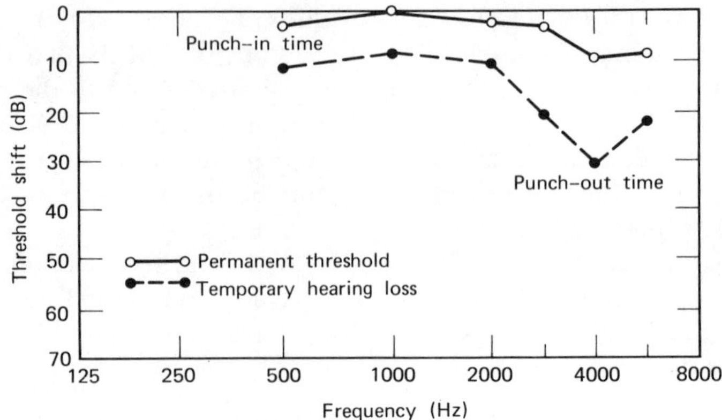

Figure 4   Temporary threshold shift found in 120 employees exposed for one day to broad band noise with an overall sound pressure of about 105 dB. Tests were made just prior to the noise exposure and immediately after work hours. (From W. Rudmose in *Handbook of Noise Control,* ed. by C. M. Harris, McGraw-Hill, 1957.)

bers of the group had already experienced some degree of permanent hearing impairment.

Recovery from a TTS after the cessation to loud sounds is highly variable. If the TTS is less than 40 dB and the exposure duration is less than 8 hr for a single event, a substantial recovery may be experienced in 50 or 100 min.[18] Recovery from temporary threshold shifts appears to be very slow when the initial threshold shift exceeds 35 to 45 dB and when the exposure lasts as long as 12 hr.[19] Even for noise levels as low as 80 dB(A), small temporary threshold shifts do not completely disappear for several days.[20] Repeated exposures to loud noise levels cause the ear to lose its ability to recover and a portion of the TTS becomes a permanent threshold shift.

Permanent Threshold Shift

Many studies have been made of permanent hearing loss induced by noise in industry.[21,22] Jute weavers were tested who had been exposed to a nearly constant noise level of 98 dB(A) for periods of 1 to 52 years in a mill in Dundee, Scotland.[23] These weavers were all women with little exposure to noise other than that received at work. Even when threshold

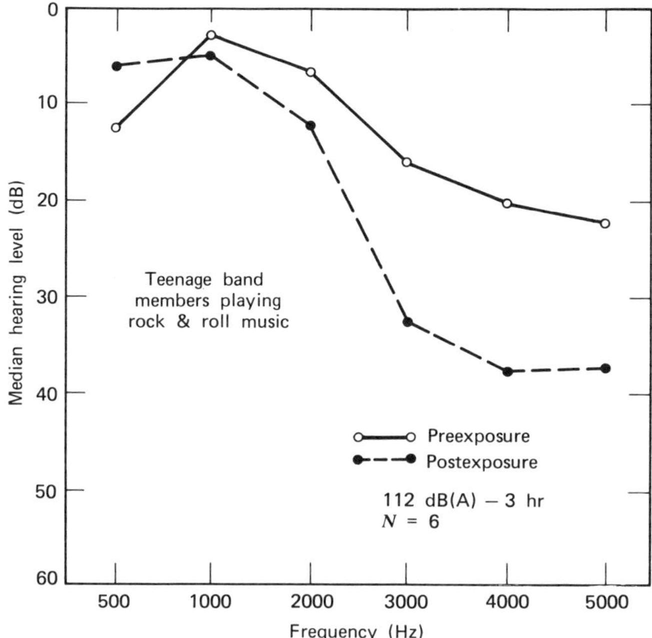

Figure 5   Temporary threshold shift observed in teen-age band players of rock and roll music. (Data by the U.S. Public Health Service.)

shifts have been corrected for presbycusis, a marked deterioration in hearing was noted. Figure 6 indicates the median noise-induced threshold shift as a function of exposure in years. After only 10 years a hearing loss of greater than 30 dB occurred at 4000 Hz. After 40 years, the loss was 50 dB at this same frequency.

Permanent threshold shifts have also been experienced by operators of heavy earth-moving machinery and farm equipment, as well as by employees in factory situations. Furthermore, based on statistical data acquired by a number of professional organizations,* the incidence of hearing impairment can be roughly predicted for groups, based upon their age and the level of their continuous exposure to noise. The upper curve of Figure 7 indicates that of 100 persons exposed to 85 dB(A), about 26 will have impaired hearing when they reach the age group of 50-59. This

* American Academy of Occupational Medicine, Academy of Ophthalmology and Otolaryngology, American Conference of Governmental Industrial Hygienists, Industrial Hygiene Association, and the Industrial Medical Association.

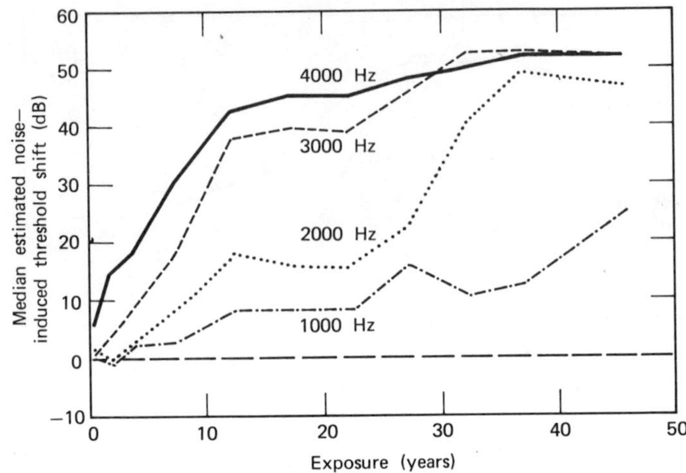

Figure 6   Permanent threshold shift observed for jute weavers with one to more than 40 years of exposure to noise at approximately 98 dB(A). Corrections have been applied for expected threshold change due to age. [From W. Taylor et al., *J. Acoust. Soc. Am.*, **38**, 113 (1965).]

compares to about 22 persons out of 100 with no occupational noise exposure, reflecting an increase of 4 persons per 100 population for the noise exposed group, or four percentage points. This difference is indeed small. However, for persons exposed to continuous occupational noise of 90 dB(A) or greater, the hearing impairment between the "non-noise" and "noise" populations is significant.

Permanent hearing loss among youth exposed to loud music, motorcycle, and other noises has also been substantial. In a recent study undertaken at the University of Tennessee, Knoxville, Dr. D. M. Lipscomb examined 3000 Knoxville high school students and 1680 incoming freshman at the university and reported that "the hearing of many of these students had already deterioriated to a level of the average 65-year-old person." At a frequency of 2000 Hz, a 15 dB loss from normal hearing was detected for the groups listed in Table 1.[24]

There now exists sufficient evidence that loud amplified music often exceeds the tolerable limits recommended for the conservation of hearing[25,26] and that sustained exposure to such environments will lead to a permanent hearing loss.

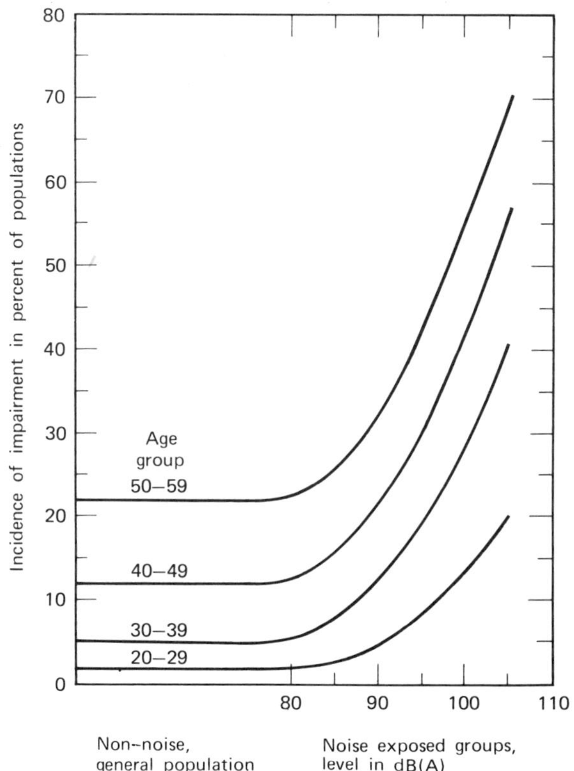

Figure 7   Incidence of hearing impairment (average threshold shift in excess of 15 dB at 500, 1000, and 2000 Hz) in the general population, and in selected populations by age groups and by occupational noise exposure.

Table 1   Percentage   of   Students   with   a 15 dB Loss at 2000 Hz[a]

Sixth grade	3.8%
Ninth grade	11.0%
Twelfth grade	12.6%
College (first year)	34.8%

[a] Data by Lipscomb (in reference 24) for 3000 high school and 1680 university students.

Acoustic Trauma

Acoustic trauma is the term used when a single exposure to an intense sound results in a permanent reduction in hearing sensitivity. Normal processes of hearing recovery may "break down" when a sound is very intense (>130 dB) even if the exposure is limited to only a few minutes or less. There is, as previously stated, a wide latitude in sound levels that an individual can tolerate,[27] but levels exceeding 130 dB should be avoided for even momentary exposures.

Figure 8 is an audiogram showing a substantial left ear hearing loss, at 4000 Hz, induced by acoustic trauma. Furthermore, when noise is concentrated in the 2400 to 4800 Hz band, the hearing damage risk is greater than at much lower or much higher frequencies.

PSYCHOLOGICAL RESPONSE TO NOISE

Man's psychological and sociological response to noise is among the most difficult to quantify. Objective measurements can be made on specific work tasks, but many reactions to noise seem to be related more to personality factors than to specific noise conditions. We should also note that perhaps humans have not had time to adapt to the large changes in noise level that have accompanied our modern era. The situation has been

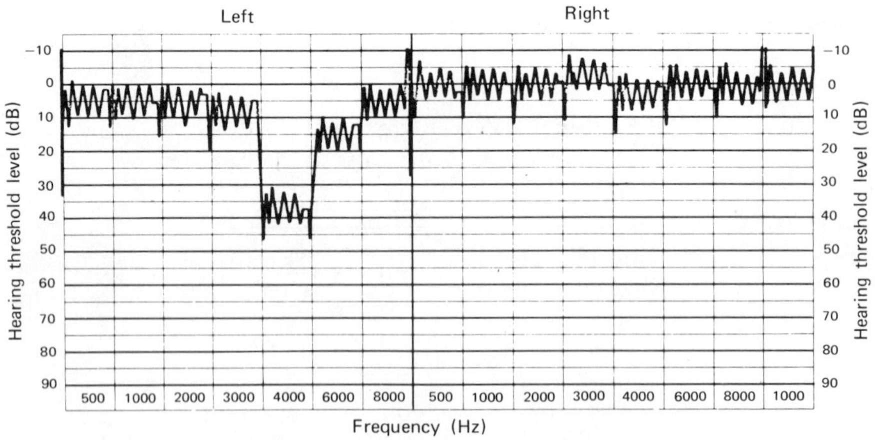

Figure 8 Audiogram showing left-ear hearing loss due to acoustic trauma. (Courtesy of Grason-Stadler Co.)

different for man's visual environment. Daily modulations in light intensity of 100 dB have taken place as long as the earth has been rotating on its axis. Moreover, man's eyes are equipped with (1) lids that are quasi-opaque, (2) pupils that respond quickly to intensity levels, and (3) a retina whose sensitivity changes with large changes in luminance.[28] One hundred decibel level sounds, aside from very infrequent loud natural sounds, have entered man's environment only within the last century. More important, the variety of lower level man-made noises has vastly increased at a time when people are living or working within audio range of each other.

In general, we are faced with a dichotomy in which noise is responsible for both positive and deleterious psychological effects. A carpenter erecting a house may indeed benefit from the noise of his own activity. The sound of hammer on nails provides the carpenter with information on whether a nail has entered or missed a joist, and the sounds of motorized drills or saws reveal substantial information about the density of material or the depth of a drill hole or saw cut. To the person generating such sounds, the effect also adds to a personal sense of skill, achievement, and accomplishment. Moderate levels of noise from the "quiet" typewriter will generate an analogous response from a secretary, that is, the noise is intimately linked with a feeling of productivity. To a neighbor or co-worker not participating in similar tasks, however, the noise is an intrusion upon privacy and, possibly, upon his own productivity and general mood. Essentially, whether a noise has a positive, negative, or null effect will be conditioned by personal factors, as well as by factors associated with the generation of the noise per se. Figure 9 suggests some, although not all, of the criteria relating to the degree of the psychological response.

Personal factors include (1) attitude or mood, (2) the personal environment, (3) whether the sound functions basically to arouse or to distract, and (4) whether the sound is sensed as a real invasion of privacy. Studies have been made that suggest that the particular mood of the individual, at any time, is very important in responding to noise; the same noise exposure levels can be either acceptable or annoying. The personal environment also plays an important role. The intrusion of noise into a person's residence is less acceptable than the same noise level in public buildings, and equivalent noise levels in outdoor environments may not be annoying. Arousal noise is sometimes beneficial; complete silence for extended periods cannot be tolerated for either mental or motor tasks. The degree of disturbance will include physical parameters such as noise level, pitch, plus the individual's particular sensitivity to noise. A per-

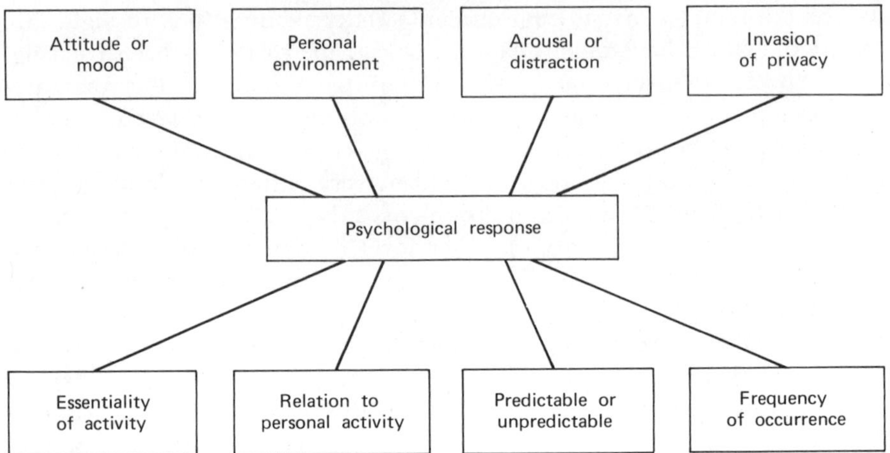

Figure 9   Factors contributing to a person's psychological response to noise.

son's psychological response will also be conditioned by whether he feels the noise is a highly personal infringement of his basic right to acoustic privacy.

Some of the other factors relate more to the noise source; that is, (1) whether the noise is essential, (2) the relationship of the noise to personal activity, (3) the predictability or unpredictability of sound, and (4) frequency of occurrence. Most persons have a much higher tolerance for noise if the activity is clearly essential; even very high noise levels will be tolerated if there is no alternative means for an important undertaking. The relationship of the noise to one's personal temporary activity is relevant. The noise at a football stadium is "shared"; so are the miscellaneous noises in a factory where many workers are participants in related tasks. If the noise is predictable, it is sometimes more acceptable than unpredictable noises, for strategies can be adopted to avoid some deleterious effects. Unpredictable sounds such as backfires, sonic booms, heavy falling objects, also connote a degree of danger or the imminence of an accident. Nevertheless, if the frequency of occurrence is very low, even sonic booms will not be a source of vigorous complaint.

EFFECT OF NOISE ON PERFORMANCE

When manual or mental tasks require the cognition of auditory signals, noise will obviously result in some impairment of the work, primarily through speech interference. What is the effect of noise, if any, when auditory signals are not involved? Because human behavior is very complex, it has been difficult to ascertain how various noises affect people undertaking different types of tasks. However, some general conclusions have emerged from studies that have attempted to provide a noise-performance correlation:[29]

1. Steady noises without special meaning do not seem to interfere with human performance unless the noise level exceeds about 90 dB(A).

2. Irregular bursts are more disruptive than steady noises; even when irregular bursts of noise are below 50 dB(A), they may sometimes interfere with the performance of a task.

3. High frequency components of noise, above 2000 Hz, may produce more interference with performance than low frequency components.

4. Noise does not seem to influence the overall rate of work, but high noise levels may increase the variability of the work rate.

5. Noise is more likely to reduce the accuracy of work than to reduce the quantity of work.

6. Complex tasks are more likely to be adversely affected by noise than are simple tasks.

Manual Tasks

Specific noise-performance data is most easily obtained for manual tasks. Kovrigin and Mikheyev[30] undertook a controlled experiment to determine the effect of noise on postal workers at mail-sorting machines. Figure 10 shows some of their results obtained in a mail-sorting room where additional noise in controlled increments was added via loudspeakers. Sorting efficiency decreased and the number of sorting errors increased with noise level. In such testing situations the results are not always conclusive because the deterioration in performance can sometimes be brief, that is, persons can adapt to adverse conditions over extended periods of time. Furthermore, many responses to noise tend to diminish when such sounds are regular or predictable. In practical long-term tests there are always factors other than noise, that make the conclusions somewhat ambiguous. Nevertheless, it has been suggested that even when a person maintains a high work output in a high noise situation, there may be a cost, in the sense that the person is left with a reduced psychological "reserve" to respond to additional demands after a given assignment is completed.[31]

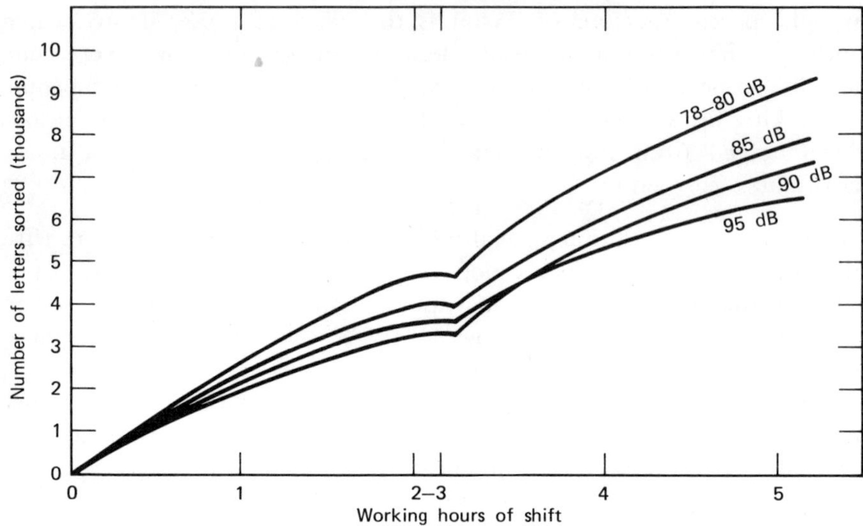

Figure 10   The effect of noise on the efficiency of postal workers. The sound pressure level readings are relative for broad band noise having a reasonably "flat" spectrum. (From S. D. Kovrigin and A. P. Mikheyev, Ref. 30, and K. D. Kryter, *The Effects of Noise on Man,* Academic Press, 1970.)

Decision Making

Tests on the decision-making ability of persons to respond to complex tasks (I.Q. tests, calculations, response to signals or instructions, etc.) have been made under a variety of conditions. Difficulties in long-term testing have included matching mental ability and personal characteristics in sufficiently large numbers for experimental and control groups. For very high noise levels, there is evidence that decision making is impaired, as exemplified in tests made by Jerison and Wing.[32] Figure 11 shows their results for nine persons who were tested in a "quiet" environment of about 83 dB(masking noise) for 0.05 hr followed by 1.5 hr of broad band noise at 114 dB. The percent of correct decisions for the group exposed to this noisy environment is seen to be significantly lower than when the same group was tested in a control test in a quiet environment. Steady-state noise more intense than 90 dB(A) has also been reported to affect judgments of time.[33] Other studies have indicated substantial increases in normal learning times in high noise-level environments.[34]

A general argument is that noise is "overstimulating" and too much stimulation means that a person is "too ready." The net result is that he

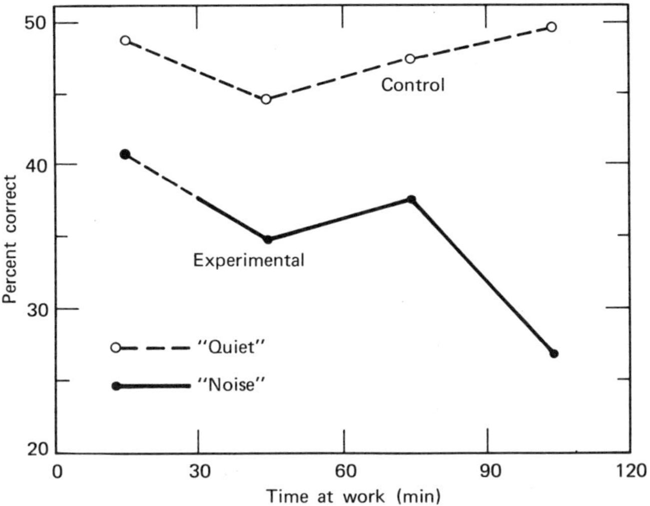

Figure 11   Effect of noise on decision making. At very high noise levels, decision making is impaired. (From H. J. Jerison and S. Wing, Ref. 32.)

may make errors that he would not otherwise make in a more relaxed steady-state condition. The effect of noise also seems to increase the frequency of momentary lapses in efficiency rather than in an overall decline of mental activity.[35,36] Nevertheless, there exists, to date, no general agreement on the magnitude of the effect that noise may have on performance, especially if the noise is anticipated, or if the noise continues over such an extended period that a person might adapt to it.[37]

EFFECT OF "STARTLE"

Several studies have been made of the effect of loud bursts of noise, lasting only a fraction of a second, on both motor control and decision making.

In one investigation,[38] 14 subjects aged 20 to 29 were exposed to pistol shots generating a peak sound pressure level of 124 dB(A). The intent was to determine the extent to which accidents and work performance changes might be attributed to involuntary muscle contraction. The pressure wave form is shown in Fig. 12; indicating that the time to decay 20 dB from the peak value was 80 msec. Background noise during the experiments was about 65 dB(A). Subjects were assigned the task of fol-

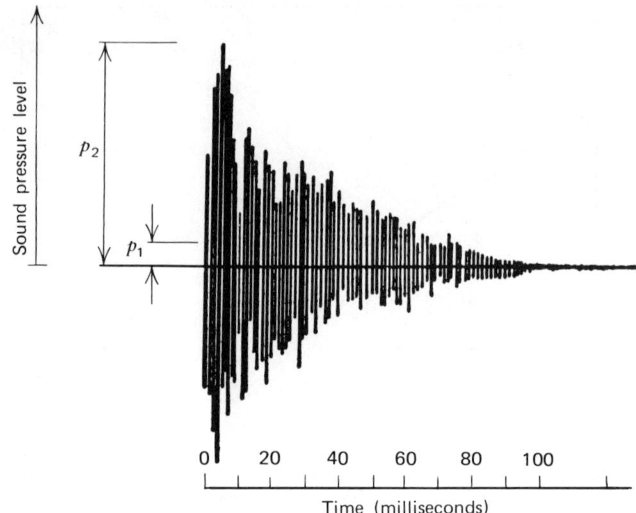

Figure 12   Sound pressure level transient produced by a 22 caliber sports start-ing pistol firing blanks through a silencer. The peak sound pressure level, p_2, was 124 dB and the decay time to p_1, where $p_1 = 0.1p_2$ was 80 msec. [From D. N. May and C. G. Rice, *J. Sound Vib.*, **15**, 197 (1971).]

lowing a probe rotating in a horizontal plane, that is, essentially a "track-ing" type of test. Results of this type of test showed that task perform-ance was significantly impaired in the two seconds immediately following each bang, and not thereafter, suggesting that this time interval follow-ing a sonic shock is a critical one with respect to accidents.

In a related study,[39] the response to short bursts of noise ($\sim$150 msec) in the range of 105 to 117 dB was found to decrease as the background noise level was increased from 45 to 84 dB(A). This result is at variance with a previous finding for animals, and the argument has thus been advanced that sonic booms are likely to be more startling in quiet environments than environments of high noise levels.

Brief loud noises also have an effect on decision making as well as upon the performance of tasks demanding good motor control. In one experiment[40] 18 subjects were exposed to short bursts (0.95 sec) of a recorded rocket firing. The noise was of low frequency and the intensity was varied within a range of 85 to 115 dB. The decision-making test consisted of matching a series of symbols that were displayed on cards moving at various speeds in front of the subject. The test results showed that performance deteriorated significantly for most of the subjects immediately after the noise bursts. The decrement in decision-making ability became especially apparent at sound levels of 95 to 115 dB.

The noxious effects of startle can also include a mild type of neurosis, especially in persons subjected to repeated startling noises,[41] (i.e., neuro-circulatory asthenia, allergic phenomena, and increased propensity to infections of the upper respiratory tract and skin). The reaction to a startling sound may, of course, level off as a person becomes familiar with a given stimulus. It has been suggested,[42] however, that the adaptation process itself involves the introduction of a novel type of stress and resultant fatigue, so that adaptation is more apparent than real.

AFTEREFFECTS OF NOISE

The concept that there is a "psychic price" paid by the individual for his adaptation to noise has been the object of a specific study by Glass and Singer.[43] They suggest that aftereffects may be as much "post-stressor phenomena as post-adaptation phenomena," and that no clear evidence exists to indicate whether "after-effects occurred in spite of adaptation or because of the effort entailed in the adaptive process." Thus, an after-effect might be thought of as a "delayed reaction" in which associated stress behavior (necessary during the stress exposure) no longer functions and is relaxed.

In any event, there is some data to indicate that an aftereffect is real, and that exposure to *unpredictable* noise leads to more adverse after-effects than exposure to *predictable* noise. In a definitive experiment involving proofreading tests,[44] personnel were subject to both periodic and random noise for about a half an hour. In the periodic noise condition, noise was presented at the end of every minute for about 9 sec. For the random or unpredictable condition, noise was presented with varying burst lengths, and with various intervals between bursts. Half of the subjects were exposed to "loud" noise [108 dB(A)], the other half heard a "soft" noise [56 dB(A)]. The total integrated time of noise exposure for both periodic and unpredictable noise was made equal. Following termination of the noise period, the test personnel were asked to perform various tasks designed to measure the after effects of adaptation.

In the postnoise proofreading test, the subjects were asked to correct numerous errors (transpositions and misspellings) that had been deliberately introduced into a seven-page manuscript. Each subject was given 15 minutes, and the quality of their performance was measured on the basis of the percentage of "errors not found" of the total number of errors that could have been detected at the point where the subject was instructed to stop. Figure 13 shows relative results for persons exposed to (1) loud and soft unpredictable noise, (2) predictable noise, and (3) no noise, that is, a control group.

No significant difference was detected in the quantity of the manuscript edited by the subjects. However, Fig. 13 shows the marked difference between the editing ability of those previously exposed to unpredictable and predictable noise, for both the loud and soft situations.

EFFECT OF MUSIC ON PERFORMANCE

Many modern offices employ music to provide an environment that is rated by employees as "favorable" or "pleasant." The Muzak Corporation of New York is engaged in the business of generating such an environ-

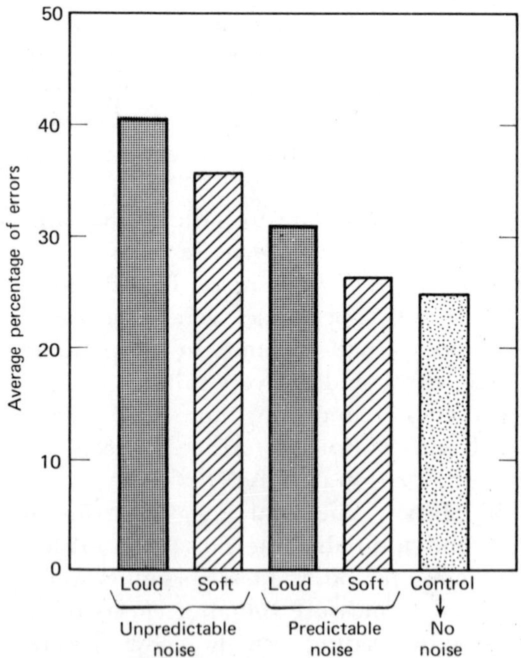

Figure 13   Average percentage of errors undetected in a proofreading test. (From D. C. Glass and J. E. Singer, Ref. 43; reprinted by permission, *American Scientist,* Journal of Sigma Xi, The Scientific Research Society of North America.)

ment, and there are many subscribers to this nationwide service. One of the early statistical studies relating to the effect of this service is cited, as it furnishes a reasonable cross-section of office-type activity. The survey,[45] conducted by an independent management firm, examined the productivity and quality of work at (1) a direct mailing operation of a publisher, (2) a major accounting office, (3) an airlines reservation office, and (4) an accounting and billing section of an electrical utility. A pre-Muzak study was made over a minimum period of several months; this was followed by an analysis after Muzak installation. A general summary of the results is presented:

Case I (Mail Operations; 24 Operators)

(a) Productivity increase approximately 8% with introduction of Muzak.
(b) Productivity improvements slightly greater for low productivity operators and older half of group.
(c) No significant change in absenteeism.
(d) Significant decrease in lateness of arrival.

Case II (Typing Groups, and Check Preparation, etc.)

(a) Error rates decreased 38% over the 6-month period for Muzak testing—some of this improvement probably attributed to music.
(b) No real correlation to tardiness, absenteeism, and employee turnover.

Case III (Airlines Reservation—Telephone Agents)

(a) Decrease in employee turnover.
(b) Improved ratings of employees work by supervisors.

Case IV (Utility and Key-Punch Operators and Bill Processing)

(a) Productivity increase of 18%.
(b) Decline in errors; 37% decrease in number of errors per thousand cards.

The *"Office Music by Muzak"* is programmed to "mirror the standard energy curve" (degrees of tiredness or boredom during the workday) and to offset the *lows*. This is accomplished by using variations in instrumentation, rhythm, tempo, and orchestra size in the different musical selections. The firm terms their programming "stimulus progression." The office music service is programmed in blocks of approximately 15 minutes duration alternated with 15 minutes of silence. To overcome the possibility of bland acceptance or possible annoyance due to sameness, programs are never repeated.

It should be emphasized that it is very difficult, even in highly controlled tests, to account for all other factors that might influence production rates, accuracy, and so on. Nevertheless, it is now generally conceded that music during working hours will often improve production, when the work is highly repetitive. Figure 14 shows the variation in the work output of a light manual task (rolling paper novelties) under alternate conditions of music and no-music.[46]

MUSIC AND EMPLOYEE ATTITUDES

While there may be some difficulty in establishing a quantitative relationship between the use of music and productivity, there seems to be little question that music is a medium for improving employee morale. Answers to extensive questionnaires bring typical responses, such as: "helps me in my work," "makes the time go faster," "makes work more enjoyable," "helps to relieve fatigue," and "you relax when you hear music."

Music probably is most effective when the individual's attention is not absorbed fully by his work; in this circumstance music appears to divert

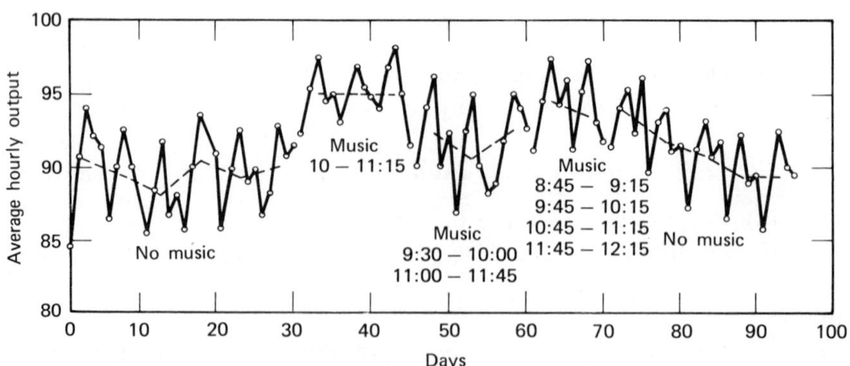

Figure 14   Work output in a light manual task (rolling paper novelties) under alternate conditions of music and no music. (From S. Wyatt and J. H. Langdon, Ref. 46, and K. D. Kryter, *The Effect of Noise on Man*, Academic Press, 1970.)

unused attention from brooding, talking, or off-the-job activities.[47] The type of music desired by employee groups will vary,[48] but the general desire for music during some portions of the working day appears to be independent of age groups. The response of 720 shop and office employees to a questionnaire on their attitudes is shown in the bar graphs of Fig. 15.[49]

It appears remarkable that such a large percentage wanted "music all the time," indicating a desire for a continually changing acoustic environment. However, many statistical studies tend to overlook individual reactions and give little consideration to the people who prefer music only infrequently, or not at all. For such persons employee morale can be adversely affected.

A further qualification is that some improvement in group morale that is often attributed to musical environments could be explained on the basis of the "Hawthorne effect."* The theory is that the motivation of workers increases with virtually *any change* made by management, provided the change is regarded by the working group as a sign that management is concerned with their well-being.[50]

ACOUSTIC OPTIONS

An ideal acoustic environment would provide the most pleasant and efficient atmosphere for every individual to work, study, play, or sleep. Such an environment might be possible for a few individuals, but it is an impossible objective when large numbers of people are involved. In most practical situations individuals are often subject to the intrusive noise of irrelevant speech, traffic, machinery, music, or other unwanted sounds. A compromise is required between a rigid "no-noise" code and a societal attitude that is highly permissive.

Sometimes simple strategies for noise reduction can be devised by advanced planning and by utilizing modern engineering methods. Airports now utilize closed-circuit television or computer-controlled display boards to transmit arrival and departure times. Prior to their use, airports were filled with loudspeaker announcements that were loud and often unintelligible. Hospital patients formerly had to endure the sound of TV and radio from fellow patients in addition to their other afflictions; today "pillow-speakers" have largely eliminated this problem. Airlines have also provided acoustic options to passengers with individual earphones

---

* An occupational psychological phenomenon so named because of its discovery at the Hawthorne plant of the Western Electric Company, in Chicago in 1947.

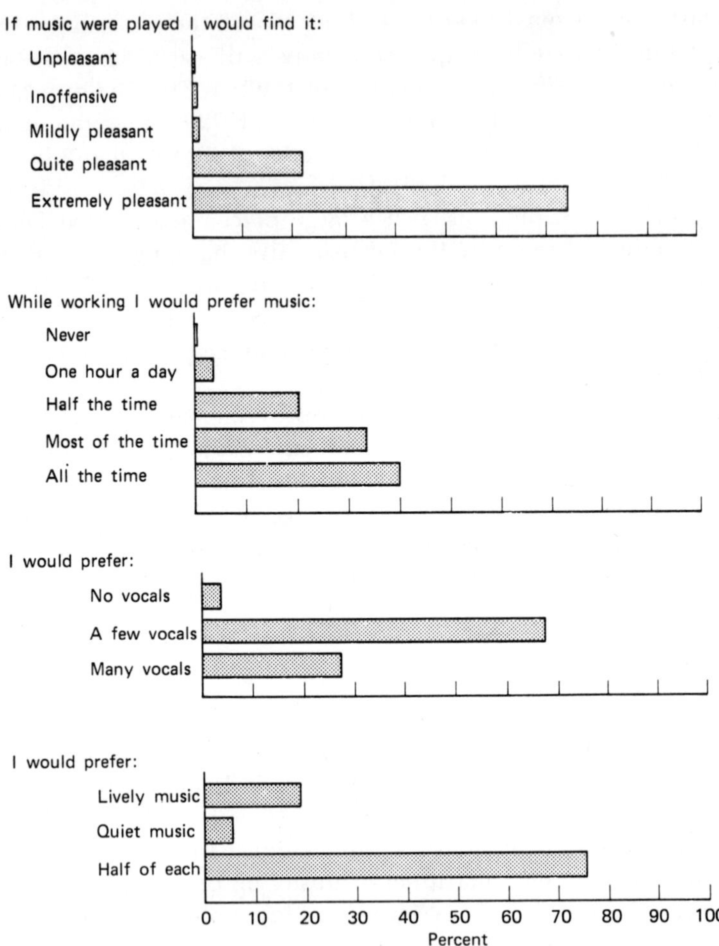

Figure 15   Response of 720 shop and office workers to a questionnaire on their attitudes toward the type of music preferred during working hours. (From *Applied Psychology Monograph,* No. 14, by H. C. Smith, Stanford University Press, 1947.)

for both TV and stereo. Options are often possible in building design and room allocation. Some buildings are now constructed with excellent acoustic isolation between dining areas, lounges, reading rooms, music listening rooms, and bowling alleys. Some libraries are essentially "zoned" for activities that generate different sound levels, and modern restaurants offer alternative rooms for dining with varying visual and acoustic decor.

But it should not be assumed that there will not be continued invasions of "air rights" that will even be upheld in the courts. For example, the U.S. Supreme Court has ruled 7 to 1 that the broadcast of radio programs and advertising on Washington buses and streetcars did not constitute a violation of freedom of speech nor did it seriously deprive people of their privacy. Furthermore, it is unrealistic to anticipate a complete cessation of noise and annoyance in an increasingly industrial society. However, substantial improvements in occupational working conditions are in evidence with incentives being provided both through federal legislation and a desire to improve employee morale.

It appears that for many years to come the most complex problem will be the abatement of noise in the residential environment. A moderate amount of noise can even be tolerated during a 40 hr work week if respite can be obtained during most of the remaining 128 hr. There is no simple solution to the problem of aircraft noise. Even if very quiet engines are engineered in future years, aerodynamic noise will still be generated by large aircraft moving through the atmosphere at high speeds. However, a substantial reduction in the intrusion of aircraft noise can be achieved over the long term by working toward several concurrent goals:[51] "(1) to make aircraft inherently quieter and to have them flown as quietly as possible, (2) to design or modify the total operating plan of an airport so as to minimize the extent of the airport noise impact zone and tailor its shape to avoid existing noise-sensitive uses, and (3) to prevent buildup of new housing or other noise-sensitive land uses in present and anticipated future noise impact zones and, where necessary, resolve by land use measures (sound proofing or conversion) those few impacted areas where the noise exposure cannot be adequately decreased by other means." For those living in close proximity to major airports, noise abatement can probably be achieved only through substantial changes in the architectural design of buildings, central air conditioning systems, and similar measures. Sound-proofed dwellings have been built in the immediate vicinity of the Heathrow Airport, London; they were costly but they are quiet.

A substantial reduction in traffic noise can be achieved for both buses and light trucks based on known technology, and manufacturers are currently developing turbocharging engines that are expected to both improve engine performance and reduce noise levels. Engine compartment shielding is also being developed using acoustically treated covers and panels. Interstate truck traffic will probably remain noisy but the next generation of buses and local service vehicles can be built with substantial noise reduction. Notwithstanding, a proper distinction must be made between (1) permissible noise levels specified for manufactured vehicles, and (2) the exposure of these vehicles within a community. Furthermore, a federal power commission predicts that millions of quiet electric vehicles (small cars and service vehicles) will be in use in the United States by 1990. Some shift to electric-powered transportation may, indeed, be needed for a national petroleum conservation effort, quite apart from noise considerations.

Community planning of a long-range nature will also be required to isolate residential from commercial and industrial areas. Building codes can be implemented to include acoustic specifications, and local ordinances with respect to noise nuisances will have to be both enacted and enforced. But in the final analysis, an individual's acoustic privacy will also depend upon good neighborliness and a genuine consideration for the acoustic sensibilities of others.

REFERENCES

1. D. E. Broadbent, *Advan. Sci.*, 406 (January 1962).

2. R. C. Davis, A. M. Buchwald, and R. W. Franklin, *Psychol. Monograph,* **69**, No. 20, 1 (1955).

3. G. Jansen, *Intern. Z. Angew. Physiol.*, **20**, 233 (1964).

4. J. D. Miller, *Effects of Noise on People,* U.S.E.P.A., National Technical Information Service, Springfield, Va., 1971, pp. 127-129.

5. G. Jansen, *Stahl Eisen,* **81**, 217 (1961).

6. K. D. Kryter, *The Effects of Noise on Man,* Academic, New York, 1970, p. 508.

7. R. J. Berger, *Sleep Physiology and Pathology, a Symposium,* edited by A. Kales, J. B. Lippincott, Philadelphia, 1969, p. 17.

8. I. Feinberg, *Sleep Physiology and Pathology, a Symposium,* edited by A. Kales, J. B. Lippincott, Philadelphia, 1969, p. 39.

9. J. Olsen and E. N. Nelson, *Minn. Med.,* **44**, 527 (1961).

10. W. P. Koella, *Sleep: Its Nature and Physiological Organization,* Thomas, Springfield, Ill., 1967.

11. Miller, *op. cit.,* p. 65.

12. A. Spoor, *International Audiology,* **6**, No. 1, 48 (1967).

13. S. Rosen, M. Bergman, D. Plestor, A. El-Mofty, and M. Satti, *Ann. Rhinol-Laryngal,* **71**, 727 (1962).

14. A. Glorig and J. Nixon, *Laryngoscope,* **72**, 1596 (1962).

15. Miller, *op. cit.,* p. 17.

16. W. Rudmose, *Hearing Loss Resulting From Noise Exposure,* Handbook of Noise Control, edited by C. M. Harris, McGraw-Hill, New York **7**, 13 (1957).

17. *Ibid.*

18. W. D. Ward, A. Glorig, and D. L. Sklar, *J. Acoust. Soc. Am.*, **31**, 522 (1959).

19. W. D. Ward, *J. Acoust. Soc. Am.*, **32**, 497 (1960).

20. J. H. Mills, R. W. Gengel, C. C. Watson, and J. D. Miller, *J. Acoust. Soc. Am.*, **48**, 524 (1970).

21. D. W. Robinson, *Occupational Hearing Loss,* Academic, London, 1971.

22. J. Sataloff, *Hearing Loss,* Lippincott, Philadelphia, 1966.

23. W. Taylor, J. Pearson, A. Mair, and W. Burns, *J. Acoust. Soc. Am.*, **38**, 113 (1965).

24. T. Berland, *The Fight for Quiet,* Prentice-Hall, Englewood Cliffs, N. J., 1970, p. 43.

25. W. F. R. Rintelmann, *J. Acoust. Soc. Am.*, **51**, 1249 (1971).

26. C. P. Lebe, K. P. Oliphant, and J. Garrett, *Calif. Med.*, **107**, 378 (1967).

27. K. D. Kryter, W. D. Ward, J. D. Miller, and D. H. Eldredge, *J. Acoust. Soc. Am.*, **39**, 451 (1966).

28. Miller, *op. cit.*, p. 2.

29. *Ibid.*, p. 118.

30. S. D. Kovrigin and A. P. Mikheyev, *The Effect of Noise Level on Working Efficiency*, Rept. N65-28, 297. Joint Publications Research Service, Washington, D. C., 1965.

31. Miller, *op. cit.*, p. 119.

32. H. J. Jerison and S. Wing, *Effects of Noise and Fatigue on a Complex Vigilance Task*, Rept. WADC-TR-57-1e, AD110.700, Wright-Air Development Center, Wright-Patterson AFB, Ohio, 1957.

33. H. J. Jerison and J. Arginteanu, Time Judgment Rept. No. WADC-TR-57-454, AD 130963, Wright Air Development Center, Wright-Patterson AFB, Ohio, 1958.

34. H. Burris-Meyer and V. Mallory, *J. Acoust. Soc. Am.*, **32**, 1568 (1960).

35. T. Berland, *op. cit.*, p. 77.

36. D. E. Broadbent, *J. Acoust. Soc. Am.*, **30**, 824 (1958).

37. Kryter, *op. cit.*, p. 546.

38. D. N. May and C. G. Rice, *J. Sound & Vib.*, **15**, 197 (1971).

39. D. N. May, *J. Sound & Vib.*, **17**, 77-81 (1971).

40. M. M. Woodhead, *J. Acoust. Soc. Am.*, **31**, 1329 (1959).

41. J. Sematan and M. Sematanova, *J. Sound & Vib.*, **10**, 480 (1969).

42. *Ibid.*

43. D. C. Glass and J. E. Singer, *Am. Sci.*, **60**, 457 (July-August 1972).

44. *Ibid.*

45. *Effects of Muzak in Office Personnel—A Survey Report;* Stevenson, Jordan and Harrison, Inc., Management Engineers and Consultants, New York, 1958.

46. S. Wyatt and J. H. Langdon, *Industrial Health Research Board Report No. 77.* Her Majesty's Stationery Office, London, 1935.

47. H. C. Smith, *Appl. Psychol. Monogr. of the Am. Psychol. Assoc.,* Stanford University Press, 14, 58 (1947).

48. A. Pepinsky, *J. Acoust. Soc. Am.,* 15, 176 (1944).

49. Smith, *op. cit.,* p. 18.

50. C. C. Homans, *Group Factors in Worker Productivity, Readings in Social Psychology,* edited by T. M. Newcomb and E. L. Hartley, Holt, Rinehart and Winston, New York, 1947.

51. E. Cuadra, *"The EPA Program for an Airport Noise Regulation,"* Office of Noise Abatement and Control, Washington, D. C. 20460, 1974.

EXPANSION OF A FUNCTION IN A FOURIER SERIES

The basic premise of the Fourier series formulation is that any function $f(x)$ defined in the interval from $-\pi$ to $+\pi$ (Fig. 1) can be expanded in a series of trigonometric functions (sines and cosines), such that

$$f(x) = \frac{a_0}{2} + (a_1 \cos x + b_1 \sin x) + (a_2 \cos 2x + b_2 \sin 2x)$$

$$+ (a_3 \cos 3x + b_3 \sin 3x) + \ldots$$

$$+ (a_n \cos nx + b_n \sin nx) + \ldots \tag{1}$$

The above can be written in the compact form

$$f(x) = \frac{a_0}{2} + \sum_{n=1}^{n=\infty} (a_n \cos nx + b_n \sin nx) \tag{2}$$

It is by no means necessary that the function represented in Fig. 1 be expressed by a single equation. Provided that $f(x)$ has only a finite number of discontinuities and a finite number of maxima and minima in the interval from $-\pi$ to $+\pi$, and provided that

$$\int_{-\pi}^{+\pi} |f(x)| \, dx$$

is finite, then a Fourier expansion is always possible. Thus a Fourier series expansion is possible for virtually all physically real situations.

The coefficients of each term in the series expansion of equation 1 are determined by evaluating the definite integrals

$$a_n = \frac{1}{\pi} \int_0^{2\pi} f(x) \cos nx \, dx \tag{3}$$

$$b_n = \frac{1}{\pi} \int_0^{2\pi} f(x) \sin nx \, dx \tag{4}$$

489

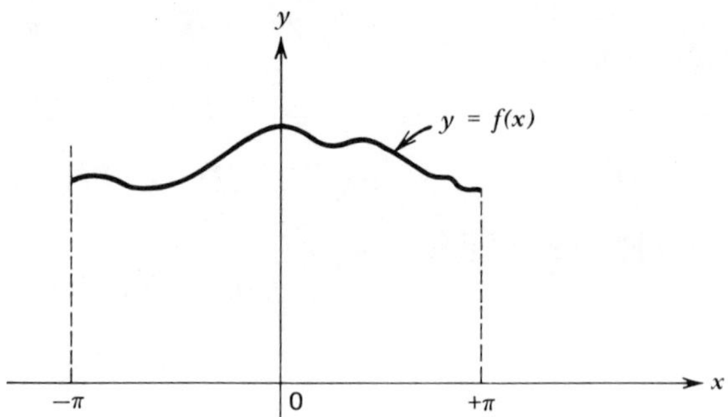

Figure 1   A Fourier series requires that a function is defined in the interval, $-\pi$ to π.

As an example, let us consider a periodic function, such as is drawn in Fig. 2. The function is indeed periodic, and it is defined with the interval $(-\pi$ to $+\pi)$ by

$$f(x) = 1 \qquad \text{from } -\pi \text{ to } 0$$

$$f(x) = -1 \qquad \text{from } 0 \text{ to } +\pi$$

According to equation 3,

$$a_0 = \frac{1}{\pi} \int_{-\pi}^{+\pi} f(x)\, dx = \frac{1}{\pi} \int_{-\pi}^{0} (+1)\, dx + \frac{1}{\pi} \int_{0}^{\pi} (-1)\, dx = \frac{\pi}{\pi} - \frac{\pi}{\pi} = 0$$

$$a_n = \frac{1}{\pi} \int_{-\pi}^{+\pi} f(x) \cos nx\, dx$$

$$= \frac{1}{\pi} \int_{-\pi}^{0} (+1) \cos nx\, dx + \frac{1}{\pi} \int_{0}^{\pi} (-1) \cos nx\, dx$$

$$= \frac{1}{\pi} \frac{\sin nx}{n} \Big|_{-\pi}^{0} - \frac{1}{\pi} \frac{\sin nx}{n} \Big|_{0}^{\pi} = 0 + 0 = 0$$

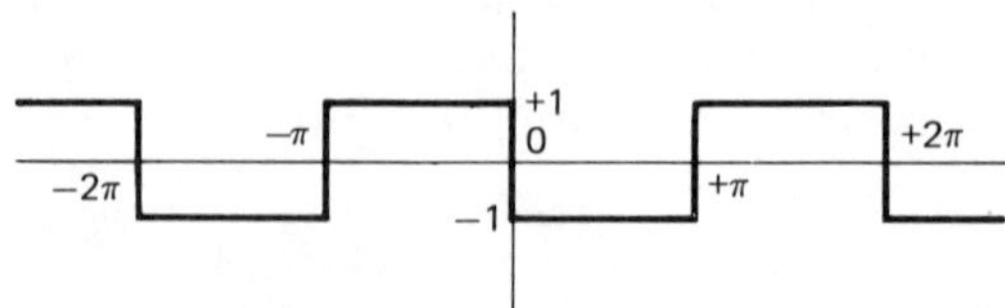

Figure 2   A square wave that is periodic and defined within the interval, $-\pi$ to π.

Thus *all* the coefficients of the cosine expansion are zero. The coefficients, b_n, are determined by equation 4

$$b_n = \frac{1}{\pi} \int_{-\pi}^{\pi} f(x) \sin nx \, dx$$

$$= \frac{1}{\pi} \int_{-\pi}^{0} (+1) \sin nx \, dx + \frac{1}{\pi} \int_{0}^{\pi} (-1) \sin nx \, dx$$

$$= \frac{-1}{n} \frac{\cos nx}{n} \Big|_{\pi}^{0} + \frac{1}{\pi} \frac{\cos nx}{n} \Big|_{0}^{\pi}$$

$$= -\frac{1}{\pi n} - \frac{1}{\pi n} - \frac{1}{\pi n} - \frac{1}{\pi n} = \frac{-4}{\pi n} \quad \text{(if } n \text{ is odd)}$$

$$= -\frac{1}{\pi n} + \frac{1}{\pi n} - \frac{1}{\pi n} + \frac{1}{\pi n} = 0 \quad \text{(if } n \text{ is even)}$$

Hence, the Fourier expansion of the function shown in Fig. 2 is

$$f(x) = \frac{-4}{\pi} \left(\sin x + \sin \frac{3x}{3} + \frac{\sin 5x}{5} + \cdots \right)$$

INDEX